Reindeer Husbandry and Global Environmental Change

This volume offers a holistic understanding of the environmental and societal challenges that affect reindeer husbandry in Fennoscandia today.

Reindeer husbandry is a livelihood with a long traditional heritage and cultural importance. Like many other pastoral societies, reindeer herders are confronted with significant challenges. Covering Norway, Sweden and Finland – three countries with many differences and similarities – this volume examines how reindeer husbandry is affected by and responds to global environmental change and resource extraction in boreal and arctic social-ecological systems. Beginning with an historical overview of reindeer husbandry, the volume analyses the realities of the present from different perspectives and disciplines. Genetics, behavioural ecology of reindeer, other forms of land use, pastoralists' norms and knowledge, bio-economy and governance structures all set the stage for the complex internal and externally imposed dynamics within reindeer husbandry. In-depth analyses are devoted to particularly urgent challenges, such as land-use conflicts, climate change and predation, identified as having a high potential to shape the future pathways of the pastoral identity and productivity. These futures, with their risks and opportunities, are explored in the final section, offering a synthesis of the comparative approach between the three countries that runs as a recurring theme through the book. With its richness and depth, this volume contributes significantly to the understanding of the substantial impacts on pastoralist communities in northernmost Europe today, while highlighting viable pathways to maintaining reindeer husbandry for the future.

This book will be of great interest to students and scholars of both the natural and social sciences who work on natural resource management, global environmental change, pastoralism, ecology, social-ecological systems, rangeland management and Indigenous studies.

Tim Horstkotte is a senior research engineer at the Department of Ecology and Environmental Science, Umeå University, Sweden.

Øystein Holand is a professor, at the Norwegian University of Life Sciences, Norway.

Jouko Kumpula is a senior research scientist at the Natural Resources Institute Finland.

Jon Moen is a professor at the Department of Ecology and Environmental Science, Umeå University, Sweden.

Earthscan Studies in Natural Resource Management

Interdisciplinary Collaboration for Water Diplomacy
A Principled and Pragmatic Approach
Edited by Shafiqul Islam and Kevin M. Smith

Peatlands
Ecology, Conservation and Heritage
Ian D. Rotherham

The Moroccan Argan Trade
Producer Networks and Human Bio-Geographies
Daniel F. Robinson

Fair and Equitable Benefit-Sharing in Agriculture
Reinventing Agrarian Justice
Elsa Tsioumani

Participatory Governance of UNESCO Biosphere Reserves in Canada and Israel
Resolving Natural Resource Conflicts
Natasha Donevska

Balancing the Commons in Switzerland
Institutional Transformations and Sustainable Innovations
Edited by Tobias Haller, Karina Liechti, Martin Stuber, François-Xavier Viallon and Rahel Wunderli

Natural Resource Sovereignty and the Right to Development in Africa
Edited by Carol Chi Ngang and Serges Djoyou Kamga

Reindeer Husbandry and Global Environmental Change
Pastoralism in Fennoscandia
Edited by Tim Horstkotte, Øystein Holand, Jouko Kumpula and Jon Moen

For more information about this series, please visit: www.routledge.com/books/series/ECNRM/

Reindeer Husbandry and Global Environmental Change

Pastoralism in Fennoscandia

Edited by
Tim Horstkotte, Øystein Holand, Jouko Kumpula and Jon Moen

Routledge
Taylor & Francis Group
LONDON AND NEW YORK
earthscan
from Routledge

First published 2022
by Routledge
4 Park Square, Milton Park, Abingdon, Oxon OX14 4RN

and by Routledge
605 Third Avenue, New York, NY 10158

Routledge is an imprint of the Taylor & Francis Group, an informa business

British Library Cataloguing-in-Publication Data
A catalogue record for this book is available from the British Library

Library of Congress Cataloging-in-Publication Data
Names: Horstkotte, Tim, editor. | Holand, Øystein, editor. | Kumpula, Jouko, editor.
Title: Reindeer husbandry and global environmental change: pastoralism in Fennoscandia / edited by Tim Horstkotte,
Øystein Holand, Jouko Kumpula and Jon Moen.
Description: Milton Park, Abingdon, Oxon; New York, NY: Routledge, 2022. |
Series: Earthscan studies in natural resource management |
Includes bibliographical references and index.
Identifiers: LCCN 2021058877 (print) | LCCN 2021058878 (ebook) |
ISBN 9780367632670 (hardback) | ISBN 9780367632687 (paperback) |
ISBN 9781003118565 (ebook)
Subjects: LCSH: Reindeer farming–Scandinavia. |
Reindeer–Climatic factors–Scandinavia.
Classification: LCC SF401.R4 R45 2022 (print) |
LCC SF401.R4 (ebook) | DDC 636.2/940948–dc23/eng/20211206
LC record available at https://lccn.loc.gov/2021058877
LC ebook record available at https://lccn.loc.gov/2021058878

ISBN: 978-0-367-63267-0 (hbk)
ISBN: 978-0-367-63268-7 (pbk)
ISBN: 978-1-003-11856-5 (ebk)

DOI: 10.4324/9781003118565

Typeset in Bembo
by Newgen Publishing UK

Contents

Contributors

Birgitta Åhman, Department of Animal Nutrition and Management, Swedish University of Agricultural Sciences, Sweden.

Bård-Jørgen Bårdsen, Norwegian Institute for Nature Research (NINA), and Arctic and Marine Biology, The Arctic University of Norway, Norway.

Svein Morten Eilertsen, Division of Forest and Forest Resources, Norwegian Institute of Bioeconomy Research, Norway.

Bruce C. Forbes, Arctic Centre, University of Lapland, Finland.

Hannu I. Heikkinen, Cultural Anthropology, University of Oulu, Finland.

Øystein Holand, Department of Animal and Aquacultural Sciences, Norwegian University of Life Sciences, Norway.

Tim Horstkotte, Department of Ecology and Environmental Science, Umeå University, Sweden.

Kathrine Ivsett Johnsen, Water and Society, Norwegian Institute for Water Management (NIVA), Norway.

Sonja Kivinen, Department of Geographical and Historical Studies, University of Eastern Finland, Finland.

Jouko Kumpula, Natural Resources Institute Finland (Luke), Finland.

Thomas Kvalnes, Department of Biology Faculty of Natural Sciences, Norwegian University of Science and Technology, Norway.

Kjersti S. Kvie, Department of Preclinical Sciences and Pathology, Norwegian University of Life Sciences, Norway.

Kristina Labba, Faculty of Law, The Arctic University of Norway, Norway.

Mia Landauer, Arctic Centre, University of Lapland, Finland.

Åsa Larsson Blind, Saami Council, Sweden.

Ilari Lehtonen, Finnish Meteorological Institute, Finland.

Élise Lépy, Cultural Anthropology, University of Oulu, Finland.

Annette Löf, Department for Urban and Rural Development, Swedish University of Agricultural Sciences, Sweden.

Jon Moen, Department of Ecology and Environmental Science, Umeå University, Sweden.

Kirsi Muuttoranta, Natural Resources Institute Finland (Luke), and Lapland University of Applied Sciences, Finland.

Asko Mäki-Tanila, Department of Agricultural Sciences, University of Helsinki, Finland.

Marius Warg Næss, Norwegian Institute for Cultural Heritage Research, Norway.

Harri Norberg, Natural Resources Institute Finland (Luke), Finland.

Amélie Paoli, Department of Biology, Concordia University, Canada.

Antti-Juhani Pekkarinen, Department of Forest Sciences, University of Helsinki and Natural Resources Institute Finland (Luke), Finland.

Jaakko Pietarinen, Department of Agricultural Sciences, University of Helsinki, Finland.

Kaisa Raitio, Department for Urban and Rural Development, Swedish University of Agricultural Sciences, Sweden.

Sirpa Rasmus, Arctic Centre, University of Lapland, Finland.

Camilla Risvoll, Nordland Research Institute, Norway.

Knut H. Røed, Department of Preclinical Sciences and Pathology, Norwegian University of Life Sciences, Norway.

Gunhild Rosqvist, Department of Physical Geography, Stockholm University, Sweden.

Javier Sánchez Romano, Department of Medical Biology, The Arctic University of Norway, Norway.

Per Sandström, Department of Forest Resource Management, Swedish University of Agricultural Sciences, Sweden.

Stefan Sandström, Department of Forest Resource Management, Swedish University of Agricultural Sciences, Sweden.

Simo Sarkki, Cultural Anthropology, University of Oulu, Finland.

Anna Skarin, Department of Animal Nutrition and Management, Swedish University of Agricultural Sciences, Sweden.

Olli Tahvonen, Department of Forest Sciences, University of Helsinki, Finland.

Morten Tryland, Department of Forestry and Wildlife Management, Inland Norway University of Applied Sciences, and Arctic and Marine Biology, The Arctic University of Norway, Norway.

Minna Turunen, Arctic Centre, University of Lapland, Finland.

Torkild Tveraa, Norwegian Institute for Nature Research (NINA), and Arctic and Marine Biology, The Arctic University of Norway, Norway.

Hans Tømmervik, Norwegian Institute for Nature Research (NINA), Norway.

Robert B. Weladji, Department of Biology, Concordia University, Canada.

Preface

The winter in 2019/2020 was extremely harsh for reindeer in northern Fennoscandia. Deep and densely packed snow made the ground vegetation inaccessible for the animals, while other areas experienced recurrent frost and thaw cycles leading to ice crusts. Both events caused starvation and high losses for the first time in decades.

Herders all across Fennoscandia mobilized to provide reindeer with emergency feeding in order to save their herds. In lack of traditionally used alternative grazing resources, emergency feeding had to be provided and hundreds of tons of concentrated feed were supplied. Despite these actions to prevent starvation, losses were high.

This provides a sobering message for the consequences of ongoing climate change, which increasingly leads to extreme winter events and worsened grazing conditions for reindeer. It warns of a potentially new trajectory for reindeer pastoralism, where supplementary feeding replaces the traditional free grazing based on natural forage resources as a regular practice to buffer extreme winter conditions. While this shift has already taken place in parts of Finland, it is increasingly becoming reality all over Sápmi, much to herders' dismay. An alternative trajectory would be to restore and protect the remaining grazing land. This would demand concerted efforts to reshape land governance in the entire reindeer herding area.

Climate change is thus only one part of the mosaic of present and future challenges to reindeer pastoralism. Grazing lands have been severely degraded for decades and are increasingly put under cumulating pressure following continued encroachment by other forms of land use. Paradoxically, the so-called green transition to carbon neutral economies and energy sector adds even more pressure as states and companies race for renewable energy and minerals in areas where reindeer pastoralism is practised. The prospect of ice-free seaways via the Northeast Passage further sparks plans to connect northernmost Fennoscandia to central Europe by railway. Such developments add to the fragmentation of reindeer habitats and collide with herders' rights and herding practices. This marks the beginning of a new area of colonialism of northern peripheries – coined 'Green colonialism'.

This is the complex political, social and ecological landscape in which the future of reindeer pastoralism is embedded: competing demands, rights and interests add up to unpredictable and cumulative challenges and place large demands on all actors involved. It demands a search for proactive strategies – that recognize the reindeer pastoralism in full – and provides herders and herding communities with real opportunities to decide over their futures. Encouragingly, the understanding of customary and Indigenous rights increased over the last decades. However, in practice many obstacles remain before a secure future of reindeer pastoralism in accordance with cultural and traditional aspirations and defined by the herders' self-determination becomes a reality.

Acknowledgements

Northern regions in Fennoscandia experience the combined impacts of accelerating environmental change and resource extraction. These transformations do not only affect the fragile ecosystem in its biotic and abiotic structure and function but also Indigenous and traditional livelihoods – including reindeer pastoralism. A multitude of adaptations to these changes is required at several fronts simultaneously – ecologically, socially and economically. At the same time, the reindeer pastoralism strives to maintain its cultural identity.

This complexity is not fully appreciated in science and policy. Too often, research within the field has been fragmented, discipline-oriented or site-specific. Likewise, among policy makers, economically important sectors are often given priority relative to immaterial but vital aspects of reindeer pastoralism such as cultural identity, ancestral heritage or attachment to places.

The Nordic Research Centre of Excellence 'Reindeer husbandry in a Globalizing North – resilience, adaptations and pathways for actions', ReiGN, funded by NordForsk, integrated perspectives from natural and social sciences to overcome this segregation. The aim of the project was to create a holistic understanding of drivers connected to globalization and climate change that affect reindeer pastoralism in Fennoscandia, and how these drivers are linked to ecological, social and political differences between Norway, Sweden and Finland.

During the funding period (2016–2020), our work in these countries resulted not only in new scientific knowledge on the challenges faced by reindeer pastoralism, but it also helped us to increase collaboration with the reindeer herding communities.

Collaborating across countries, and involving reindeer herders and researchers, senior scientists with long experience and early career scientists, our ambition in ReiGN was to evaluate the adaptive capacity of reindeer pastoralism in a natural and social environment that is experiencing the consequences of climate change and globalization, and how these challenges can best be met. The multisite and multidiscipline centre spans 13 universities and research institutions in the 3 Fennoscandian countries, as well as many international collaborators.

During the process we began to realize that knowledge gaps existed about aspects in the respective other countries that had not been the focus of our previous work. Yet, a holistic understanding – across scientific disciplines and across country borders – is essential to fully grasp the challenges towards reindeer pastoralism as a livelihood and social-ecological system. The diversity of reindeer pastoralism at the local and regional level across Fennoscandia regarding the natural environment, herding practices, traditions and history, as well as the shared challenges and ambitions, inspired us to write this book.

This work would not have been possible without the commitment and patience of all involved authors – chapters kept being changed and updated in response to each other as the book progressed. We therefore would like to express our gratitude to all contributors for realizing this project, including the Vice President of the Saami Council, a Sámi non-governmental organization with member organizations in Finland, Norway, Russia and Sweden for her insights into urgent policy tasks in particular on Sámi matters.

Out thanks also extend to the many reviewers, internal as well as external, that helped us to sharpen the chapters' arguments and clarity. Importantly, we are indebted to the reindeer herders who have collaborated with us during the project – by sharing their knowledge and experiences of their livelihood, and their hopes and concerns for the future.

Furthermore, we thank NordForsk and the Nordic Council of Ministers for funding such an ambitious project as ReiGN and supporting us during the course of our work. We further appreciate the collaboration with the three other centres founded by NordForsk within the Arctic umbrella – ArcPath, CLINF and REXSAC. Some authors in this book work within one or both of the latter two centres. Indeed, this collaboration has contributed to added Nordic values. To strengthen the position of Arctic peoples' involvement in decision making and policy making, as well as ensuring integrity of intact ecosystems for the delivery of ecosystem services, including cultural services, is a major keystone for sustainable development of Northern regions.

We believe that our project has contributed with scientific insights to address these challenges.

Lastly, we would like to express our hope that the present book will make an impact not only within the scientific community but also within policy making for increased well-being of people and reindeer in the North.

Introduction

This book offers a holistic understanding of the environmental and societal challenges that affect reindeer pastoralism in Fennoscandia today. These challenges may be external and/or internal to the livelihood, spanning from the animal itself, to the herders and their cultural norms, to interactions with surrounding society, including resource exploitation on traditional grazing lands. All chapters have a strong forward-looking perspective towards the changes taking place in Northern Fennoscandia.

Addressing these present-day challenges to reindeer pastoralism in Fennoscandia, this book consists of five parts. Each part is devoted to a particular component of the social-ecological system dynamics inherent in reindeer pastoralism. Originating from different disciplines, the chapters reflect the methodological heterogeneity and different epistemological backgrounds and traditions of the different disciplines involved in the book. However, the dynamics described in these five sections are interdependent, and clear linkages between different chapters exist.

Part I introduces **reindeer pastoralism as social-ecological system** and sets the stage for the book as a whole.

In Chapter 1, **'Reindeer pastoralism in Fennoscandia'**, we define important concepts and describe the natural and social environment of reindeer pastoralism. This includes a short presentation of the herding year with season-specific key events. Further, we provide a historical overview of how reindeer pastoralism developed in Finland, Norway and Sweden until the present. Presenting maps and statistics of the present situation, this chapter provides a basis for all the following chapters.

Chapter 2, **'Genetic structure and origin of semi-domesticated reindeer'**, analyses the effects of the transition from hunting to herding on genetic variation and structure in reindeer. Differences in the genetic structure of semi-domesticated reindeer between the countries are substantial and necessary to understand when and how domestication occurred, as well as what implications it has for future adaptation to a changing environment.

Part II revolves around **reindeer in their environment**. Reindeer affect and are affected by a multitude of ecological interactions in their arctic and

DOI: 10.4324/9781003118565-1

boreal habitats, with considerable consequences for the herders' herding practices and interaction with the surrounding society.

Chapter 3, **'Reindeer behavioural ecology and use of pastures in pastoral livelihoods'**, focuses on habitat selection and foraging behaviour of reindeer during the different seasons. Different migration patterns throughout Norway, Sweden and Finland that determine the use of pastures during different seasons are presented. These differences also affect the interaction with other forms of land use.

These different forms of land use and their effects on grazing resources of reindeer are reviewed in Chapter 4, **'Pastures under pressure: effects of other land users and the environment'**. By competing for space, industrial resource developments affect where, when and how the landscape can be used for reindeer grazing. These cumulative impacts can reduce pastures directly or indirectly by either increasing landscape fragmentation or changing reindeer behaviour. By comparing the different land use pressures and environmental drivers across the countries, we analyse the consequences of reduced pasture availability on reindeer pastoralism. Strategies for pasture restoration are outlined, as well as we emphasize the need to include the herders' traditional knowledge in land use planning for coexistence.

Such a coexistence is increasingly challenged by climate change. Therefore, Chapter 5 is devoted to **'Reindeer husbandry and climate change: challenges for adaptation'**. As the consequences of climate change affect the natural environment, herding practices need to be changed and adapted to these new conditions. However, adaptations to climate change also have an inherent socio-political dimension regarding the interaction between different actors in society, which are contextualized in the chapter.

Finally, the conservation and return of large predators is in many aspects a success story in the Fennoscandian countries. However, an increased number of predators cause losses for reindeer pastoralism. Chapter 6, **'Large predators and their impact on reindeer husbandry'**, compares the abundance of predators, the losses they cause to reindeer pastoralism and management strategies of large predators, including different economic compensation schemes for reindeer losses between the countries. Furthermore, the challenge of knowledge integration between reindeer herders and state authorities with regard to predators is addressed.

Part III connects to the **governance of reindeer pastoralism** – how the livelihood with cultural embedded traditions and aspirations interacts with the nation states and what institutional frictions can occur.

The reindeer herders' own customary institutions, including laws, norms, rights and traditional knowledge are explored in Chapter 7, **'Implications of norms and knowledge in customary reindeer herding units for resource governance'**. We analyse how reindeer herders' customary institutions are integrated into state governance of natural resources or recognized in national legislation. Central to the chapter is the Sámi *siida* and the corresponding Finnish *tokkakunta* – both represent customary herding groups that seek to balance the relationship between human-reindeer units to the spatial and temporal availability of grazing resources.

Chapter 8, **'Unpacking reindeer husbandry governance in Sweden, Norway and Finland: a political discursive perspective'** dissects for each country how governance of reindeer pastoralism by state policies might conflict with the herders' livelihood or rights to land. Where such governance challenges persist, they limit the capacity to identify and agree on common solutions and visions to ensure a sustainable reindeer pastoralism.

One example of such a major challenge of reindeer pastoralism is addressed in Chapter 9, **'Governing maximum reindeer numbers in Fennoscandia'**. Affecting the state of pastures, being a social marker and being influenced by weather and predator abundance, governance of reindeer herd size is a complex task for herders and state authorities alike. Governing systems in setting an upper limit of reindeer numbers therefore need to consider whose knowledge is used, and whether and how much and in what phases herders can influence the process.

Part IV progresses from questions of reindeer herd sizes to **challenges for productivity, health and adaption of reindeer**. Herd size is one factor that relates to the productivity of reindeer pastoralism. Chapter 10, **'The productive herd: past, present and perspectives'**, investigates how reindeer pastoralism developed from the early livelihood to today's rationalized meat production. This includes effects on productivity, herd structure, slaughter strategies that can differ between counties. The chapter also addresses challenges and strategies for maintaining viable and productive herds in climate change context.

Effects of herd productivity have implications for the herders' economy. These are explored in Chapter 11, **'Bioeconomics of reindeer husbandry in Fennoscandia'**. By applying bio-economic optimization models, this chapter approaches economically reasonable adaptations and herding practises, such as optimal herd structure and slaughtering strategy – all dependent on environmental conditions that vary between and within Fennoscandian countries. Furthermore, the chapter shortly reviews economic incentives, pasture conditions and government regulation in Nordic countries.

As an important factor both for herd productivity and bioeconomics, providing reindeer with supplementary feeding in times of forage shortages is identified. The causes and consequences, benefits and disadvantages, as well as practices with this herding strategy are discussed in Chapter 12, **'Role of supplementary feeding in reindeer husbandry'**. Despite being more common in Finland, the practice of supplementary feeding has increased in the last decades also in the other countries for several reasons. This may cause concern for the herders, e.g., for economic reasons, threats to traditional knowledge or traditional practices, as well as on animal health and welfare.

Therefore, Chapter 13 investigates **'Health and diseases of semi-domesticated reindeer in a climate change perspective'**, in relation to supplementary feeding and to other factors, such as the role of insects and other arthropods as vectors for pathogens and other disease outbreaks. Their role is of particular concern in a predicted warmer climate in the Arctic and sub-Arctic regions.

Finally, **Part V** concludes with **prospects and synthesis** based on the results presented in previous chapters.

Departing from a perspective of system dynamics and regime shifts, that is, interactions that create interconnected patterns at higher and lower levels of organization, Chapter 14 investigates **'Tipping points and regime shifts in reindeer husbandry: a systems approach'**. Based on this framework, the chapter assesses the future risk of reindeer pastoralism to cross a tipping point, beyond which the current system changes considerably due to new feedbacks. Such a risk will hinge to a large degree on the continued loss of natural pastures, conflicts with other land users and the effects of climate change. The chapter concludes that system qualities that provide resilience need to be strengthened, including structural, institutional and legislative changes.

These necessary changes are also emphasized in Chapter 15, presenting **'Pathways for action: the need for Sámi self-determination'**. The vice president of the Saami council shows why reindeer herders' rights to participate in decision making and their self-determination need to be strengthened for full and effective participation in decision-making processes. This includes the embracement of Indigenous knowledge in these decisions, as well as a creative engagement between Indigenous knowledge holders and academia to create the best available knowledge to address the multifaceted challenges for reindeer pastoralism.

Closing the book, Chapter 16 will end with some **'Final reflections'** on the future of reindeer pastoralism in Fennoscandia, especially in relation to national and international policies and strategies to reach a sustainable future.

Part I

Reindeer pastoralism as social-ecological system

1 Reindeer pastoralism in Fennoscandia

Øystein Holand, Tim Horstkotte, Jouko Kumpula and Jon Moen

Reindeer herders – the pastoralists of the north

Pastoralism is a livelihood where animals are reared for their products for consumption and trade. It is a dominating land use where harsh and unpredictable environmental conditions do not allow productive crop cultivation. Seasonal patterns of forage and water availability therefore often require movements of people and their herds across the landscape. This mobility across ecological gradients makes pastoralism a livelihood with significant social, cultural, economic and political ramifications.

Pastoralism is found in all parts of the globe. Several hundred million people (the exact estimate is uncertain), particularly in developing countries, are dependent on this form of livelihood (McGahey et al. 2012). Contrastingly, pastoralism in western societies is marginal from a food production and economic perspective. Competition from other land users, often the economic stronger competitors, limits the pastoralists' physical and adaptive space. Yet, pastoral systems still exist in several western countries, such as in the European Alps, central Spain and Northern Europe. Fennoscandian reindeer pastoralism is today one of the most viable forms of pastoralism in the western world.

The Association of World Reindeer Herders is a representative organ for 24 ethnic groups who practice reindeer pastoralism in the northern taiga and tundra of the Eurasian continent – from Norway to Chukotka in the Russian Far East. Also in northern America, one Indigenous group in Alaska and one group in Canada have adopted reindeer pastoralism as part of their livelihood.

Spanning about 3,700,000 km^2, reindeer pastoralism outcompetes all other pastoral systems areawise. Today, the shared life between people and reindeer encompasses a gradient of interdependence: from subsistence hunting-based societies where reindeer serve as means of transportation to large, more or less free-ranging, herds for market-based meat production and to sedentary farming. The Association of World Reindeer Herders (2021) estimates that around 100,000 people, including herders and their families, are engaged in this production system. In many households, reindeer pastoralism is one part of a multifaceted livelihood. The Eurasian winter population of semi-domesticated reindeer (see Box 1.1) is estimated at around 2.3 million. In Russia, most of

DOI: 10.4324/9781003118565-3

the about 1.6 million semi-domesticated reindeer are found east of the Ural, centred in and around Yamal Peninsula.

In Fennoscandia, excluding the Kola Peninsula, the size of the winter population at present is around 650,000 animals, rather equally distributed between Finland, Norway and Sweden. The reindeer herding area covers around 35%–50% of each countries' land area (Figure 1.1), mainly practised in the northern

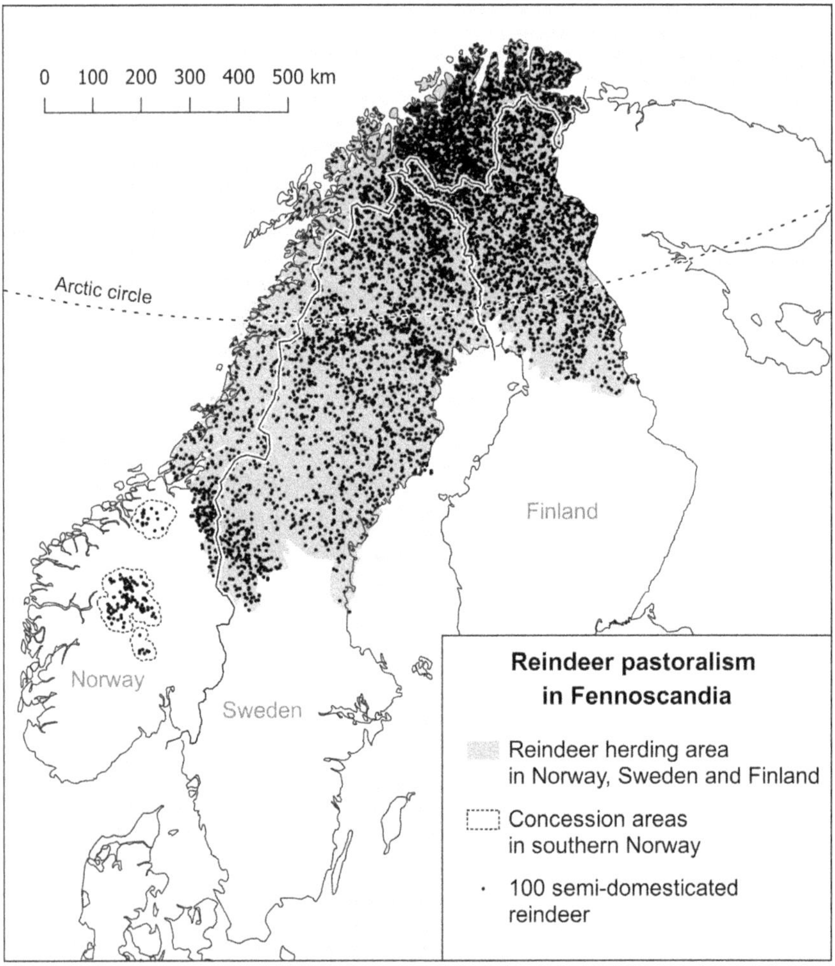

Figure 1.1 The reindeer herding area in Finland, Norway and Sweden covers approximately 500,000 km², today stocked with a winter population of around 650,000 animals. The size of the reindeer population within each herding district is represented by randomly distributed points derived from the average population size in each herding district between the years 2010 and 2020.

Source: Data from Landbruksdirektoratet (Norway), Sametinget (Sweden) and Paliskuntain yhdistys (Finland).

alpine and northern boreal regions. On the Kola Peninsula, Sámi, Komi and ethnic Russians are engaged in reindeer pastoralism, encompassing around 80,000 km^2 with around 60,000 animals.

Reindeer pastoralism is integrated with the countries' market economies. The annual meat production from Finland, Norway and Sweden combined at present is around 5,000 tons, mainly for domestic consumption. Not included in these numbers is private consumption of meat and by-products. In a national economy context, reindeer meat production is of minor importance. However, reindeer pastoralism provides an array of ecological services and is of social and cultural importance. Most notably, reindeer pastoralism in Finland, Norway and Sweden is a cultural keystone for the Indigenous Sámi. In Finland and locally also in Norway and Sweden, it is tightly connected to local non-Sámi people livelihood, culture and way of living.

It is estimated that about 3,000 persons have their primary income from reindeer pastoralism in Finland, Norway and Sweden (Vistnes et al. 2009). However, the number of reindeer owners, that is, having their own earmark that documents animal ownership, exceeds 12,000 people, as many family members are associated with the livelihood, even if their main income comes from other sources.

Box 1.1 Concepts and definitions

Throughout the book, we use the following key concepts that are essential to understand reindeer pastoralism as a social-ecological production system.

Reindeer husbandry

Reindeer husbandry, if not specified otherwise in particular chapters, is a general term to describe the livelihood of reindeer pastoralism. It encompasses the social-ecological relation between people, animals and the natural environment, as well as the economic and cultural dimension inherent in the livelihood. Political and legal regulations are important societal components. In this book, reindeer husbandry and reindeer pastoralism are used interchangeably.

Reindeer herding

Reindeer herding, if not specified otherwise, is defined as the practical work with the herd or individual animals to secure their well-being. This includes, e.g., migration between seasonal grazing grounds or directed movements between different grazing sites, gathering or separation of several herds based on ownership or herding groups, calf marking, slaughter and guarding the herd against disturbances, such as predators.

Semi-domesticated reindeer

The interactions between herders and reindeer, their frequency and intensity have been changing throughout the history of reindeer pastoralism. Even today, it varies between and within countries. A diverse terminology therefore exists to describe the degree of reindeer domestication. We define Fennoscandian reindeer in a pastoral relationship with humans as "*semi-domesticated*", to distinguish their less pronounced degree of domestication from other livestock, which often live in a stronger controlled environment. This terminology is a qualitative description, but no scientific definition of the term has been established. Other authors may use the term "domesticated reindeer", arguing that the degree of domestication is sufficiently advanced as it needs to be for the herders.

Reindeer herding area (RHA)

The *RHA* includes all the areas where the right to herd reindeer exists. This can refer either to the area of all countries combined or to a particular country. Based on the right to pursue reindeer herding, the specific extent of this geographical area can be disputed and change over time. In some areas, the RHA can extend beyond national borders.

Herding district

The spatial, ecological, historical, legal and economic attributes of distinct units where reindeer herding is practiced differ between the countries, as well as the terminology to describe them. This diversity is summarized as a *herding district*, irrespective of the country. Although the terminology implies a geographical unit, the social dynamics within the district, with neighbouring districts and the surrounding society, are equally important.

Siida

A *siida* (Northern Sámi) or *sijdda* (Lule and Pite Sámi) or *sijte* (South Sámi) is a customary group that shares the work of reindeer herding within a designated area, working together for the benefit of its members. *Siida* is normally used in the text as a common denominator. The corresponding Finnish term, used for both Finnish and Sámi herding units, is *tokkakunta*.

The reindeer pastoralists' landscape

From the reindeer pastoralists' perspective, the landscape is characterized by the dynamic relationship between the pasture resources, the herd, other pastoralists and themselves (Figure 1.2). The balance in this pastoral system is kept by

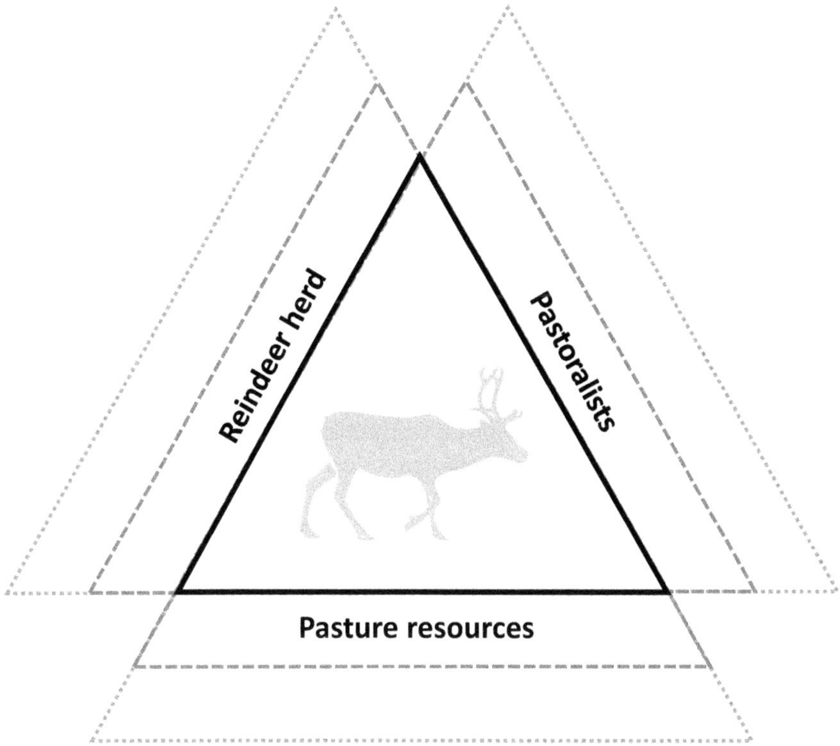

Figure 1.2 A conceptual framework of the dynamic interaction and self-adapting relationship between the herd, the herders and the pasture (inspired by Ingold 1980 and Paine 1994). The interaction between these varies over time, depending on their relative size as indicated by the grey lines.

adjusting herd size and structure, dependent on production goals, to the pasture and labour resources available. Reindeer pastoralism is both a competitive and co-operative endeavour, and the composition and size of herds, as well as of the size of collaborating herding groups, will vary seasonally based on available pasture resources and social setting (Chapter 7).

The environment is not only a bio-physical but also a cultural landscape shaped by herders and their herds, where both play a role in assessing range quality and risks (Ingold 1980). The pastoralists' perception of their landscape is described by the Sámi artist Nils-Aslak Valkeapää in his poetry collection: *Beaivi, áhčážan* (*The sun, my father*, 1997):

the land
is different
when you have lived there,
wandered

sweated,
frozen,
seen the sun
set, rise
disappear, return

the land is different
when you know
here are
roots,
ancestors

Climate and physical geography

Natural and climatic conditions differ between and within the three Fennoscandian countries. Geological, latitudinal, longitudinal, altitudinal and topographic gradients shape pasture conditions and resources accessible to reindeer.

Topography in Fennoscandia is diverse. Along the western coast of Norway, the terrain is rugged, climbing into the Scandinavian mountain range. East of the mountain range, the landscape falls gently with undulating moraine hills with lakes, streams and peatlands. The short time span, in geological terms, since the last glaciation and the cold climate, has contributed to slow soil formation.

Climate in Fennoscandia is strongly influenced by the Gulf Stream, moving enormous amounts of heat into the North Atlantic. The North Atlantic Oscillation (NAO), a climatic pattern dependent on atmospheric pressure across the North Atlantic, influences winter climate in Northern Europe: in its positive phase, winters are wet and warm due to westerly winds across the Atlantic Ocean, whereas a negative NAO results in cold and dry air transported into Fennoscandia from the east. By influencing snow conditions and thus winter grazing conditions for reindeer, the NAO during winter affects reindeer population dynamics in Fennoscandia (Helle & Kojola 2006). The NAO also influences to a lesser degree the summer climate in northwest Europe, where a positive NAO during summer leads to warm and dry summers (Folland et al. 2009).

The Scandinavian mountain range influences patterns of precipitation and temperature, and thus the gradient between oceanity and continentality in Fennoscandia (Seppälä 2005). The heaviest precipitation falls on the western coastal mountain slopes in Norway, with decreasing amount in the eastern rain shadow in Sweden and Finland. Here, an interior climate dominates with relatively warm summers and cold and dry winters, as compared to the coastal climate with mild and wet winters and cool and moist summers. These climatic gradients form the basis for the seasonal migration pattern characteristic for a large part of Fennoscandian reindeer pastoralism (see Figure 1.4).

The temperature is highly influenced by altitude and latitude as well as oceanity. Accordingly, the length of the growing season spans between 80 and

160 days in northern Fennoscandia (Karlsen et al. 2008). At the local level, the vegetation period depends on topography and its influences on snow accumulation in depressions, or snow-free patches on exposed ridges. A diverse topography therefore contributes to a pasture mosaic that enables reindeer to select favourable vegetation patches and types on fine spatial and temporal levels.

Summer is a period of rapid vegetation growth due to the long availability of daylight, increasing towards northern latitudes. The almost unlimited photosynthetic activity generates highly digestible plants that reindeer forage on intensively throughout the arctic summer. Contrastingly, winters are long with snow cover lasting approximately half the year in the north and in the mountains. However, ongoing climate change is affecting the seasonality in Fennoscandia, decreasing winter duration but with frequent extreme events, with important consequences for reindeer pastoralism (Chapter 5).

The alpine region of Fennoscandia covers approximately 250,000 km^2, most of it within the reindeer herding area (Seppälä 2005, Figure 1.3). The subalpine zone encompasses belts of mountain birch (*Betula pubescens ssp. czerepanovii*), whereas the alpine vegetation at lower altitudes is dominated by shrubs, dwarf birch (*Betula nana*), willows (*Salix ssp.*), meadows with a varied species composition of forbs and graminoids, as well as bogs (Moen 1999). Alpine vegetation at high altitudes is sparse, and boulders, bare rocks and glaciers dominate.

In Sweden and especially in Finland, the reindeer herding area is largely found within the boreal region (Figure 1.3), dominated by Norway spruce (*Picea abies*) on moist sites and Scots pine (*Pinus sylvestris*) on dry and oligotrophic heaths. The field layer is mainly composed of ericaceous shrubs (*Vaccinium ssp.*), with a ground layer of mosses on moist sites and lichens on dry sites (Esseen et al. 1997). Productive riparian zones along streams, lakes and mires can be important grazing areas in mountainous and forest habitats.

The primary production is low and declines with increasing altitude and latitude. In Utsjoki, northernmost Finland, the biomass production of the vascular ground layer in alpine birch forest is estimated at 110 g dry mass per m^2 per year on average (Kjelvik & Kärenlampi 1975). In the alpine region, the yearly above-ground production is even lower (Wielgolaski 1975). However, production varies greatly depending on vegetation type and growing conditions. In the interior alpine and boreal regions, well developed lichen-rich types may reach lichen biomass of over 1,000g dry mass per m^2 (Kumpula et al. 2014).

The reindeer

Reindeer and the North American caribou (*Rangifer tarandus* spp.) are grouped into three major ecotypes: the tundra/barren ground ecotype with extensive migration between winter and summer habitat, the forest/woodland ecotype that spends all year in the boreal forest and the high arctic type. The total population of wild *Rangifer* is estimated at ca. 2.9 million (Gunn 2016) and covers the largest range of any large terrestrial mammal. Only remnants of wild reindeer populations are found in Fennoscandia. European tundra reindeer (*R.t.*

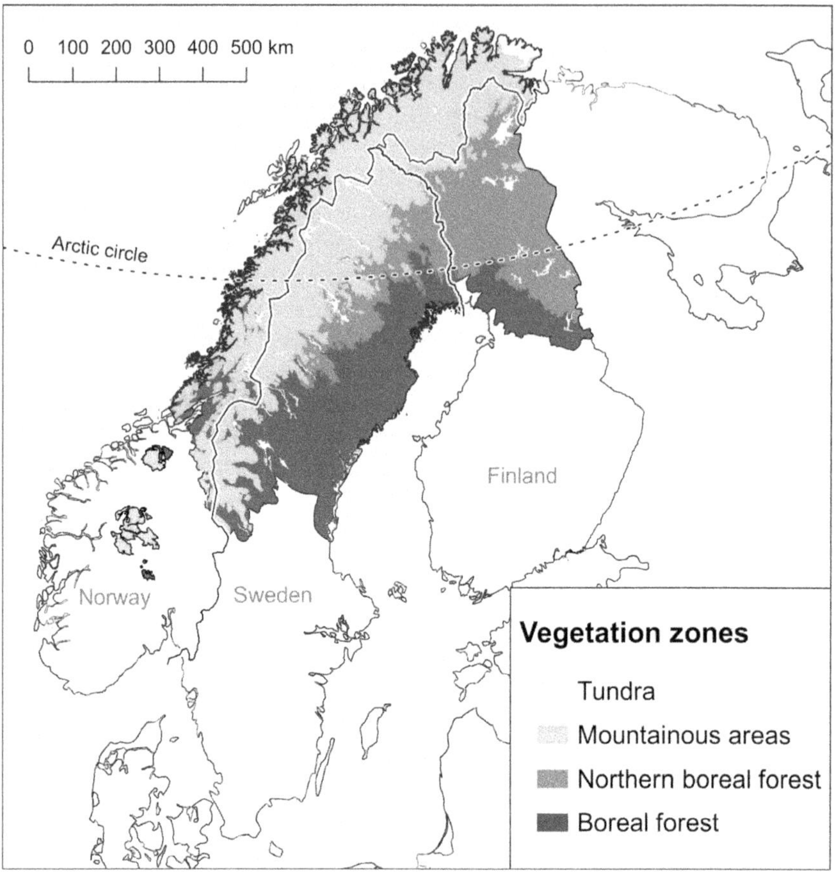

Figure 1.3 The vegetation zonation in Finland, Norway and Sweden where reindeer pastoralism is practised. Major lakes are shown in white. Vegetation zones adapted after Dinerstein et al. (2017) and Moen (1999).

tarandus) are found in small pockets in south-central Norway, estimated at around 35,000 animals (Gunn 2016). Their viability rests on a strict management regime to protect their ranges and to control the populations by hunting regulations. Indeed, Norway has acknowledged its special responsibility to safeguard the European wild tundra reindeer. Approximately 2,300 forest reindeer (*R.t. fennicus*) are found in eastern and middle parts of Finland as estimated by the Natural Resources Institute of Finland in 2018. They are a high-profile species and actions are taken to protect their sub-populations and to prevent crossbreeding with semi-domesticated reindeer.

Rangifer has been described as an Ice Age relict due to morphological and behavioural adaptations to cold climate and snow. Outstanding morphological adaptations to snow are broad flexible hooves that provide a large surface in relation to body weight and reduce energy expenditures when moving on snow (Telfer & Kesall 1984). Sharp edges of the hooves serve as an optimal tool for digging through snow to reach the forage beneath. The dense fur consisting of hollow and stiff hair provides optimal insulation (Geist 1999). In comparison to summer, the basal metabolic rate in winter is 20%–30% lower in non-activity functions (Russel & Martell 1984).

Their highly gregarious behaviour is a response to their coevolution with predators and pests, especially the wolf and parasitic insects (e.g., *Oestridae*) (Geist 1999). Behavioural adaptations include trail-following behaviour to save energy, e.g., during spring migration when the snow becomes soft. Communication is well developed, such as alarms or threats and display signals. Individual recognition by smell and sound is important for mother–young communication (Espmark 1971).

Reindeer are highly sexual dimorphic, with adult males weighing up to two-fold of females (Geist 1999), suggesting a sexual selected species with a highly polygynous mating system (Holand et al. 2019). It is the only deer species where both sexes wear antlers. These are functional weaponry for males during competition for females and probably an honest signal of quality. The females' weaponry is primarily used for defending winter food resources when males have dropped their antlers after rutting in late autumn.

Reindeer are seasonal breeders with a rut spanning from end-September to late-October, depending on latitude as well as body condition. This leads to a synchronized parturition in May after a gestation period of ca. 220 days (Geist 1999).

The short summer season implies selection for early parturition and transfer of reserves to offspring during early lactation, important for their calves' winter survival and build-up of own reserves, typical for capital breeders. The females' risk-aversive life history strategy is essential to understand *Rangifer* population dynamic (Bårdsen et al. 2008).

Reindeer are classified as an intermediate forager (Hofmann 1989). The varied summer diet is composed of forbs, sedges, grasses and shrubs. Undisturbed, they are able to realize their selective grazing potential, alternating between grazing, while steadily moving and resting/ruminating. High nutrient acquisition leads to rapid body growth and accumulation of body reserves.

The winter diet in many areas consists largely of terrestrial lichens (*Cladonia sp.*), however, also dwarf shrubs and grasses/sedges may contribute substantially to their winter diet (Åhman & White 2019). Lichens are rich in digestible carbohydrates but low in protein. Hence lichens are well-suited forage for supporting winter survival. Slow growth of lichens, and consequently a slow regeneration, may lead to heavy grazing pressures at patches with suitable snow conditions, especially at high reindeer densities. Therefore, the *Rangifer*–lichens interaction is a key for understanding the species' winter ecology and population dynamics and fluctuations (Skogland 1990). The amplitude and wavelength of the population cycles will vary according to lichen distributions, accessibilities and summer growth conditions, as well as stochastic weather events.

Many scholars have described reared reindeer as semi-domesticated (Clutton-Brock 1999, see Box 1.1). This refers to a rather short domestication process but also to specific needs placed on the reindeer–herder relationship compared to most other livestock species (Chapter 2). Indeed, the ecology, life history and behaviour resemble their wild ancestors and they are reproductively compatible with their wild counterparts. This underlines the herders' conscious selection for strong, free-ranging but controllable herds (Chapter 10). The highly gregarious behaviour of the ancestors was a prerequisite for the domestication process and has been further selected for by the herders. Therefore, anxious and straying animals are deliberately culled.

The herders

When the relationship between people and reindeer changed from hunting to herding, protection of animals from predators and securing the herd's well-being became important and required cooperation between herders. The Sámi *sijte/sijdda/siida* (South-Sámi/Lule and Pite Sámi/North-Sámi), earlier based on the hunting and fishing livelihood, evolved as a social network to distribute the herding work often between close relatives, based on mutual agreements (Sara 2009). Its main task was to balance the relationship between herders and herd size according to pasture resources and to protect their range and grazing rights (Figure 1.2). This means mutual agreements with and respect of other *siidas'* territory (Chapter 7). Other organization regimes evolved within non-Sámi reindeer pastoralism, today spanning from community-based herding districts as seen in Finland to private enterprises with hired professional herders in some of the Norwegian reindeer herding companies outside the Sámi herding area.

The herders' social-ecological knowledge and mutual understanding between them and their herd is the essence of their livelihood and is transferred between generations through practise. A well-tuned balance between pasture resources and herd size generates a calm herd easy to control. This underlines the dynamic self-regulation of the pastoral relationships (Figure 1.2). If the herd increased beyond its pasture capacity, the range had to expand. Otherwise, food shortage will downregulate herd size. The herders' traditional ambition for herd accumulation therefore often materialized in fluctuations of herd sizes (Ingold 1980).

Main forms of reindeer pastoralism today

The degree to which reindeer migrate between their seasonal grazing grounds, the distances covered and means of migration or transportation vary significantly throughout Fennoscandia (Chapter 3). Three main forms can be distinguished, depending on the biogeography of the pastures, as well as on historical roots and traditions (Riseth et al. 2019; Figure 1.4).

The dominant form in Norway and Sweden is characterized by seasonal migrations between summer grazing grounds in the mountains or at the Atlantic

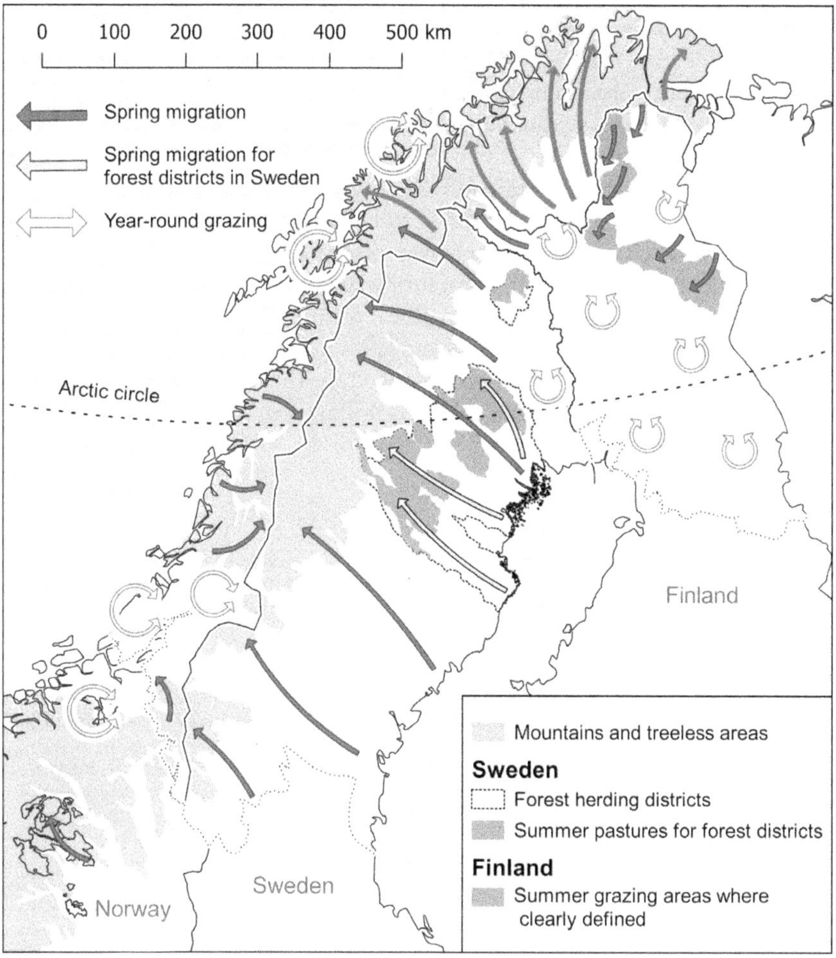

Figure 1.4 The three main forms of seasonal pasture use in Fennoscandia reflect the biogeographical variation and zonation constrained by historical and political events.

coast and winter grazing grounds in the drier and more continental inland. A second, but limited form in the middle part of Norway utilizes the Atlantic coast for winter grazing due to snow-free winters, with short migrations into summer pastures further inland. Some herds spend the whole year on islands and peninsulas. A third form remains year-round in the boreal forest, primarily in Finland and parts of Sweden. Seasonal migrations are limited, as this form is confined to relatively small grazing areas, especially in the southern parts of reindeer herding area in Finland.

The yearly cycle

The pronounced seasonality in Fennoscandia, and its effects on reindeer ecology, divide the herding year into eight seasons (Figure 1.5). However, the timing of particular events strongly depends on variation in, e.g., weather conditions between and within years. We can therefore only describe the major generalized characteristics of the yearly cycle, emphasizing the animals' ecology and the herders' main operations of the migratory form today (Figure 1.5). Similar strategies are found in the forest and coastal form, but covering smaller distances. Climate change increasingly disrupts the seasonal pattern (Chapter 5).

In late winter (March–April), migration starts towards the calving grounds, often following rivers or other topographical features. To avoid soft and energy-demanding snow conditions, migration takes place preferably during early morning on ice-crusted snow. During the days, the reindeer rest and graze on emerging snow-free patches. Herders follow on snowmobiles and may

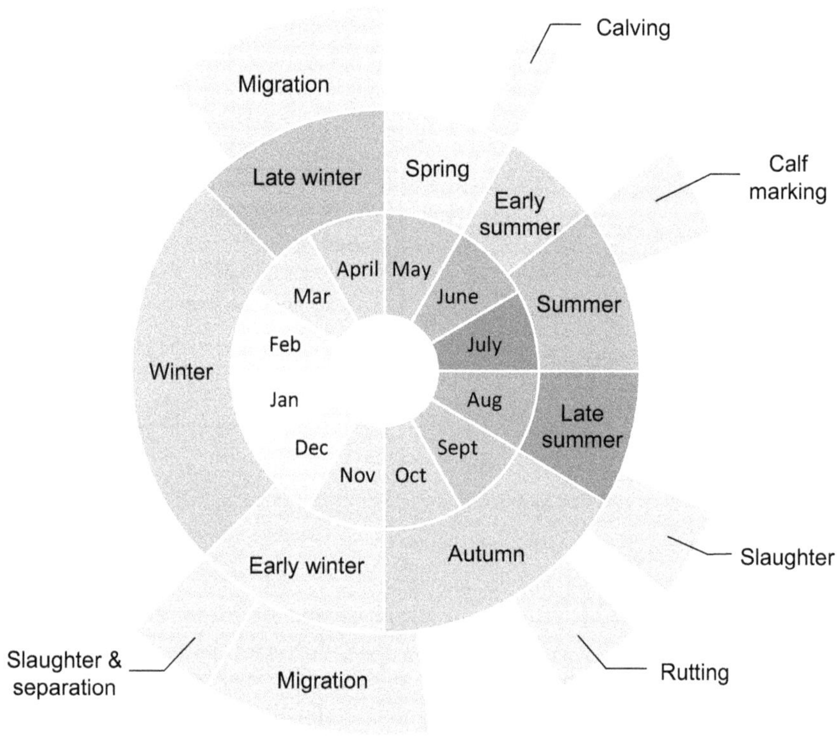

Figure 1.5 The herding year reflects the animals' life history, behaviour and spatial utilization of the land and the herders' main operations. This generalized pattern may vary between countries, as well as between years, mainly due to weather variation.

use supplementary feeding to guide the herd and keep it gathered. Animals in bad condition are moved separately or transported by trucks. Where migration routes have been destroyed, e.g., by hydropower development or other infrastructure (Chapter 4), reindeer are transported by trucks. At the coast of Finnmark, some herds used to swim to their calving grounds on islands, but today they are mostly transported by ferries.

Upon arrival at the calving grounds in early spring (late April to early May), pregnant females disperse for calving. The herders avoid disturbing the animals but protect the herd from predators and other disturbances. In areas with high predation pressure, some herding groups have started to gather the pregnant females into calving enclosures to minimize losses. After calving, the females and their calves congregate into fostering herds.

In the early summer (June), the herd follows and selectively feeds on the emerging vegetation. Calf marking starts in late June. The animals are gathered in corrals, with individual marks cut into the animals' ears by their respective owner. The operations are executed carefully as all handling and stress weaken the calves. Herders may work at nights to relieve the animals from extra heat stress.

After calf marking, the animals are released back on the summer range. Blood-sucking insects may hamper the grazing activity and the animals seek relief on wind-exposed habitats or snow patches, an important component of a functional summer land. As a time of minimal herd surveillance, herders may allocate time to other activities to supplement their income, such as fishing, handicraft (*duodji*) and tourism.

Late summer (August) is characterized by grazing and reduced lactation burden for the females, leaving surplus for the build-up of body reserves for the winter, and in case of males, for the rut in autumn. The males form separate groups during the summer.

During autumn (September–October), migration to the early winter ranges starts. The animals are corralled for the traditional bull slaughter before the rutting season. Today many calves are trucked to slaughterhouses, as state subsidies favour this slaughter strategy (see Chapter 10). Mushrooms become important to build-up body reserves, and the animals may disperse widely for foraging. Their movements can become hard to control and herds belonging to different herding groups often mix. The rut starts in late September and lasts for a month.

After the arrival onto the early winter range in November, the herds are gathered. Snowmobiles are used for rounding up the herds. Advanced corral systems (Figure 1.6) ease the separation of mixed herds, selection of animals for slaughter and division into smaller winter herds. Reindeer may also receive medication and anti-parasite treatment.

Around Christmas, the animals enter the winter pastures. The herders divide their herding activity into shifts and survey the animals' grazing behaviour and the snow conditions carefully. Good winter pasture conditions with easy snow conditions induce a relaxed herd and the herders have time to maintain and

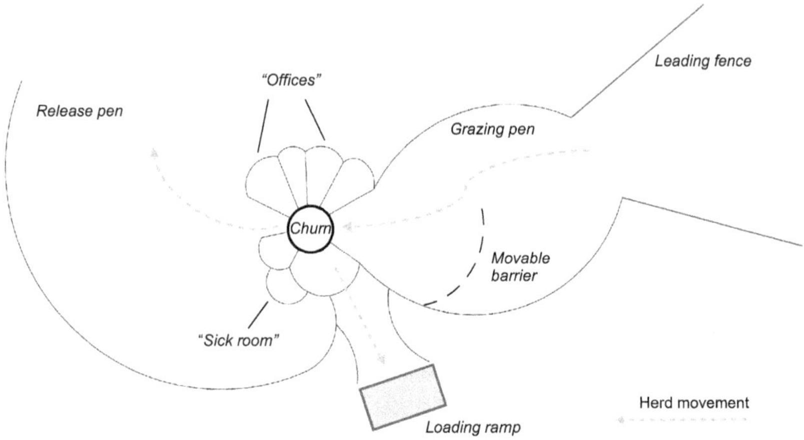

Figure 1.6 An effective round-up infrastructure eases the working operations and improves the animals' welfare. The illustration shows a combined separation and slaughter system. In the churn, animals are separated according to the herders' management plan: released back into the living herd or loaded for slaughter. The movable barrier, often of fabric, is used to separate small groups from the herd to guide them into the churn. Specific animals can be put into corrals for medical treatment or observations or into "offices", e.g., corrals for animals belonging to other herds.

repair their gears. Under difficult winter conditions, such as hard and deep snow, the herds may need to be split into smaller units with frequent movements according to local variations in grazing conditions. If reindeer cannot access their natural grazing resources due to ice-formation for instance, supplementary feeding becomes necessary (Chapter 12). Alternatively, the herds may be allowed to spread in search of suitable grazing on their own to prevent starvation, inducing mixing and extra separation, which entail a heavy workload for the herders and a high risk of losing animals, e.g., to predators. However, in many areas supplementary feeding is used to some extent in each winter for keeping reindeer gathered and maintaining their good body condition. Before starting their migration back to their calving area in late winter, herds are often gathered and separated into functional groups that are easier to control during migration.

This migratory pattern in large parts of Sweden and Norway may extend up to several hundred kilometres annually. Being labour intensive for the herders and energy consuming for the animals, it involves risks for both animals and herders. However, it is regarded as an optimal utilization of the seasonal variation of forage availability, generally enhancing the animals' and the herders' well-being.

Where reindeer stay in the boreal forest year-round, these seasonal migrations are less extensive. In the southern reindeer herding area of Finland, the limited size of districts and scattered lichen pastures have led to heavy grazing and trampling, reducing the terrestrial lichen biomass. Furthermore, the long-term forest harvesting in the commercial forest area which covers nearly two-thirds of the reindeer herding area of Finland and large parts of Swedish winter pastures has contributed to the loss of ground lichens and reduced the arboreal lichen resources (Chapter 4). Therefore, winter feeding in pens has become common in the southern part of the herding area in Finland (Chapter 12).

The coastal form reverses the migration pattern between summer and winter grazing areas. The animals stay along the Atlantic coast during winter on ranges poor in lichens, but normally with shallow snow. Here they survive on shrubs and graminoids. Animals are confined on isolated islands and peninsulas. Migration during spring towards the mountain areas needs continuous surveillance due to a high predation pressure (Chapter 6). Though short, these migrations may cover several biogeographical zones owing to the steep elevation gradients.

A brief history of reindeer pastoralism in Fennoscandia

Ecological, social, and political processes are path-dependent: past events affect present and future dynamics. Understanding the past may therefore offer insights to identify drivers and trends and prepare for future challenges.

Wild reindeer – an important prey for hunters

Wild reindeer have been a vital resource for humans since the Pleistocene period. Pitfall systems, strategically placed along migration routes, are found all over Fennoscandia (Myrvoll et al. 2011), and rock carvings document an advanced hunting culture (Figure 1.7).

Odner (1985) argues that the Sámi culture and identity was formed as their hunting, fishing and gathering culture met the settlers with their agricultural practices expanding northward in the second half of the first millennium BC. This mutual relationship developed and is reflected in the written records of the Norwegian chieftain Ottar during his visit to King Alfred the Great in England around 890 AD. Ottar's description is also the first written source of an early phase of domestication of reindeer by the Sámi and dates the use of tame reindeer for transportation and as decoys in hunting operations back to the Iron Age.

During the Medieval Period, northern Fennoscandia was colonized and influenced from several directions. This resulted in conflicts between the kingdoms of Norway-Denmark, Sweden (then including Finland) and the Novgorod Republic (conquered by pre-Tsarist Russia in 1478). Agreements between these emerging states were established in the 1320s, as they defined their areas of influence and overlapping interests. Northern Fennoscandia became a melting pot of cultures, languages and livelihoods, but embedded in

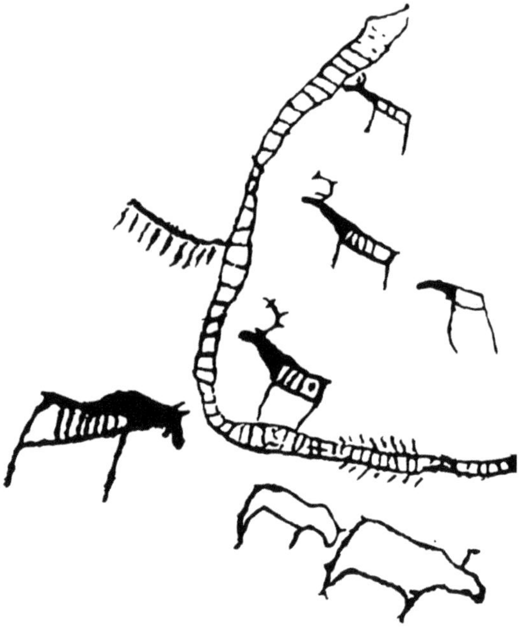

Figure 1.7 The Alta carvings in Northern Norway, spanning 6500–2000 BC depict an array of hunting and fishing scenes, among them well-organized reindeer hunting operations.

a growing colonial matrix. Sámi goods, including pelts, fish and transportation services, were highly priced and traded through the Hanseatic as well as the Novgorod networks. However, the Plague shattered the social structure and the states' central and local power bases in the mid-1300s. The Plague, particularly along the Norwegian coast, implied a setback for the agricultural settlements and the colonization of the north. This was amplified by worsening climatic conditions during the Little Ice Age, commencing in the 15th century (Nesje & Dahl 2003).

In the Kalmar Union (1397–1523), Denmark engulfed Norway into its political and economic sphere. Sweden, reigning Finland, eventually broke out of the Union in 1523 and started to challenge Denmark's/Norway's sovereignty along the North Atlantic coast (Hansen & Olsen 2014). The Sámi were heavily taxed, as the European market for fur, hides and pelts soared, to finance the states' expanding warfare. This induced decimation of wild reindeer populations and fur species (Wallerström 2000). Advanced pitfall systems using long fences with converging "arms" leading the animals into small enclosures contributed to their decimation (Hansen & Olsen 2014). This system was later refined by the Sámi as an essential infrastructure for handling their semi-domesticated herds (Fjellheim 2005).

Reindeer pastoralism evolves, expands and diversifies

In a primarily hunting-based economy, reindeer were a collective resource. In the gradually emerging social, cultural and economic transition to reindeer pastoralism, protection from predators, herd accumulation and privatization became important. In Sámi communities, social prestige and wealth became increasingly connected to ownership of large herds (Ingold 1980).

The transformation probably started in south-central Sámi areas in the boreal forest. Small herds were kept under close surveillance for transportation purposes during hunting and fishing expeditions. This practice spread to the north and east (Vorren 1973). It diversified into four main forms in the 1600s and early 1700s, depending on different natural and social conditions. At the north Atlantic coast, small-scale pastoralism in combination with farming and fishing remained stationary, while another form stayed for winter grazing at the coast, moving into the mountains for summer herding. The migratory form with long seasonal migrations between summer pastures in the mountains and winter grazing in the boreal lowlands predominated in Sweden and northernmost Norway. Lastly, forest Sámi practiced small-scale pastoralism with short seasonal migrations found in parts of the boreal zone in Sweden and Finland. Hunting, fishing and herding were an integrated part of a multifaceted livelihood (Bjørklund 2013). The forms partly overlapped spatially as well as temporally and were integrated ecologically, economically and socially. In Finland, settlers moved to the Sámi area in the 17th and 18th centuries and soon adopted the forest Sámi livelihoods, including reindeer herding, and further developed it for transportation and meat production. A custodial reindeer husbandry system developed between farmers (including stationary Sámi) and the herders, which was practised till the mid-1900s in many places (Nordin 2002; Evjen 2007). This reciprocal exchange diversified the involved livelihoods, and thus was economically beneficial. Further, it strengthened the social bonds between the pastoralists and the farmers.

During the 16th and 17th centuries, Sámi reindeer pastoralism grew in economic and cultural importance, thus strengthening the pastoralists' position. In Sweden, this was partly driven by new taxes placed on the Sámi in 1602. These taxes as a means to secure land ownership (*lappskatteland*) had previously been paid in meat and fish rather than in meat and furs. This increased pressure lead to a decrease in the wild reindeer population and an increase in the semi-domesticated herds (Lundmark 2006). In the 17th century, a clearer colonial agenda shaped the kingdoms' policy in northern Fennoscandia (Kuusela et al. 2020). The Settlement Bill of Lapland in 1673 stimulated settlers to move north by exemption from military service and taxes for 15 years. Mineral resources and metals, especially copper, iron and silver, were extracted on Sámi land on the so-called Crowns' land. The Great Northern War (1700–1721) was partly related to the increased exploitation and intensified overlapping taxation in some of the northern border areas.

In 1751, a formal border agreement between Denmark/Norway and Sweden/Finland was signed. It included a Codicil, granting the Sámi reindeer herders the right to cross borders during their seasonal migrations. This so-called Lapp Codicil also, for the first time, officially incorporated Sámi as part of the nations. As such, they were obliged to acquire citizenship, pay taxes and hold land in only one kingdom. This indicates that the states regarded migratory reindeer pastoralism as essential for "the Conservation of the Sámi Nation". This was followed up in 1766 by the Swedish Parliament which drew the so-called *Lappmarksgränsen* (Lapland border) to protect the herders' interests and land use from the expanding agricultural activities. In the Kalix and Torne valley, Swedish and Finnish farmers started to practice small-scale pastoralism in the forest (Åkerman 1990). In the southern Norwegian mountains, outside the Sámi area, reindeer pastoralism was also adapted by farmers in the late 1700s and expanded its range in the early 1800s (Bitustøyl & Mossing 2019).

The Sámi migratory pastoralism with large herds expanded towards the east into Northern Finland. It is estimated that in the early 1800s about 100,000 reindeer belonging to Sámi who paid taxes to the Swedish Kingdom had their summer pastures in Norway, whereas about 50,000 reindeer belonging to Sámi who paid taxes to the Kingdom of Denmark-Norway wintered in Northern Finland (Aarseth 1989). The stationary pastoralism in the boreal forests and along parts of the Atlantic coast diversified, where milking and transportation were key elements, often in combination with hunting, fishing and farming.

The Napoleonic Wars (1803–1815) induced a domino effect in Fennoscandia. Russia annexed Finland from Sweden in 1809, while Denmark surrendered Norway to Sweden in 1814. A deliberate colonization by the states followed to secure their territorial interests. Settlers were stimulated to cultivate the north. Conflicts between farmers and herders intensified, fuelled by the dramatic population growth. The herders were increasingly pushed back from their traditional grazing areas. Expanding wolf populations put the herders under extra pressure.

Border closures and regulations

In 1826, the border between Russia and Norway was settled, impeding the seasonal cross-border migration of the reindeer herding Skolt Sámi, who lived in the borderland between Norway, Russia and Finland (Niemi 2007). The Union between Norway and Sweden unilaterally issued a law to restrict reindeer grazing across the border. They also restricted the crossing between Norway and Finland, then a Grand Duchy under the Russian Empire (Strøm Bull 1997). Negotiations to reach an agreement regarding access to fishing grounds at the Norwegian coast for Finnish Sámi and access to winter grazing grounds in Finland for Norwegian Sámi failed. In 1852, Russia in agreement with Finland therefore closed the border for Norwegian herders. In 1859, the Finnish authorities wanted to renegotiate the cross-border migration and proposed to return to the practice based on the 1751 Codicil. However, Norway refused (Pedersen 2007). The unrest was amplified by population growth in the area and northward influx of farmers which both contributed to increased natural resource competition. A temporary cultivation boundary ("*Odlingsgräns*") was drawn in 1867 in northern Sweden and later finalized to shield the herders from further agricultural expansion (Lundmark 2006).

The herders adapted several strategies to cope with the border closure between Finland and Norway. Some herders in Finnmark and Northern Finland registered as Swedish citizens. This enabled them to access their traditional winter grazing areas in Finland and summer grazing areas in Norway, respectively, as migration between Norway and Finland via Sweden was still an option, as the Lapp Codicil of 1751 was still in force. The Russian/Finnish authorities regarded this as a circumvention and in 1889, they also closed the border between Finland and Sweden.

Following the turbulences in the traditional migratory pattern, the states intensified their institutional grip and governance of reindeer pastoralism, and partly dismantled the reindeer *siida* organization (Labba 2016). The Norwegian/Swedish Common Grazing Act of 1883 (excluding Finnmark in Norway) confirmed the Sámi pastoralists' rights to cross the borders. However, it enabled authorities to regulate the number of animals, divide the pastures into herding districts and enforce joint responsibility by the members of a herding district for reindeer damage on agricultural land (Strøm Bull 1997). This was immediately implemented by the Norwegian authorities in some regions to constrain the Swedish herders' activities. In 1888, a law for reindeer pastoralism in Finnmark was issued. The law regulated and partitioned the summer grazing areas, whereas the winter ranges were regarded as commons for the herders and no collective liability for damages on farmland was enforced (Strøm Bull 1997).

The blocking and restrictions of migration across borders led to an overcrowding of herders and reindeer in northernmost Sweden and northwestern Finland. In combination with several severe winters in the early 1890s, this resulted in strong declines in the reindeer population (Riseth et al. 2019). The Swedish regional authorities therefore enforced migrations of Northern Sámi families southwards along the Fennoscandian mountain chain (Aarseth 1989). These relocations created conflicts between and among Sámi reindeer herding communities, as their practices came into conflict with the local practices by southern Sámi, focused on milk and cheese production in closely tended, smaller herds. Both forms could not coexist, and the meat-producing livelihood became the dominant form even in the South (Lundmark 2006). The Swedish Sámi had to carry most of the burden for Norway's hostile policy towards the cross-border pastoralism.

In Finland, many Sámi families moved east, e.g., to the current Municipality of Sodankylä. They brought with them their herds and forms of herding, including an extensive milking practice that was adopted locally. Other locals, traditionally using reindeer for transportation, adopted an extensive form of herding largely focused on meat production in comparatively large herds left unattended for extended periods during the summer.

In Finland, settlers moved further north, and land use conflicts emerged between them and Sámi herders, as well as between Sámi and Finnish herders, amplified by the border closures (Kortesalmi 2008). Gradually settlers and Sámi mixed in many areas and a multi-ethnic culture emerged based on small-scale farming, reindeer herding, hunting and fishing. For this reason, reindeer pastoralism is practiced today not only by Sámi but also by ethnic Finns, many of them with Sámi roots. In 1898, the Finnish Senate ordered reindeer owners to establish geographically defined herding districts, based on the more sedentary herding practices associated with agriculture. Many Sámi herders claimed it unsuitable for their migratory way of life and customary rules, arguing that the system favoured farmers' livelihoods (Pennanen & Näkkäläjärvi 2002). The Sámi reindeer herders were still to some degree able to continue their *siida*-based pastoralism, however, operating still within the official herding districts (Chapter 7).

Social Darwinists claim Sámi pastoralism a backward way of life

In the late 1800s, Social Darwinism gained momentum. A patronizing attitude towards reindeer pastoralism was a characteristic driving element in Sweden, separating reindeer herders from modern society, which was seen as a threat to Sámi culture, also known as the "*Lapp shall remain Lapp*" ideology (Lantto 2000). The Norwegian assimilation policy was motivated by Norwegian nationalism (Minde 2003) and was deliberately used to weaken the Sámi pastoralist rights in favour of the farmers. Several Norwegian Supreme Court trials in the late 1800s resulted in loss of grazing land in the south (Fjellheim 1999). In Finland, the institutional and cultural assimilation was similar, but evolved in its own pace. The states' policies were also driven by an expanding natural resource extraction, such as for iron, copper and timber as part of the industrial revolution. Indeed, the industries profited from the marginalizing of the rights to land formerly held by reindeer pastoralists.

During the de-union negotiations in 1905 between Norway and Sweden, Norway aimed to terminate the Lapp Codicil but had to accept the immemorial grazing rights embedded in the Codicil. However, attempts to limit the Codicil are ongoing to this day. The first Norwegian-Swedish Reindeer Pasture Convention, signed in 1919, deprived Swedish herders of large summer pasture areas in Norway. The reduced access contributed to a new over-accumulation of animals in northernmost Sweden and enforced relocations southwards. In parallel, a growing Sámi self-esteem surfaced. The first Norwegian Sámi herders' national gathering in 1917 and in Sweden in 1918 aimed to counter the devastating policies towards reindeer pastoralism and the language assimilation (Andresen et al. 2021).

In the early 1900s, Swedish businessmen in Arjeplog and Arvidsjaur owned big reindeer herds and hired Sámi as herders (Jernsletten & Beach 2006). The Swedish Reindeer Act, passed in 1928, put an end to this practise to protect the Sámi herder livelihood. Only in the Concession area in the Torne and Kalix valley, local landowners were allowed to own a small number of reindeer if herded by Sámi. The 1928 law removed the last pieces of earlier rights to land ownership by the Sámi (Cramér & Ryd 2012). They were subjugated to a Sámi bailiff system with close control over reindeer herding. This system continued the segregation policy established in the late 1800s. The first Reindeer Herding Act in Finland was enacted in 1932 based on the earlier established system of herding districts (*paliskunta*), which poorly paid attention to the traditional *siida*-based herding system. The Norwegian Reindeer Act passed in 1933 underlined that the Sámi grazing rights were "tolerated use" that had to yield for other interests (Strøm Bull 1997), imposed detailed regulations of where and when to graze and stressed the collective liability for damages on farmland caused by reindeer. This was executed by special bailiffs. The regulations in Finnmark were less firm, e.g., the contested liability paragraph was not enforced that strictly (Strøm Bull 1997).

In Fennoscandia, the Nordic welfare model gradually developed, from a partly subsistence-based economy where over half of the population was engaged in farming to an engulfing industrialization and urbanization. In the first half of the 1900s, reindeer pastoralists were not a part of this gradual transition and practised largely a subsistence-based economy. The 1920s and 1930s were continuously fuelled by Social Darwinism, arguing that Sámi pastoralism was a dying culture, which was also reflected in legislation.

The Second World War (WWII, 1939–1945) affected the Fennoscandian countries differently, so also the herders. The reindeer herding Skolt Sámi were hit hard as they were trapped in the warfare at the Kola flank. In autumn 1944, the retreating German troops burned almost all settlements in Finnmark and northern Finnish Lapland and destroyed the infrastructure. Reindeer numbers collapsed (Figure 1.8) and many herders lost control of their remaining herds. The Swedish pastoralists' migration to their summer pastures in Norway did not fully re-establish after the war, as they had partly changed their seasonal grazing pattern during the war and was further excluded from parts of their old grazing grounds. In southern Norway, Sámi herders had deliberately let the herds loose to deter forced slaughter by German troops. This led to the feralization of the reindeer and difficulties to re-establish herding practices after WWII (Fjellheim 1999).

Post war – a gradual change

The post-war period constitutes a gradual change in public opinion towards the Sámi in general and specifically towards reindeer pastoralism. This may be seen as a response to the 1948 UN Human Rights Declaration and the international decolonization starting after WWII. Sámi herders organized and the herders were gradually included in the welfare states. Sámi organizations grew stronger and claimed their propagated and legitimate rights. The responsibility to "dismantle oppressive, hegemonic structures" was also starting to surface on the academic agenda (Kuokkanen 2010). The Nordic Sámi Council was established in 1956 and has worked actively to promote Sámi "cultural, political, economic, civil, social and spiritual rights and interests" but also engaged in international processes related to Indigenous peoples. In 1992, the Kola Sámi were fully taken on board and the name changed accordingly to the Saami Council. The Council now holds status as a permanent participant to the Arctic Council, established in 1996.

At the same time, resource extraction on reindeer grazing lands intensified, leading to functional deteriorations of pastures (Chapter 4). A strengthening of Sámi grazing rights started to emerge in the late 1960s. The Altevatn verdict in 1968 compensated Swedish herders for the loss of summer pastures in Norway caused by the construction of a hydroelectric power plant. This is regarded as a watershed regarding the relevance of Sámi immemorial grazing rights, which later have been included in the legislation in Norway and Sweden. In Finland, the construction of large reservoirs for hydropower in 1967 in Lappi herding district (Chapter 4) sparked political agency by the Sámi to improve their rights and status in land use issues. However, the revisions of the Reindeer Husbandry Acts, in Sweden in 1971, in Norway in 1978 and in Finland in 1990, were primarily targeted to modernize and rationalize the sector. The focus on production was strengthened by formalizing production units and strengthening the authorities' regulation of the number of reindeer per district, hence weakening the herders', the *siidas*' and the districts' self-determination and management of their herds (Chapters 7 and 9).

The revisions coincided with technical innovations adopted into herding practices, with the introduction of the snowmobile in the 1960s. Snowmobiles, and later the introduction of the all-terrain vehicles, changed the mode of mobility, eased the control of the herds and as a consequence altered the herd structure and production goals (Chapter 10) and influenced the internal organization. The authorities induced incentives to modernize the production. Reindeer pastoralism was modernized, and the herder's organizations were interwoven into the states' network through negotiations and agreements about incentives and subsidies. The increased use of technology led to a higher demand for cash flow, partly made available by governmental subsidies, and amplified the inclusion into a market economy (Riseth et al. 2019). Herd sizes started to increase in the late 1970s in all three countries, especially in the northern parts, accompanied by favourable winter conditions in the 1980s, and peaked around 1990 (Figure 1.8). In Finland, the maximum permitted number of reindeer was surpassed in the 1980s, and state authorities required both forced slaughter and a reduction of the maximum permitted numbers in particular in the northernmost herding districts.

The Chernobyl accident in 1986 contaminated grazing resources in the middle and southern parts of the RHA in Norway and Sweden. Especially lichens and mushrooms were polluted by radioactive Caesium. This led to forced slaughter of contaminated reindeer and mitigation actions, including supplementary feeding (Chapter 12).

The high grazing pressure combined with several difficult winters in the late 1990s reduced population sizes in all countries (Figure 1.8). Herd sizes increased again in the early 2000s, especially in Norway (Figure 1.8). In Finnmark, the

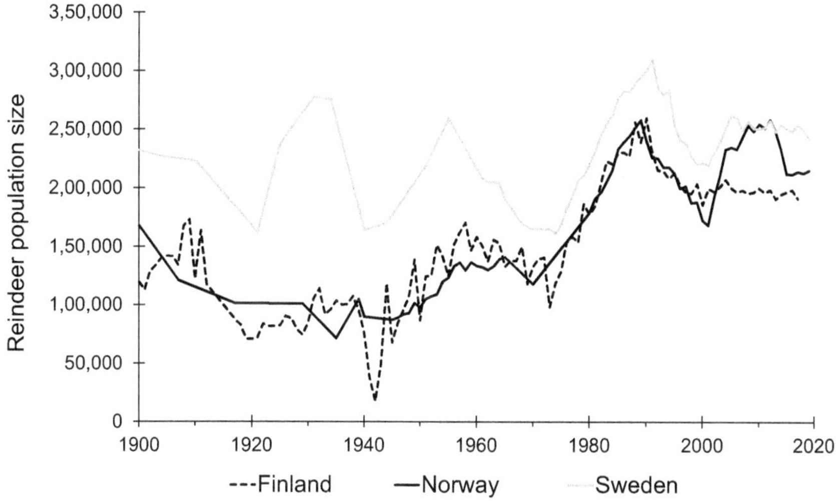

Figure 1.8 Population fluctuation of semi-domesticated reindeer in Finland, Norway and Sweden.

Source: Data compiled from Tømmervik & Riseth (2011) and Landbruksdirektoratet (Norway), Statistiska centralbyrån (1999) and Sametinget (Sweden), Kortesalmi (2008), A. Ermala (pers. comm) and Paliskuntain yhdistys (Finland).

northernmost region in Norway, a strict regulation regime was enforced around 2010 to reduce the herd sizes. Since 2012, population sizes have remained rather stable in all three countries – one reason being the increased use of supplementary feeding (Chapter 12).

At the same time, encroachment and fragmentation of pastures, in particular by growing infrastructure and other forms of land use, increased in all countries. In Sweden and Finland, forest exploitation intensified with the introduction of intensive forestry methods in the 1950s (Chapter 4), and in Finnmark the hydropower development raised awareness of Sámi herders' rights, leading to the establishment of Sámi Parliaments: 1989 in Norway, 1993 in Sweden and 1996 in Finland. Though their political power remains limited, the Parliaments' management responsibilities and rights to have their say in, e.g., land use planning and resource extraction issues have been extended. They were, indeed, instrumental for advocating the Indigenous rights and culture. The Sámi Parliamentary Council, a co-operative body between the Sámi Parliaments in Norway, Sweden and Finland, applied in 2014 for full membership in the Nordic Council. They were granted extended observer status in 2016.

While several international agreements and treaties exist to protect Indigenous cultures and develop their self-determination, not all of these are equally ratified in the three countries. The International Labour Organization's Indigenous and Tribal Peoples Convention (ILO 169), established in 1989, emphasizes the importance of Indigenous cultures and their access to traditional land, though without establishing new rights. It was signed by Norway in 1990 but has not been ratified by Sweden and Finland. However, it has had an impact on recent court cases in Sweden and strengthened Sámi land rights (Allard & Brännström 2021). It is a predecessor to the United Nations Declaration on the Rights of Indigenous People (UNDRIP), passed in 2007, which aims to protect Indigenous cultures and develop their self-determination. Contrary to ILO 169, the UNDRIP is not legally binding. Indeed, the Sámi have played an active international role in the Indigenous rights issue. The herders have worked actively to enter the international scene. In 2000, the Association of World Reindeer Herders, established in 1997, was granted observer status in the Arctic Council, which brought reindeer pastoralism onto the Arctic agenda.

A Nordic Sámi Convention has been on the agenda since the late 1990s. The three states agreed upon a draft in 2005, negotiated by a commission composed of governmental and Sámi representatives. The document recognized the right of the Sámi people to self-determination and acknowledged reindeer herding as a keystone in their livelihood. Further, the draft underlined the Sámi reindeer herding grazing rights, also across national borders. In 2016, a revised draft was agreed upon. The Convention underlined not only the importance of traditional rights and livelihoods but advocated support to develop new areas based on their traditional way of life, tourism and other businesses. Yet the Convention is not finally ratified and hence not put into action, partly because of incongruence between countries in the functional definition of Sámi.

The herders' grazing rights, including the right to graze on private land, were still disputed during the 20th century, such as in the Korsjøfjell case (Norway, 1988), the Selbu case (Norway, 2001), the Nellim case (Finland, 2010), the Skattefjällsmålet (Sweden, 1981) and the Nordmaling case (Sweden, 2011), as the herders were not able to document their immemorial use due to lack of written sources or archaeological remains. However, with the Selbu verdict in 2001, the Norwegian Supreme Court clearly stated that the right to practice reindeer herding is a legal right, rather than a tolerated use, even on private land (Strøm Bull 2010). Several cases in the Swedish Supreme court, e.g., the Nordmaling case (2011), ruled that the Sámi herders have customary rights (*sedvanerätt*) to graze their reindeer on privately owned land (Allard 2011). In Finland, general grazing rights of reindeer herders have been accepted, while some exceptions are found in southeast where debates and tensions with private forest landowners still do occur. Although the reindeer herding laws underscore the Sámi rights to land based on use since time immemorial, future challenges of access to grazing lands might develop from the increased claims for land due to the transition towards so-called green energy, such as wind power development or mining for minerals (Normann 2021).

The "Green transition" challenges reindeer pastoralism

The UN 2020 Agenda for Sustainable Development, accentuated by the recently released IPCC report (August 2021), puts pressure on natural resource extraction for green energy development. In several court cases related to the establishment of wind power development or in relation to mining sites, it has been argued by the herders that their Indigenous rights are being violated. Examples include the herding district Fosen (Storheia and Roan) in Norway and Stekenjokk in the herding district Vilhelmina södra in Sweden with regard to wind power, and Repparfjord in the herding district Fiettar in Norway, Gallók in the herding district Jåhkågasska tjiellde in Sweden, and Lätas in the herding district Käsivarsi in Sámi Homeland, Finland with regard to mining. The companies on the other hand argue that the Environmental Impact Assessments (EIAs) conducted conclude that the establishments will not impede their herding practice. This has been disputed by alternative assessments and points to the lack of consistency in such assessments. Indeed, social and community assessments could be integrated into the process (Chapter 8). Larsen et al. (2017) argued that Cumulative Effects Assessment governance is of particular importance on Indigenous territory. They concluded that governance tends to limit itself to impact mitigations of new development projects while ignoring the structural barriers to securing Indigenous peoples' rights. This struggle is amplified by an imbalance in power and resources between the involved actors, draining the herders' time and economy and hence negatively impacting their socio-cultural well-being. However, the recent verdict by the Norwegian Supreme Court (11 October 2021, Norges Høyesterett. 2021) concludes that the planned wind power development in Storheia and Roan violates the Sámi herders' Indigenous rights. This verdict may represent a turning point to secure Sámi herders' rights in Norway (Chapter 15).

Synopsis

The transitions from reindeer hunting to reindeer herding in Fennoscandia towards today's reindeer pastoralism were driven by interdependent changes in the ecological, social-economic and cultural evolution of Fennoscandia, including drivers from local to global levels. Figure 1.9 schematically summarizes important events and processes. With the increasing control of the Nordic states over natural resources and Sámi affairs, Indigenous rights were violated and have regained some recognition only in recent years. The future trajectory of reindeer husbandry will considerably depend on the degree to which these rights are implemented and strengthened and thus empower reindeer herders to shape their strategies to meet the challenge of climate change and continued resource extraction in the reindeer herding area.

Present

We present briefly the most recent (2019/2020) national statistics on production and economy based on official records from the Finnish Reindeer Herders' Association, the Norwegian Government (*Landbruksdirektoratet*) and the Swedish Sámi Parliament. We further outline the legislation and the administrative organization concerning reindeer pastoralism in the three countries.

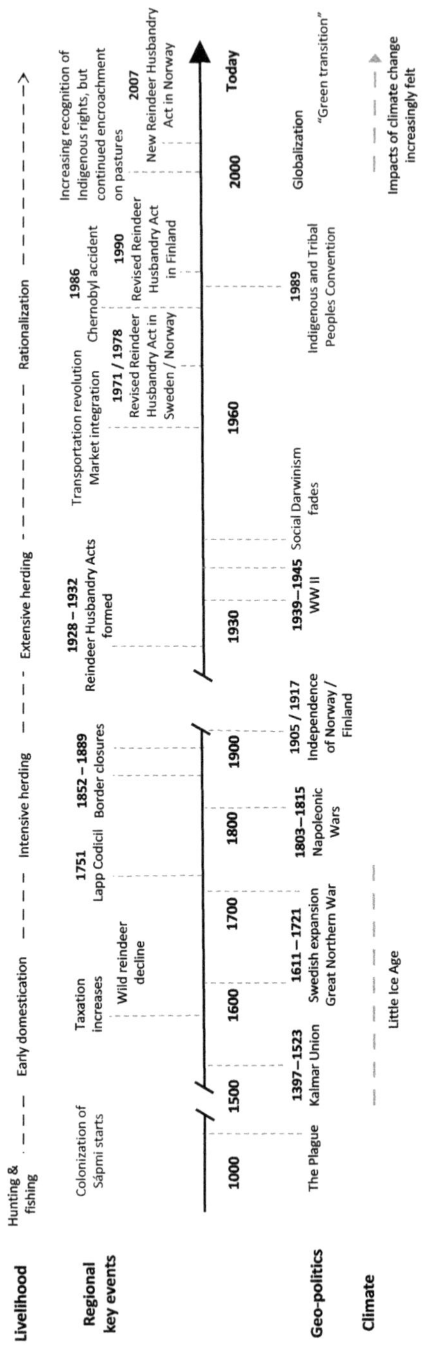

Figure 1.9 Timeline of major transitions of the livelihood revolving around reindeer, and important key events at the regional and global level.

Owners, production and economy

Basic statistics for Finland, Norway and Sweden are presented in Table 1.1. In all three countries, about 75% of the owners own less than 50 animals each. These owners often are household members or close family members who do not take active part in the daily herding activities. Between 30% and 40% of the owners are women. However, their share as active herders is low, but they often have important administrative parts in the livelihood. The number of herders who have their main income from reindeer pastoralism is estimated to around 1,000 in each country, but difficult to define and in flux.

For an equal amount, it is a significant secondary income. However, the multiplicator effect is large, e.g., by creating jobs indirectly related to reindeer pastoralism, and reindeer pastoralism is essential for securing a viable and culturally diverse economy in northern Fennoscandia.

Fennoscandian reindeer pastoralism generates a value of sold products to the herders and their households estimated to around 50 million €. The meat production output has stabilized in the last decade and is highest in Finland (Table 1.1, Chapter 10). This is probably the result of an extensive winter feeding practice (Chapter 12). However, feeding implies extra costs and the high productivity does not necessarily generate high revenues.

Reindeer meat provides about 90% of the production income from the animal in all three countries. Hides, antlers and other by-products contribute around 10% of the cash income. Other sources of income integrated into reindeer pastoralism include hunting, fishing, tourism and handicraft (*duodji*). However, these are hard to quantify and to compare across countries. In Norway, this accessory income is estimated to be around 8% of the production income. Indeed, the income spectrum varies greatly between households and underlines the multifaceted livelihood and way of life.

Table 1.1 Key statistics for 2019/2020 of Fennoscandian reindeer pastoralism. Reindeer numbers are for the winter herds after slaughter

Country	*Total area of RHA (km²)*	*# Reindeer*	*# Owners*	*Meat Prod. (tons)*[a]
Finland	123,000	188,000	4,300	2,000
Norway	145,000	213,000	3,300[b]	1,600
Sweden	226,000[c]	241,000	4,600[d]	1,260

Sources: Data from Landbruksdirektoratet (2020a, 2020b) in Norway, Swedish Sámi Parliament (2021) and Paliskuntain yhdistys (2020) in Finland.

Notes:
a Registered by slaughterhouses.
b Only Sámi owners.
c The exact boundaries of reindeer herding areas remain largely undefined in Sweden. Sandström's (2015) estimate is presented.
d Concession owners included.

During the last decades, production income has shown a falling proportion of the total income, as both subsidies and compensation have increased their shares. An array of state subsidies contributes to the herders' economy (Chapter 11). Especially in Norway, the scheme is intricate and extensive (Hausner et al. 2011, Chapter 11). In 2019/2020, the economic direct support ("*ordinære tilskudd*" in Norwegian) to the herders and herding groups amounted to around 8 million €, which represented around half of the first-hand production value. In Sweden, a total of 1.48 million € was granted in 2019/20, accounting for 18% of the slaughter income. The Finnish support system is primarily based on animal heads in the winter herd and amounted to 4.5 million € in 2020. Norway, not being a member of EU, can form its own reindeer pastoralism policy, including economic support schemes. This opens also for the protection of its reindeer meat production by import and toll regulations of venison.

The states provide subsidies to professionalize reindeer pastoralism and to drive the ongoing rationalization. To receive support in Finland, it is required that the owners have at least 80 reindeer and 500 at most. In Norway, a minimum production income per *siida* share (the functional production unit) from meat must be documented for being eligible support. There is an upper roof for this support at 600 animals in the winter herd per *siida* share. Indeed, many small owners, especially in Finland, do not receive financial animal-based state support. In Sweden, subsidies are not related to upper or lower limits in reindeer herd size. In all three countries, general support for establishing rural businesses, e.g., connected to local meat processing and tourism related to reindeer pastoralism, is available although limited.

The investment costs are primarily related to transportation, mainly cars, snowmobiles and ATWs and infrastructure (fences, huts and corrals), as well as communication equipment and other technical gear. The main running costs are related to herding activity (maintenance of carriers, infrastructure and fuel). The active herders are private business enterprises and the profitability is generally low, however, with variation between as well as variation within each country. For many herders, maintaining a good way of life seems more important than optimizing the economic output. Regionally hard competition to stay in business also contributes to the low profitability and many herder households are dependent on income from other sources.

In recent decades, the share of compensations in the herders' total income has increased, mainly for losses to predators, traffic accidents and to some degree degraded pastures by other land users, while the income from meat production has declined. The compensation for losses caused by predators equals around 90%–95% of the total compensation in Sweden and Norway (Chapter 6).

Tenure and legislation

About 48% of the Swedish and 64% of the Finnish reindeer herding area is classified as state-owned land, with the highest share in the northern part. In Norway, the percentage is around 58, including the Finnmark Estate which was established in 2005 (Broderstad et al. 2020). Also here the highest share is found in the northern part.

Reindeer herders have the right to graze reindeer on private lands within the RHA. Privately owned plots of land are generally small in Norway. In Sweden, large forest companies, including state-owned companies, own about 50% of the productive forestland within the reindeer herding area. In the southern part of the herding area in Finland, small private forest landowners dominate, some of them also owning reindeer. In Northern Finland, in Sámi Homeland, in particular, large state areas are set aside as National Parks or wilderness areas where reindeer herding is permitted. Indeed, in all three countries reindeer pastoralism is practiced in National Parks and large conservation areas.

The legal structures for reindeer pastoralism in Fennoscandia are partly interwoven, reflecting the states' shared history. Sámi grazing rights are acknowledged in all three countries as a form of use since time immemorial, i.e., established by the long-term use of a given area (*alders tids bruk* in Norway and *urminnes hävd* in Sweden and Finland) (Allard 2011). In Finland, the herders' rights to use the land for reindeer pastoralism are protected as a civil right in national legislation, and no distinction is made between Sámi and ethnic Finnish herders. However, Sámi herders have stronger rights to practice their culture, including reindeer pastoralism, which may give them a stronger position, e.g., in land use conflicts.

The reindeer herding right also includes rights to construct infrastructures (huts, corrals and fences), necessary for herding practices. In addition, resource extraction such as small game hunting, in Sweden also moose hunting, fishing, collection of firewood and construction materials are included. In Norway, the herders' rights to fish and hunt on private land are disputed (Strøm Bull 2010). In Sweden and Norway, pasture use is regulated by seasonal zonation. In the reindeer herding area of Finland, the right to reindeer grazing is given both on the state and private land for reindeer herders according to the Reindeer Husbandry Act.

There is a clear distinction between the right to own reindeer and perform reindeer pastoralism in Norway and Sweden as compared to Finland. In Norway and Sweden, it is an exclusive right of the Sámi. Only in the Concession areas in Sweden, reindeer may be owned by ethnic Swedes, but all herding work is done by Sámi herders (Figure 1.10b). In Finland, all citizens and EU members living in the reindeer herding area can own reindeer and take part in herding if approved as a member by a herding district. Also, in Norway and Sweden approved membership is required.

Although all three countries have a rather sophisticated and, on paper, strict legislation which regulates land use issues, the herders regard encroachment and fragmentation of their ranges as the biggest threat to their livelihood (Chapter 4). Large projects that result in considerable land use changes, such as wind parks, mining sites, power lines and main roads, must in connection with their application implement an EIA. If a planned activity will affect reindeer herding, the company/body responsible for the project must describe its effects

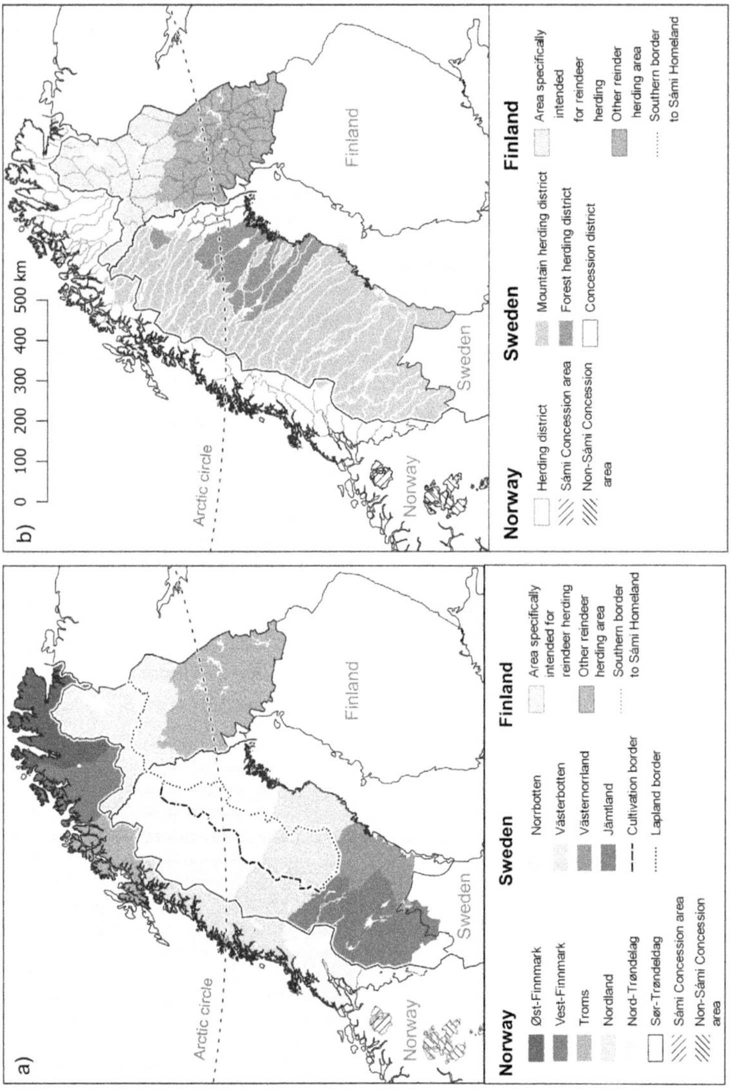

Figure 1.10 (a) Main administrative borders and (b) reindeer herding districts in Fennoscandia.

for pursuing the activity before the deciding body decides to give permission or not to. In special cases, the decision can be appealed to the central Authority and may also be tried in the court system. Recently Indigenous rights have moved to the forefront (Chapter 8). Larsen et al. (2017) argued that Cumulative Effects Assessment governance is of particular significance on Indigenous territory. However, they concluded "that governance tends to limit itself to mitigating the impacts while ignoring the structural barriers to securing indigenous peoples' rights". Compensation regimes are intricate and are partly negotiable. Often the claims end up in court. In Sweden, an environmental court has been established and is actively involved in the process.

However, local but more frequent land use changes and encroachments are probably a bigger threat than large projects. Often the herders' right to be heard and consulted is overridden in such cases.

Administrative organization

Reindeer pastoralism sorts under the Agricultural Ministry in all three countries, representing the highest administrative level. The central level in Norway also includes a national Reindeer Husbandry Board appointed by the government, the Sámi Parliament and a directorate (*Landbruksdirektoratet*). In Finland, the Reindeer Herders' Association (*Paliskuntain yhdistys*), where all herding districts are members, is engaged in the state administration at several levels. A general assembly meets yearly and elects a Board with representatives from reindeer herding co-operatives including also representatives from Ministry and a member from the Sámi Parliament in Finland. In Sweden, the Swedish Sámi Parliament has a more central role in reindeer management as compared to Norway and Finland. They act as an advisory body for the Ministry and administer the regulations and the funding for the predator survey and compensation.

Administration at the regional and local levels deviates slightly between the countries. In Norway and Sweden, the county administration plays an important role in following up the herding districts. In Norway, each herding district has a board and may be regarded as the lowest level of the state administration, although their major role is played out in the internal management. The herding districts in Sweden also have a board elected by the members to administer their internal management and defend their interest in land use and other conflicts. In Finland, the county borders do not align with the administrative regional borders. The three northernmost counties are involved in administrative issues relevant for reindeer husbandry. In addition, regional advisers working at *Paliskuntain yhdistys* assist reindeer herders and herding districts with administrative issues. The members' influence within the district is partly based on the number of animals they own.

Reindeer pastoralism within the three countries

There is considerable variation within the countries with regard to many aspects of reindeer pastoralism. This section therefore shortly portrays the reindeer pastoralism within each country.

Finland

In Finland, the reindeer herding area covers about 36% of the surface of the country. According to the reindeer herding law, the area is divided into (i) the area where the state land is specifically intended for reindeer herding in the North and (ii) the remaining reindeer herding area (Figure 1.10a). In the former, encompassing 20 districts (Figure 1.10b), reindeer herding may not significantly be hindered on state land by other forms of land use. Even though, this area is practically affected by several industrial developments (Chapter 4). Reindeer pastoralism is practiced in 54 reindeer herding districts (*paliskunta* in Finnish, *bálgosat* in Sámi) (Figure 1.10b). The 13 northernmost districts are located in Sámi Homeland (Figure 1.10b) and are granted a high degree of autonomy on issues relating to the Sámi culture and language. In Sámi Homeland, reindeer owners predominantly are Sámi.

About 38% of the semi-domesticated reindeer population in Finland are found in Sámi Homeland (Figure 1.10a), and also the number of reindeer owners is lower than in the southern herding area (Table 1.2). However, herd size per individual owner is higher in Sámi Homeland compared to the south, where reindeer herding forms less commonly is the main income for herders. In Sámi Homeland, around 92% of the land is state-owned with large areas set aside for nature conservation. In the other districts of the area specifically intended for reindeer husbandry, 79% of the land is state-owned, and 44% in the remaining reindeer herding area. More than 75% of the reindeer herding

Table 1.2 Key statistics for 2019/2020 for reindeer husbandry in Finland. Reindeer numbers are for the winter herd after slaughter

Main regions	*Herding area (km²)*	*# Reindeer*	*# Owners*	*# Districts*
Sámi Homeland area	35,600	38,997	1,220	13
Special Reindeer Herding Area except for the Sámi Homeland	22,100	76,447	916	7
Southern part of the reindeer herding area	65,300	72,528	2,218	34
Total	123,000	187,972	4,354	54

Source: Paliskuntain yhdistys (2020) in Finland.

area in Finland is located within the boreal region (Figure 1.3). The potential conflict with the forest industry is therefore pronounced (Chapter 4).

The reindeer population has stabilized in the two last decades (Figure 1.8). Many districts are facing challenges with high grazing pressures on limited lichen pastures, partly mitigated by the extensive use of supplementary feeding (Pekkarinen et al. 2015). This challenge seems most pronounced in tundra areas with limited summer pastures. In the boreal areas, establishing a reasonable system of pasture rotation is starting to emerge.

Norway

In Norway, the Sámi reindeer herding area stretches from the middle to the northernmost part of the country (Figure 1.10a), which is divided into 89 districts (Figure 1.10b). Approximately 40% of the country's land area is used for reindeer herding. In the southern part, most of the range is privately owned, whereas state-owned land and commons dominate in the North. In Finnmark, almost all land (95%), encompassing 46,000 km^2, was defined as state land until 2005 when it was collectively transferred by law to all inhabitants in Finnmark (Broderstad et al. 2020). The dynamic land tenure situation in Finnmark seems not to influence the herders' grazing rights directly. However, land claims by local villages and municipalities (Broderstad et al. 2020) as well as by different *siidas* (Marin & Bjørklund 2015) may in the long run reduce the *siidas*' flexibility to access different winter pastures.

In south-central Norway, the Concession areas with non-Sámi reindeer herding encompasses about 6,100 km^2 of alpine pastures (Figure 1.10b). Concessions are based on access to well defined reindeer pastures and granted for a certain period and the management can be subject to specific conditions. The herding in Trollheimen is enforced by a special law where the state has expropriated and rented land for Sámi herders.

Around 70% of the semi-domesticated reindeer and 75% of reindeer owners are found in Finnmark (Table 1.3), consisting of two management regions: Øst-Finnmark and Vest-Finnmark (Figure 1.10a). Herd size in the four remaining Sámi herding regions, Troms, Nordland, Nord-Trøndelag and Sør-Trøndelag/Hedmark, is around 12,000 animals in the winter herd, although the potential herding area varies up to fourfold between them (Table 1.3). The average number of reindeer per owner is about 65, but 80% of the owners possess less than 50 animals. The highest proportion of herders with small herd sizes is found in Vest-Finnmark. The reindeer number per *siida* share varies between 243 in Troms and 458 in Sør-Trøndelag/Hedmark.

In addition, four non-Sámi companies run by local farmers and community members herd 10,800 reindeer in the Concession area. Some of them border on areas with wild reindeer. It is challenging to avoid mixing between the wild and semi-domesticated herds. The Chronic Wasting Disease (CWD) outbreak in 2016 in one of the neighbouring wild reindeer areas has increased the tension, and actions are taken to reduce the risks of mixing of populations.

Table 1.3 Key statistics for 2019/2020 for reindeer husbandry in Norway. Reindeer numbers are for the winter herd after slaughter. Figures of non-Sámi herding also included

Regions	*Herding area (km²)*[a]	*# Reindeer*	*# Districts*	*# Siida shares*	*# Owners*
Øst-Finnmark	30,800	69,000	15	164	979
Vest-Finnmark	25,900	78,910	32	212	1,535
Troms	18,300	12,410	19	51	214
Nordland	32,600	13,880	12	39	239
Nord-Trøndelag	22,300	13,860	6	39	198
Sør-Trøndelag/ Hedmark	8,600	13,760	5	30	164
Non-Sámi	6,100	10,780	-	-	-
Total	144,600	212,600	89	535	3,329

Note:
a Including all potential land area.

The population size of reindeer has varied greatly in the last decades, particularly in parts of Finnmark (Tømmervik & Riseth 2011). Riseth and Vatn (2009) ascribed the increase in herd size in Finnmark in the 1980s (Figure 1.8) to increased economic support, failed governmental actions, as well as favourable winter conditions. Næss (2020) argued that internal competition and lack of trust, within and between districts, have contributed to an "assurance game" resulting in periods of herd accumulation (Chapter 7). During the 1990s, herd sizes declined in many parts of Finnmark due to a series of severe winters in combination with deteriorated lichens ranges (Tømmervik et al. 2012) but increased again after the turn of the millennium. Strict herd regulations have been enforced by the authorities in the last decade. These regulations induced tensions internally in and between herding districts, as well as between authorities and herding districts (Chapter 7). Since 2012, population sizes have remained rather stable in Finnmark. In all other regions, the populations have been comparatively stable for decades (Tømmervik & Riseth 2011).

A new Reindeer Husbandry Act came into force in Norway in 2007 and underlines that reindeer pastoralism must be economically, ecologically and culturally viable based on Sámi culture, traditions and customary practices. The law underscored the principle of *use since time immemorial*. It re-introduced the *siida* concept more in line with the Sámi tradition but with several serious misconceptions that can create frictions between *siidas* today (Chapter 7). The *siida* share is put at the centre. A *siida* share is licensed to a given person. To be entitled and to lead a *siida* share, a person must have reindeer pastoralism as the main profession. The share can be passed on to the next generation. The Act has retained the system of districts, defined as a geographical specific administrative and management unit. This seemingly self-determination according to Sámi customs at the local level is, however, constrained by the maximum number of animals within a summer *siida* which is finally determined by the central administration (Chapter 9).

Meat production on average provides ca. 50% of the income in most regions. In Troms and Nordland, predation has been very high in the last decades and hence deteriorated the production output and increased compensation for predator damage (Chapter 6). The profitability varies greatly between regions and *siida* shares. The differences are clearest between the two Finnmark regions, Troms and Nordland, compared to the two southernmost regions and the non-Sámi Concession herding. However, big variations in Finnmark are found.

Sweden

In Sweden, the reindeer herding area covers about 50% of the land area, i.e., approximately 226,000 km^2 (Sandström 2015). The estimates vary, and borders are not always clearly defined. As in Norway, the right to reindeer husbandry in Sweden is an exclusive right to the Sámi people. Only Sámi who are members of a reindeer herding district (*Sameby* in Swedish) have the right to pursue reindeer husbandry. Only in the Concession districts, a restricted area at the border to Finland (Figure 1.10b), special legislation based on customary reindeer ownership by local farmers, allows non-Sámi ownership of reindeer. However, herding practices are always carried out by Sámi herders even in these Concession districts.

There are 51 reindeer herding districts in Sweden. A reindeer herding district constitutes a geographic, economic and legal entity. The members of a herding district are usually organized in winter groups (*siida*), which may consist of one or several reindeer herding enterprises.

In Sweden, reindeer pastoralism is practiced in two distinct forms (Figure 1.10b, Chapter 3). Mountain herding districts migrate between summer pastures in the mountains at the border to Norway and winter pastures in the lowland forest. Some mountain districts migrate to Norway for summer grazing or have parts of the summer pasture in both countries. In forest herding districts, reindeer stay in the boreal forest throughout the year. Thirty-three districts practice mountain herding, while forest herding is practiced in ten districts (Table 1.4). The remaining eight districts are Concession districts, where reindeer herding resembles forest herding.

The Reindeer Husbandry Act (1971) divided the reindeer herding area into year-round grazing areas and winter grazing areas. Year-round areas for mountain herding districts in the two northernmost counties Norrbotten and Västerbotten are located west of the cultivation border, and for forest herding districts west of the Lapland border (Figure 1.10a), as well as in the mountains in Jämtland and Dalarna. Contrary to the year-round areas, reindeer grazing is allowed on the winter grazing areas only between October 1st and April 30th. The winter grazing area is located east of the year-round grazing areas.

There is a large variation between reindeer herding districts in both herd size (Chapter 10), number of reindeer owners and production (Table 1.3). More

Table 1.4 Key statistics for 2019/2020 for reindeer husbandry in Sweden. Reindeer numbers are for the winter herd after slaughter

Herding districts with county	*Total area (km²)*[a]	*# Reindeer*	*# Owners*	*# Districts*
Norrbotten, excl. Concession districts	104,000	131,121	3,160	15 mountain districts, 9 forest districts
Västerbotten	65,500	51,748	321	6 mountain districts, 1 forest district
Jämtland[b]	80,000	47,116	396	12 mountain districts
Concession districts	16,500	11,069	765	8 districts
Total[c]	226,000	241,054	4,642	51 districts

Source: Swedish Sámi Parliament (2020).

Notes:
a Area of the individual districts, some of which do overlap and partly transgress and include pastures in Norway.
b Includes the area of pastures in the counties of Västernorrland and Dalarna.
c The total area is an approximation, accounting for overlap between districts.

herders and reindeer, but lower reindeer densities, characterize the northern herding districts compared to the south. Northern herding districts therefore consist of many small enterprises with fewer reindeer per herder. Eighty per cent of reindeer owners own less than 50 reindeer. This is in particular true for Norrbotten, while in Västerbotten and Jämtland 50%–55% of owners have herds of less than 50 reindeer. The southern herding districts seem more oriented towards productivity (Chapter 10). In the Concession districts, non-Sámi landowners are allowed to graze a maximum of 30 reindeer on common land. Therefore, the number of herders in Concessions districts is high in relation to animal numbers.

Perspectives

This chapter conceptualized reindeer pastoralism within a socio-ecological framework. The relationship between the reindeer herd, pasture resources and pastoralists is the most fundamental component, delivering an array of ecosystem services. However, reindeer pastoralism has experienced environmental, socio-political and socio-cultural change since its origin. These complex and dynamic interactions reach from the local to the regional (national) and to the global level, including cross-scale linkages between them (*sensu* Cash et al. 2006; Figure 1.11).

In the present era of globalization and climate change, these linkages are reinforced. They present formidable challenges for the resilience of reindeer pastoralism and its capacity to maintain its cultural identity while adapting to the multifaceted changes. The combined impacts of environmental change and resource extraction exert pressure on the remaining land and intensify

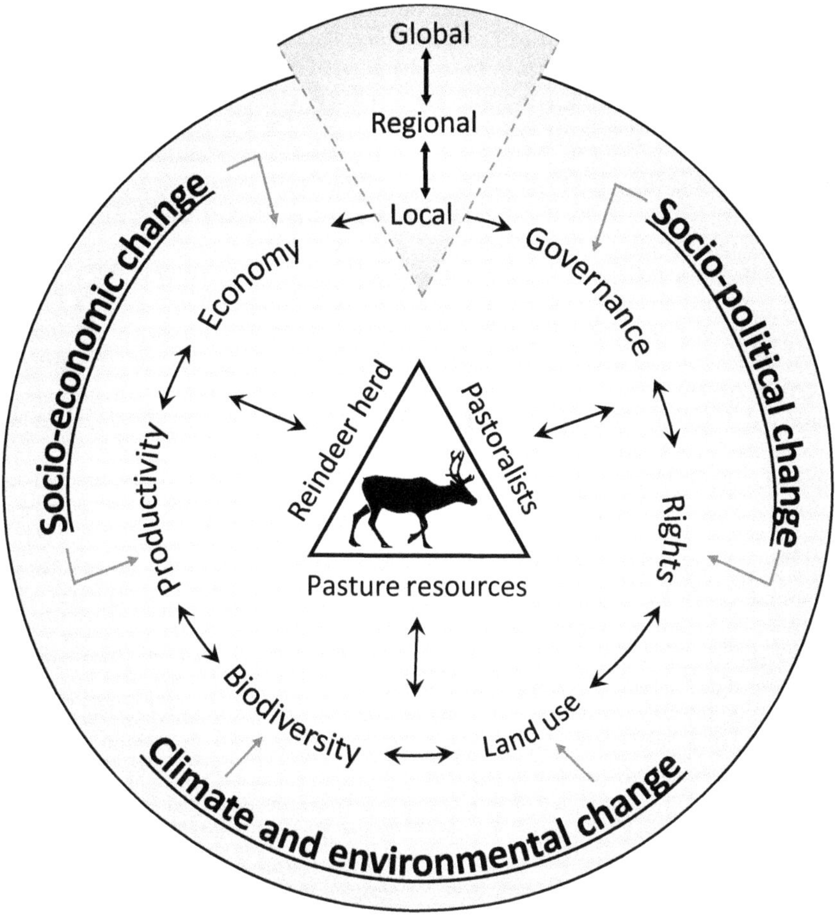

Figure 1.11 External drivers at the global, regional (national) and local level interact with reindeer pastoralism and the ecosystem services the livelihood provides.

the struggle for strengthening the herders' rights to land and resources. The following chapters will address these challenges from different angles.

References

Aarseth, B. (1989). Grenseoppgjørene og konsekvensene av disse for den nordsamiske bosetting i Norge. (In Norwegian). In: Aarseth, B. (ed) *Grenser i Sameland*. Oslo: Samiske samlinger, 13. 43–82.

Åhman, B. & White R.G. (2019). Rangifer Diet and Nutritional Needs. In: Tryland, M. & Kutz, S.J. (eds) *Reindeer and Caribou: Health and Disease*. New York: CRC Press. 107–134.

Åkerman, S. (1990). Ett ekosystem och dess förvandling: Storfamilj och storhushåll vid övergången mellan medeltid och nyare tid i två norrländska älvdalar. In: Åkerman, S. & Lundholm, K. (eds) *Älvdal i norr: människor och resurser i Luledalen 1300–1800.* Luleå: Umeå university studies in the humanities, 91. 29–31.

Allard, C. (2011). The Nordic Countries' Law on Sámi Territorial Rights. *Arctic Review.* 2(2), 159–182.

Allard, C. & Brännström, M. (2021). Girjas Reindeer Herding Community v. Sweden: Analysing the Merits of the Girjas Case. *Arctic Review.* 12, 56–79.

Andresen, A., Evjen, B. & Ryymin, T. (2021). *Samenes historie fra 1751 til 2010.* Oslo: Cappelen Damm Akademisk.

Bårdsen, B.-J., Fauchald, P., Tveraa, T., Langeland, K., Yoccoz, N.G. & Ims, R.A. (2008). Experimental Evidence of a Risk-sensitive Reproductive Allocation in a Long-lived Mammal. *Ecology.* 89(3), 829–837.Bitustøyl, K. & Mossing, A. (2019). *Reinen på Hardangervidda-sterkt blanda med tamrein.* www.villrein.no/aktuelt/tag/tinn [20.08.2021].

Bjørklund, I. (2013). Domestication, Reindeer Husbandry and the Development of Sámi Pastoralism. *Acta Borealia.* 30(2), 174–189.

Broderstad, E.G., Hausner, V., Josefsen, E. & Søreng, S.U. (2020). Local support among arctic residents to a land tenure reform in Finnmark, Norway. *Land Use Policy*. 91: e101326.

Cash, D.W., Adger, W.N., Berkes, F., Garden, P., Lebel, L., Olsson, P., Pritchard, L. & Young, O. (2006). Scale and Cross-scale Dynamics: Governance and Information in a Multilevel World. *Ecology and Society*, *11*(2): 8.

Clutton-Brock, J. (1999). *A Natural History of Domesticated Mammals, 2nd edition.* Cambridge: Cambridge University Press.

Cramér, T. & Ryd, L. (2012). *Tusen år i Lappmarken.* Skellefteå: Ord & visor forlag.

Dinerstein, E et al. (2017). An Ecoregion-Based Approach to Protecting Half the Terrestrial Realm. *BioScience*. 67(6), 534–545.

Espmark, Y. (1971). Individual Recognition by Voice in Reindeer Mother-Young Relationship. Field Observations and Playback Experiments. *Behaviour.* 40(3–4), 295–301.

Esseen, P.A., Ehnström, B., Ericson, L. & Sjöberg, K. (1997). Boreal Forests. *Ecological Bulletins*. 46, 16–47.

Evjen, B. (2007). Custodial Reindeer and Custodial Goats – Part of Reindeer Herding and Animal Husbandry. *Rangifer.* 27(2), 79–91.

Fjellheim, S. (1999). *Samer i Rørostraktene* (in Norwegian). Snåsa: Fjellheim.

Fjellheim, S. (2005). Fra fangstbasert til nomadisk reindrift i Rørostraktene (in Norwegian, english summary). *Rangifer Report*. 10, 21–30.

Folland, C.K., Knight, J., Linderholm, H.W., Fereday, D., Ineson, S. & Hurrell J.W. (2009). The Summer North Atlantic Oscillation: Past, Present, and Future. *Journal of Climate*. 22(5), 1082–1103.

Geist, V. (1999). *Deer of the World: Their Evolution, Behaviour, and Ecology*. Shrewsbury: Swan Hill Press.

Gunn, A. (2016). *Rangifer tarandus (Linnaeus, 1758).* The IUCN Red List of Threatened Species 2016.

Hansen, L.I. & Olsen, B. (2014). *Hunters in Transition: An Outline of Early Sámi History.* Leiden: Brill Academic Publisher.

Hausner, V.H., Fauchald, P., Tveraa, T., Pedersen, E., Jernsletten, J.-L. L., Uivevadet, B., Ims, R.A., Yoccoz, N.G. & Bråthen, K.A. (2011). The Ghost of Development Past: the

Impact of Economic Security Policies on Saami Pastoral Ecosystems. *Ecology and Society*. 16(3), 4.

Helle T. & Kojola I. (2006). Population Trends of Semi-Domesticated Reindeer in Fennoscandia – Evaluation of Explanations. In: Forbes, B.C., Bölter M., Müller-Wille, L., Hukkinen, J., Müller, F., Gunslay, N. & Konstantinov, Y. (eds) *Reindeer Management in Northernmost Europe. Ecological Studies*, 184. Berlin: Springer. 291–339.

Hofmann, R.R. (1989). Evolutionary Steps of Ecophysiological Adaptation and Diversification of Ruminants: A Comparative View of Their Digestive System. *Oecologia*. 78(4), 443–457.

Holand, Ø. & Weladji, R.B. (2019). Rangifer mating strategies. 79–89. In: Tryland, M. & Kutz, S. J. (eds). *Reindeer and Caribou: Health and Disease*. New York: CRC Press. 23–43.

Ingold, T. (1980). *Hunters, Pastoralists and Ranchers: Reindeer Economics and Their Transformations*. New York: Cambridge University Press.

Jernsletten, J.-L. & Beach, H. (2006). The Challenges and Dilemmas of Concession Reindeer Management in Sweden. In: Forbes, B.C., Bölter, M., Müller-Wille, L., Hukkinen, J., Müller, F., Gunslay, N. & Konstatinov, Y. (eds) *Reindeer Management in Northernmost Europe. Ecological Studies, 184*. Berlin: Springer. 95–116.

Karlsen, S.R. Tolvanen, A., Kubin, E., Poikolainen, J., Høgda, K.A., Johansen, B., Danks, F.S., Aspholm, P., Wielgolaski, F., E. & Makarova, O. (2008). MODIS-NDVI-based Mapping of the Length of the Growing Season in Northern Fennoscandia. *International Journal of Applied Earth Observation and Geoinformation*. 10(3), 253–266.

Kjelvik, S. & Kärenlampi, L. (1975). Plant biomass and primary production of Fennoscandiàn subarctic and subalpine forests and of alpine willow and heath ecosystems. In: Wielgolaski, F. E. (ed.) *Fennoscandian Tundra Ecosystems. Part 1: Plants and Microorganisms*. Berlin: Springer. 111–120.

Kortesalmi, J. (2008). *Poronhoidon synty ja kehitys Suomessa. Suomalaisen Kirjallisuuden Seuran toimituksia*. Helsinki: Tiede.

Kumpula, J., Kurkilahti, M., Helle, T. & Colpaert, A. (2014). Both Reindeer Management and Several Other Land Use Factors Explain the Reduction in Ground Lichens (Cladonia Spp.) in Pastures Grazed by Semi-Domesticated Reindeer in Finland. *Regional Environmental Change*. 14(2), 541–559.

Kuokkanen, R. (2010). The Responsibility of the Academy: A Call for Doing Homework. *Journal of Curriculum Theorizing*. 26(3), 61–74.

Kuusela, J.M., Salmi A. K. & Äikäs, T. (2020). Hunters, Fishers, Traders – an Archaeological and Zooarchaeological Perspective on the Development of the Late Iron Age and the Medieval Northern Fennoscandia Trade Network. *Estonian Journal of Archaeology*. 24(2), 141–160.

Labba, K. (2016). The Legal Organization of Sami Reindeer Herding and the Role of the Siida. In: Allard, C. & Skogvang, S.F. (eds) *Indigenous Rights in Scandinavia*. New York: Routledge. 141–153.

Landbruksdirektoratet. (2020a). *Ressursregnskapet for reindriften 2019/2020*. www.landbruksdirektoratet.no/nb/nyhetsrom/aktuelle-tema/ressursregnskapet-for-reindriftsnaeringen [20.08.2021]

Landbruksdirektoratet. (2020b). *Totalregnskap for reindrifts-næringen. Regnskap 2019 og budsjett 2020*. www.landbruksdirektoratet.no/nb/nyhetsrom/aktuelle-tema/totalregnskap-for-reindriftsnaeringen [20.08.2021]

Lantto, P. (2000). Tiden börjar på nytt. En analys av samernas etnopolitiska mobilisering i Sverige 1900–1950. (In Swedish, English abstract). PhD. diss. Umeå University.

Larsen, R.K., Raito, K., Stinnebom, M. & Wik-Karlson, J. (2017). Sami-state Collaboration in the Governance of Cumulative Effects Assessment: A critical Action Research Approach. *Environmental Impact Assessment Review.* 64, 67–76.

Lundmark, L. (2006). *Samernas skatteland i Norr- och Västerbotten under 300 år.* (in Swedish). Stockholm: Institutet för rättshistorisk forskning.

Marin, A. & Bjørklund, I. (2015). A Tragedy of Errors? Institutional Dynamics and Land Tenure in Finnmark, Norway. *International Journal of the Commons.* 9(1), 19–40.

McGahey, D., Davies, J., Hagelberg, N. & Ouedraogo, R. (2012). *Pastoralism and the Green Economy – A Natural Nexus? Status, Challenges and Policy Implications.* Nairobi: IUCN and UNEP.

Minde, H. (2003). Assimilation of the Sami – Implementation and Consequences. *Acta Borealia.* 20(2), 121–146.

Moen, A. (1999). *National Atlas of Norway: Vegetation.* Hønefoss: Norwegian Mapping Authority.

Myrvoll, E.R., Thuestad, A. & Holm-Olsen, I.M. (2011). Wild Reindeer Hunting in Arctic Norway: Landscape, Reindeer Migration Patterns and the Distribution of Hunting Pits in Finnmark. *Fennoscandia Archaeologica.* 28, 3–17.

Natural Resources Institute of Finland. (2018). *Finnish Forest Reindeer.* www.luke.fi/en/natural-resources/game-and-hunting/finnish-forest-reindeer/ [22.08.2021]

Nesje, A. & Dahl, S. O. (2003). The "Little Ice Age" – Only Temperature? *The Holocene.* 13(1), 139–145.

Niemi, E. (2007). Østsamene – grenseminoritet og nasjonal problem (in Norwegian). In: Broderstad. E. G., Niemei, E. & Sommerseth, I. (eds) *Grenseoverskridene reindrift før og etter 1905* (in Norwegian). Tromsø: Universitetet i Tromsø. 71–86.

Nordin, Å. (2002). *Relationer i ett samiskt samhälle. En studie av skötesrensystemet i Gällivare socken under första hälften av 1900-talet.* Samiska studier 87. Umeå universitet.

Norges Høyesterett. (2021). *Vedtak om konsesjon til vindkraftutbygging på Fosen kjent ugyldig fordi utbyggingen krenker reindriftssamenes rett til kulturutøvelse.* www.domstol.no/globalassets/upload/hret/avgjorelser/2021/oktober-2021/hr-2021-1975-s.pdf [13.10.2021]

Normann, S. (2021). Green Colonialism in the Nordic Context: Exploring Southern Saami Representations of Wind Energy Development. *Journal of Community Psychology.* 49(1), 77–94.

Næss, M.W. (2020). Cultural Group Selection and the Evolution of Reindeer Herding in Norway. *Human Ecology.* 48, 279–291.

Odner, K. (1985). Saamis (Lapps), Finns and Scandinavians in History and Prehistory. Ethnic Origins and Ethnic Processes in Fenno-Scandinavia. *Norwegian Archaeological Review.* 18(1–2), 1–12.

Paine, R. (1994). *Herds of the Tundra: A Portrait of Saami Reindeer Pastoralism.* London: Smithsonian Institution Press.

Paliskuntain yhdistys. (2020). Tilastoja poronhoitovuodelta (in Finnish). Extracted from Poromies.

Pedersen, S. (2007). Lappekodisillen av 1751 – Samene – det grenseløse folket. In: Broderstad. E.G., Niemei, E. & Sommerseth, I. (eds) *Grenseoverskridene reindrift før og etter 1905.* Tromsø: Universitetet i Tromsø. 9–20.

Pekkarinen, A.-J., Kumpula, J. & Tahvonen, O. (2015). Reindeer Management and Winter Pastures in the Presence of Supplementary Feeding and Government Subsidies. *Ecological Modelling.* 312, 256–271.Pennanen, J. & Näkkäläjärvi, K. (2002). *Siiddastallan: From Lapp Communities to Modern Sami Life.* Inari: Sami museum foundation.

Riseth, J.Å. & Vatn, A. (2009). Modernization and Pasture Degradation. A Comparative Study of Two Sámi Reindeer Pasture Regions in Norway. *Land Economics*. 85(1), 87–106.

Riseth, J.Å., Tømmervik, H. & Forbes, B.C. (2019). Sustainable and resilient reindeer herding. In: Tryland, M. & Kutz, S.J. (eds) *Reindeer and Caribou: Health and Disease*. New York: CRC Press. 23–43.

Rusell, D.E. & Martell, A.M. (1984). Winter Ecology of caribou (Rangifer tarandus). In: Olson, R., Hastings, R. & Geddes, F. (eds) *Northern Ecology and Resource Management*. Alberta: University of Alberta Press. 117–144.

Sandström, P. (2015). *A Toolbox for Co-production of Knowledge and Improved Land Use Dialogues*. PhD. diss. Swedish University of Agricultural Sciences. Umeå.

Sara, M.N. (2009). Siida and Traditional Sámi Reindeer Herding Knowledge. *Northern Review*. 30, 153–178.

Seppälä, M. (ed.) (2005). *The Physical Geography of Fennoscandia*. Oxford: Oxford University Press.

Skogland, T. (1990). Density Dependence in a Fluctuating Wild Reindeer Herd; Maternal vs. Offspring Effects. *Oecologia*. 84, 442–450.

Statistiska centralbyrån. (1999). *Svensk rennäring*. Halmstad.

Strøm Bull, K. (1997). *Studier i reindriftsrett* (in Norwegian). Tano: Aschehoug.

Strøm Bull, K. (2010). Reindriftens bruksrettigheter. (In Norwegian, English Summary). *Kart og Plan*. 70, 174–181.

Swedish Sámi Parliament (2020). *Statistik rennnäringen*. www.sametinget.se/renstatistik [20.08.2021]

Telfer, E.S. and Kelsall, J.P. (1984). Adaptations of Some Large North American Mammals for Survival in Snow. *Ecology*. 65(6), 1828–1834.

The Association of World Reindeer Herders. https://reindeerherding.org/world-reindeer-herders [28.8.2021]

Tømmervik, H. & Riseth, J.Å. (2011). *Historiske tamreintall i Norge fra 1800-tallet fram til i dag*. NINA Rapport 672. Tromsø.

Tømmervik, H., Bjerke, J.W., Gaare, E., Johansen, B. & Thannheiser, D. (2012). Rapid Recovery of Recently Overexploited Winter Grazing Pastures for Reindeer in Northern Norway. *Fungal Ecology*. 5, 3–15.

Valkeapää N.-A. (1997). *The Sun, My Father*. (First published in 1988 titled *Beaivi, áhčážan*). Vasa: Dat.

Vistnes, I. I., Burgess, P., Mathiesen, S. D., Nellemann, C., Oskal, A., & Turi, J. M. (2009). *Reindeer Husbandry and Barents 2030: Impacts of Future Petroleum Development on Reindeer Husbandry in the Barents Region*. Alta: Fagtrykk Idé AS.

Vorren, Ø. (1973). Some trends of the transition from hunting to nomadic economy in Finnmark. In: Berg, G. (ed.) *Circumpolar Problems: Habitat, Economy and Social Relations in the Arctic*. Oxford: Pergamon Press. 185–194.

Wallerström, T. (2000). The Saami between East and West in the Middle Ages: An Archaeological Contribution to the History of Reindeer Breeding. *Acta Borealia*. 17(1), 3–39.

Wielgolaski, F.E. (1975). Primary production of alpine meadow communities. In: Wiegolaski, F.E. (ed) *Fennoscandian Tundra Ecosystems. Part 1: Plants and Microorganisms*. Berlin: Springer. 121–128.

2 Genetic structure and origin of semi-domesticated reindeer

Knut H. Røed, Kjersti S. Kvie and Bård-Jørgen Bårdsen

Introduction

Domestication of reindeer, *Rangifer tarandus*, and the emergence of large-scale reindeer herding were fundamental social transformations for the Indigenous people of the circumpolar north (Krupnik 1993; Kofinas et al. 2000; Jernsletten & Klokov 2002; Bjørklund 2013; Hansen & Olsen 2013). This pastoral transition established new relationships between humans and animals and led to new settlement and land use patterns across large portions of northern Eurasia. Today, reindeer husbandry as a livelihood is increasingly threatened by changes in the climate (Vors & Boyce 2009; Pape & Löffler 2012) and in land use (Horstkotte 2013; Skarin & Åhman 2014; Sandström 2015). Future adaptation, selection and improvement of reindeer herding are dependent on the genetic variability of the animal populations (Groeneveld et al. 2010; FAO 2015).

The pastoral transition

Today's large-scale and extensive reindeer husbandry in the Nordic countries is essentially linked to animal numbers, with the pastoral economy associated with the sale of meat from slaughtered calves (Næss 2010). However, long before large-scale reindeer herding developed, the first reindeer herders primarily lived in a hunter-gatherer economy and domesticated reindeer were mainly used for transportation (Bjørklund 2013) and as decoy animals to attract wild reindeer (Tornæus & Wiklund 1900). The early use of reindeer for transport purposes is illustrated by the 3500-year-old remains of Sámi-type sledges from the burial site in the Murmansk Fjord in northwest Russia (Murashkin et al. 2016). This early phase of reindeer domestication resulted in only small changes in local modes of subsistence, which remained focused on hunting, fishing and gathering. During the 16th and 17th centuries there was development towards increasing small-scale intensive herding, usually following a nomadic lifestyle, based on the provision of transport and food products including milk (Nieminen 1992). However, driven by diverse economic, social and ecosystem forces during the 18th and 19th centuries, there was a change towards increasingly large-scale,

DOI: 10.4324/9781003118565-4

extensive herding with herders subsisting primarily on domesticated animals (Bjørklund 2013).

In Fennoscandia, reindeer pastoralism has been associated with the Indigenous Sámi population, and the questions related to when, why and how it came about have been the focus of recurrent scientific debates. Vorren (1973) argued that Sámi communities shifted from hunting to reindeer pastoralism during the period 1550–1750 due to increased taxation, expanding trade relations and the introduction of firearms. Others have emphasized social tensions already inherent in Sámi hunting societies, which favoured a pastoral system with individual ownership over a hunting economy based on sharing (Hansen & Olsen 2014). Some archaeologists, on the other hand, have argued that the emergence of pastoralism can be dated back to the Viking age or the 9th to 13th century (Storli 1993), or even as early as the beginning of the Christian era (Aronsson 2009). Whatever the cause of the pastoral transition, the debate continues about whether the rapid growth of semi-domesticated reindeer herds actually involved the importation of a new domesticated type of reindeer not native to Fennoscandia (Røed et al. 2018), or if it primarily involved the adoption of husbandry techniques enabling different societies to domesticate wild stocks locally (Vorren 1973).

Genetic change associated with pastoral transition

Recently, examining DNA variation in archaeological specimens of various farm animals and comparing this with present-day material has helped to reveal the origin and spread of the domestication process. In reindeer from Finnmark county in northern Norway, such studies have shown genetic changes associated with the transition from a predominantly hunting economy to reindeer pastoralism (Bjørnstad et al. 2012; Røed et al. 2018). These studies analysed the control region of the mitochondrial DNA (mtDNA). The mtDNA is non-recombining and maternally inherited and therefore suitable as a genetic marker to study preserved demographic processes; it is, therefore, able to provide clues about the early history of reindeer husbandry. Different clusters of closely related mtDNA variants (haplotypes) represent genetic lineages preserved in the maternal lines through generations. Comparing the mtDNA in reindeer from 5000-year-old archaeological sites with those from the Medieval and more recent sites up to present-day semi-domesticated reindeer reveals that reindeer in Finnmark have gone through massive genetic replacement since Medieval times. This genetic transition is characterized by a significant loss of native haplotypes, together with the introduction of new ones (Figure 2.1). Out of a total of 62 mtDNA haplotypes identified in both the modern and archaeological samples, only 14 appear among samples known to represent semi-domesticated reindeer. This implies that the transition from the historical wild reindeer to today's semi-domesticated reindeer involved a significant bottleneck with massive loss of genetic variation. The pairwise genetic differences between sample sites also show a clear pattern of

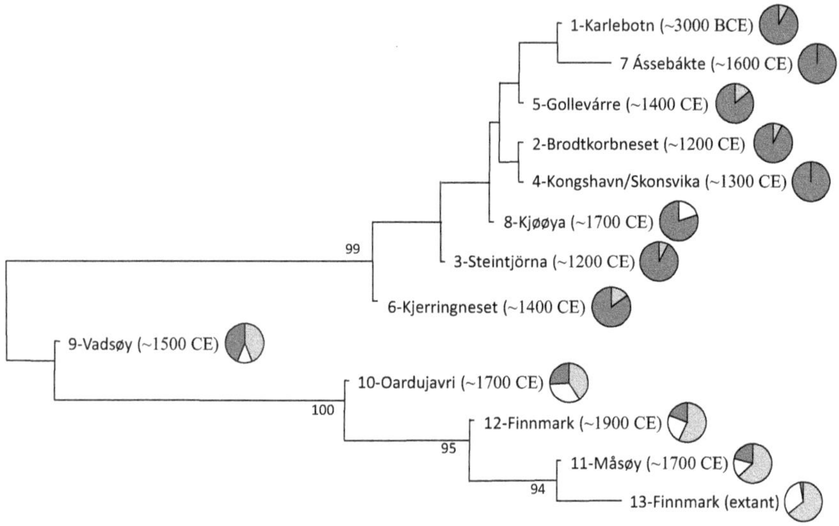

Figure 2.1 Population dendrogram based on mtDNA genetic distances and haplotype cluster distribution (pie charts) in prehistoric and historic reindeer from Finnmark county, northern Norway. In the dendrogram the support values from bootstrap replicates are given as percentages at the nodes. Number and age of sampling sites are given beside the pie charts, where dark grey represents mtDNA haplotype cluster I, light grey represents cluster Ib and white represents cluster II.

Source: (Modified from Røed et al. 2018).

low, or no, genetic differentiation between the ancient sites (1–8), while low or no genetic differentiation is found at sites 9–13, representing relatively more recent periods. However, between these two sets there is a substantial genetic difference, as illustrated by the high bootstrap values in the cluster analysis (Figure 2.1), suggesting a sudden genetic change over a relatively short period. This coincides with the transition of the Sámi economy from mainly hunting and gathering to a reliance on extensive reindeer pastoralism, indicating that the distinct genetic shift was closely associated with the beginning of pastoralism (Bjørnstad et al. 2012; Røed et al. 2018).

The fact that the archaeological samples show evidence of a reindeer population with high genetic variation and a homogeneous genetic structure up to the Late Medieval Period suggests relatively large wild reindeer herds in the region at that time. The detection of a major subsequent genetic bottleneck makes it likely that the wild reindeer populations were reduced and fragmented prior to, or during, the first phases of the pastoral transition. The putative population reduction and fragmentation of the wild reindeer herds may have allowed the domesticated type to expand rapidly. Notably, several other reindeer herding peoples across Russia experienced rapid growth in the herd size of domesticated reindeer in the 18th and 19th centuries (Krupnik 1993). This may indicate a more general drive for pastoralism, such as the onset of the Little Ice Age with the coldest interval occurring between the 17th and mid-19th centuries (Kaufman et al. 2009). Reindeer are well-adapted to cool summers and cold winters, and increasing domesticated herd sizes may have led to increased human mobility that greatly facilitated the hunting of wild reindeer. The decline of wild reindeer populations may not have preceded but followed the increase in number of semi-domesticated reindeer, although local political and economic factors may have been influential in encouraging herders to keep large herds (Krupnik 1993; Stépanoff 2017). The pastoralism was followed by a shift towards subsisting primarily on domesticated animals. Further reduction and eventual depletion of the wild reindeer may have taken place due to the challenges associated with the co-existence of both large wild and semi-domesticated herds.

The maternal genetic shift in Finnmark reindeer was not only characterized by loss of genetic variation but also by the replacement of haplotype clusters assumed to represent different maternal lineages. The lineage characterized by the cluster **I** haplotypes dominated the ancient material but became rare and is almost absent in extant semi-domesticated reindeer, while the opposite trend was seen for the lineages characterized by haplotype cluster **II**. The most common cluster **II** haplotypes among extant domesticated reindeer were completely absent in the more ancient specimens (Figure 2.1). The pastoral transition appears, therefore, to have been founded on a limited number of individuals with a maternal ancestry of partly non-native origin. The rapid growth in herd sizes from the 17th to 19th centuries may have facilitated the development of a unique reindeer type based on a small number of imported pioneers. Where these animals came from is an intriguing but genetically still unresolved issue.

Possible eastern influence on the early history of reindeer husbandry

The dominance of the genetic lineage characterized by mtDNA cluster **II** haplotypes among extant Nordic domesticated reindeer may imply a common ancestry. It has been suggested that cluster **II** evolved during the last glaciation period in a few refugia in southern Europe, isolated from the general Euro-Beringia lineage that survived the glaciation as part of the large Beringia refugium encompassing a major part of north-eastern Russia and parts of north-western America (Flagstad & Røed 2003). The cluster **II** lineage then migrated northwards as the ice retreated. Among Nordic semi-domesticated reindeer, cluster **II** haplotypes consisted of one haplotype at high frequency, with all other haplotypes radiating from this by one or two mutations, a pattern suggesting a sudden population expansion 2500–6000 years ago (Røed et al. 2021). This is well before the rapid growth associated with the pastoral transition and points towards the possibility of the modern semi-domesticated reindeer originating from one or a few rapidly growing wild populations.

The absence of the characteristic cluster **II** haplotypes in ancient reindeer from both Medieval and earlier sites in southern and northern Norway (Røed et al. 2011; 2014; 2018) suggests colonization of these haplotypes from the east. This is also in accordance with the decreasing gradient for this lineage from east to west among the modern semi-domesticated reindeer population in the Nordic countries (Røed et al. 2021). Probable refuge areas for these ancestral wild populations could have been in the current taiga areas in Fennoscandia or western Russia. Today, this is the habitat of the wild Finnish forest reindeer living in Finland and north-western Russia (Banfield 1961). Based on morphological data, the origin of the Scandinavian semi-domesticated reindeer from this population was first suggested by Lönneberg (1909). The present-day Finnish forest reindeer population is descended from a previously large population with a geographical distribution that probably covered the northern part of Finland and Sweden as well as western Russia. The population became nearly extinct in Finland and Sweden in the early 20th century but then recovered as some herds migrated from Russia to the Kainuu district in Finland during the 1950s (Nieminen 2013). The mtDNA haplotype cluster dominating in the Nordic semi-domesticated population (cluster **II**) has been reported to be present in this population, although at low frequency (Røed et al. 2008). More common distribution of this haplotype that has changed during the population decline in the early 20th century might be the case.

The present-day domesticated reindeer within the Nordic countries

Today, there are approximately 200,000 semi-domesticated reindeer in each of Norway, Sweden and Finland, and the husbandry area covers approximately 30%–50% of the area of each country. The number of animals kept on winter pastures is regulated by administrative units (Reinbeitedistrikter in Norway, Samebyar in Sweden and Paliskunta in Finland) and varies substantially from about 500 reindeer in Ikonen paliskunta in Finland to just above 20,000 reindeer in Karasjok reinbeitedistrikt in Norway. Reindeer husbandry represents a socio-ecological system with considerable cultural and ecological variation (Holand et al. 2021). This herding is traditionally associated with the Indigenous Sámi and has evolved as an adaptation to natural conditions, being moulded by history, competing land use and legal rights (Käyhkö & Horstkotte 2017). In contrast to Norway and Sweden where Sámi are the only people permitted to practice reindeer husbandry, all local citizens are entitled to own reindeer in Finland, with Sámi husbandry mainly confined to northern Lapland. In the mountain areas of south-central Norway, which is outside the Sámi herding areas, there are local farmers practising reindeer husbandry.

Genetic structure of Nordic semi-domesticated reindeer

To better understand both ancient and more recent processes that affect the genetic structure and variation in the Nordic semi-domesticated reindeer,

Røed et al. (2021) analysed variation in both the control region of mtDNA and in 18 DNA microsatellites in reindeer from 31 reindeer herding districts in Norway, Sweden and Finland (Figure 2.2). Microsatellites are highly variable nuclear markers, inherited by both males and females, and are thus particularly appropriate to reveal demographic processes related to both sexes. The characterizing of the genetic structure in the Nordic herds revealed genetic variation both within and between reindeer husbandry areas. In extant reindeer in Finland, compared to Sweden and Norway, there is distinct differentiation with respect to the nuclear markers, but less so in the maternal marker (mtDNA). The general pattern of relatively high levels of microsatellite variation within all husbandry areas indicates relatively large effective population sizes (i.e. with limited effects of ongoing inbreeding and genetic drift). Compared to the microsatellites, levels of mtDNA variation were greater between reindeer herding areas, with several herds showing highly reduced levels of genetic variation indicating previous bottlenecks (small effective population sizes) in the number of reproducing females. The discrepancy between the microsatellite and mtDNA variation can be explained by mtDNA being more prone to genetic loss and fixation, since its effective population size is only a quarter that of microsatellites (Moore 1995). Alternatively, the discrepancy may reflect the greater mobility of males compared to females both at present and historically, as well as introgression within the Nordic reindeer husbandry areas.

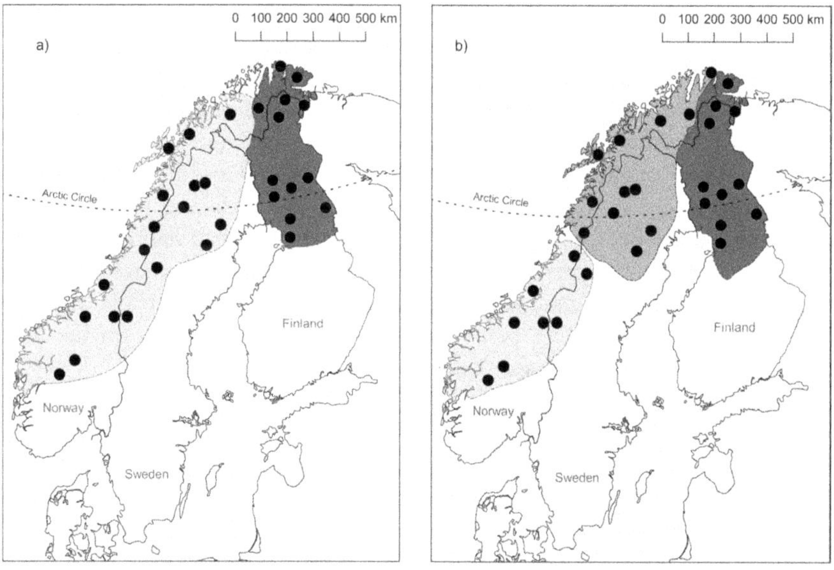

Figure 2.2 Spatial interference of the microsatellite ancestry distribution of semi-domesticated reindeer in the Nordic countries for (a) the main structure of two gene pools and (b) the three-part structure. Black dots indicate locations of husbandry areas sampled.

Source: (Modified from Røed et al. 2021).

Despite the relatively high amount of genetic variation in microsatellites, there were also substantial genetic differences between the Nordic populations. Analysing microsatellite genetic structure, based on the assignment of individual reindeer to genetic clusters, revealed two distinct gene pools, one dominating in Finland and one in Norway and Sweden combined, with the exception of the most northern herds in Norway which shared a gene pool with the Finnish reindeer (Figure 2.2a). In addition to this main structure, there was a substructure in which the joint Swedish and Norwegian gene pool could be divided into southern and northern parts, again except for the small area in northern Norway which was similar to the Finnish reindeer (Figure 2.2b).

In Finland, early nomadic reindeer husbandry is believed to have spread from the mountain areas of northern Sweden and Norway to the Käsivarsi area in north-western Finland early in the 17th century, from where large-scale reindeer herding gradually spread to other areas (Kortesalmi 2008). During the 17th and 18th centuries there was also frequent trade and transport of animals between the Finnish herders and the Indigenous reindeer herding people in the neighbouring eastern region (Kuusela et al. 2016). At that time, the taiga type of reindeer husbandry was common in both present Arkhangelsk Oblast and the Republic of Karelia on the eastern side of the current border between Finland and Russia (Koz'min 2003), implying a possible transfer of both knowledge and animals from these areas into the Finnish reindeer herding culture. The importation of particularly strong male reindeer with an eastern ancestry from Finland mostly to northern Norway for transport purposes could explain the present shared genomic pattern between these areas (Figure 2.2). Elsewhere more persistent maintenance of female-based herds primarily used for subsistence, including milking, may have been an important factor for the Swedish and Norwegian gene pool.

The distinct genetic separation of reindeer herds in Finland versus herds in Sweden and Norway combined has probably also been influenced by isolation after the closure of the border between Russia–Finland and both Norway in 1852 and Sweden in 1889. At the beginning of the 19th century, the yearly movements of reindeer were huge, with tens of thousands of reindeer crossing what have become the present borders between the four countries. After the border closures, these crossings became impossible, and this greatly affected the Nordic reindeer herders' pastoral livelihood. The restrictions on the seasonal migration pattern probably caused diversification of the reindeer genetic resources in Finland and the combined Sweden and Norway area. Notably, dissolution of the Swedish–Norwegian Union in 1905 appears not to have had similar genetic consequences, despite the resulting loss of access of particularly Swedish herders to important grazing and fishing grounds along the Norwegian coast (Riseth et al. 2016).

Social-ecological drivers of genetic structure

Rather than following the national borders, the genetic resources of reindeer in Norway and Sweden separated into southern and northern clusters, reflecting social-ecological processes across the countries. Possible relationships between the three-part genetic structure of the semi-domesticated reindeer in the Nordic countries and different social-cultural factors have been examined by Røed et al. (2021). Besides the national and ethnic contrasts, there is social-cultural variation within Sámi society as represented by the many Sámi languages (Sammalahti 1998). The cultural boundaries do not follow the national borders as most of the Sámi languages are spoken in more than one country. The language variants belong to the Fenno-Ugrian languages and can be divided into three main types: Eastern Sámi spoken mainly on the Kola Peninsula and in some parts of north-eastern Finland; Central Sámi spoken mainly in northern Finland, Sweden and Norway; and Southern Sámi spoken in southern areas of Norway and Sweden (Hermanstrand et al. 1919). Northern Sámi belongs to the Central Sámi language and is the most common Sámi language spoken in northern Finland, Sweden and Norway.

The three-part genetic structure of the Nordic semi-domesticated reindeer is strongly associated with nation states (Norway, Sweden and Finland), ethnicity (Sámi versus non-Sámi) and languages that characterize the different herding areas (Norwegian, Finnish, South Sámi and Central Sámi), with language group being clearly the best factor to explain the genetic clustering shown in Figure 2.2b. Particularly in Norway, the distribution of the genetic clusters follows the traditional language borders, with South Sámi dominating the southern sub-cluster and Central Sámi the more northern genetic sub-cluster. This illustrates that the different reindeer husbandry systems within the Nordic countries are closely associated with the socio-cultural gradients within Sámi society. Like other traditional subsistence uses of natural resources, reindeer husbandry is based on cultural transmission of traditional ecological knowledge to exploit and adapt to environmental changes. Both the long-term pattern of reindeer migration and awareness of preserving the cultural identities including the exchange of animals as part of dowries and friendship have probably contributed to the genetic structure seen today.

The characterization of the genetic structure among the Nordic domesticated reindeer herds revealed no obvious differences in herding practices between the two genetic sub-clusters in Sweden and Norway. Within the same genetic cluster, the different herding types from the coastal climate in northern Norway across the alpine area in inland Sweden to more coastal and forested areas in north-eastern Sweden reflect ecological plasticity rather than genetic adaptations. Although one would expect some adaptive processes responding to the ecological diversity, the genetic structure detected by neutral genetic markers appears primarily to reflect history and spread of the pastoralist culture, while social-cultural variation across the Nordic reindeer husbandry areas is a secondary influence.

The distinct genetic structure of Eurasian semi-domesticated reindeer

At least two hypotheses regarding when and where reindeer herding originated in Eurasia have been debated (Gordon 2003). One theory, the diffusion theory or monocentric hypothesis, suggests that semi-domesticated reindeer first appeared a few thousand years ago east of the Urals in the southern part of the

Siberian taiga, when they spread to other regions. A second theory, the evolutionary theory or polycentric hypothesis, suggests that domestication of reindeer occurred independently multiple times in different parts of Eurasia. Studies based on both mtDNA and microsatellites show distinct genetic differentiation between Nordic and Russian semi-domesticated reindeer (Røed et al. 2008; Kvie et al. 2016) – a pattern also reported when using whole-genome DNA sequencing (Weldenegodguad et al. 2020). The differentiation probably reflects historic and evolutionary events and further implies that semi-domesticated reindeer in the two regions have different domestication origins.

A separate analysis of both microsatellites and the mitochondrial control region of semi-domesticated reindeer from 25 sites across Eurasia yielded a similar distinct genetic differentiation between Nordic and Russian reindeer (Kvie et al. unpublished), adding more support to the polycentric hypothesis (Figure 2.3a).

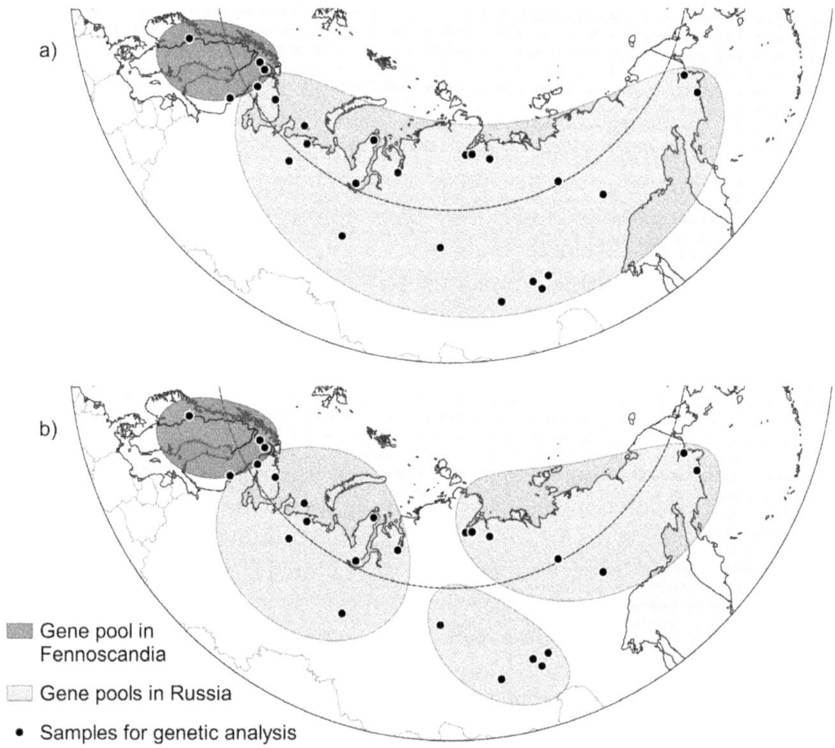

Figure 2.3 Genetic separation of Eurasian semi-domesticated reindeer into two and four groups based on individual assignment analyses in the software STRUCTURE. (a) The main structure with a separation between the Nordic and the Russian semi-domesticated reindeer. (b) Further sub-structuring divides the Russian semi-domesticated reindeer into north-western, north-eastern and southern genetic groups.

From the analysis, further sub-structuring was apparent, supporting the presence of as many as four genetic clusters within Eurasian semi-domesticated reindeer (Figure 2.3b). Among them, the semi-domesticated reindeer in the Nordic countries still comprise a distinct cluster, while those in Russia could be separated into north-western, north-eastern and southern clusters. This separation was evident only from the microsatellite data and could not be confirmed from mtDNA analyses – implying that these clusters probably reflect more recent demographic processes, rather than different domestication origins within Russia. Reindeer husbandries in Russia are usually separated into tundra and taiga forms. Beyond this general distinction, they are subdivided into four official breeds: Nenets, Even, Evenkiyskaya and Chukot (hargin), which are named after the ethnic groups assumed to have established these breeds showing particular traits and adaptations to their own environment (Zabrodin & Borozdin 1989). Analyses of single nucleotide polymorphisms (SNPs) in Russian reindeer by use of a Bovine 700K SNP Chip have revealed the genetic uniqueness of each of these breeds; it has been suggested that this reflects ecological processes, internal gene flow, breeding practices and geographical features (Kharzinova et al. 2020). Gene flow and breeding practices are also emphasized by Kvie et al. (unpublished) to explain the three-part genetic structure revealed by microsatellites, which appear to overlap with established Russian husbandry zones, namely the north-western, north-eastern and southern (Siberian taiga) zones (Klokov 2012).

However, Russian reindeer herding is without doubt very diverse and includes several different economic strategies associated with the natural and social environment (Klokov 2012). Hence, future studies on population genomics may be able to provide clearer answers about the extent to which human impact and adaptation to different environments drove the differentiation within Russia. Nevertheless, analyses based on neutral markers, e.g., microsatellites and mtDNA, have revealed that the semi-domesticated reindeer in the Nordic countries form a distinct group, probably originating from a different source population than the Russian domesticated reindeer. This further implies that Nordic domesticated reindeer should probably be considered a separate genetic type within Eurasia, with the conservation and management responsibilities that follow such a status.

References

Aronsson, K-Å. (2009). Relations between man and reindeer – traces of reindeer herding. In Recent perspectives on Sámi archaeology in Fennoscandia and North-West Russia. *ISKOS.* 17, 17–24.

Banfield, A. W. F. (1961). *A revision of the reindeer and caribou, genus Rangifer.* National Museum of Canada. Bull. No. 177. Biol. Ser.66.

Bjørklund, I. (2013). Domestication, reindeer husbandry and the development of pastoralism. *Acta Borealia.* 30(2), 174–189.

Bjørnstad, G., Flagstad, Ø., Hufthammer, A. K. & Røed, K. H. (2012). Ancient DNA reveals a major genetic change during the transition from hunting economy to reindeer husbandry in northern Scandinavia. *Journal of Archaeological Science*. 39(1), 102–108.

FAO (2015). *Coping with climate change – the roles of genetic resources for food and agriculture*, Rome, Italy.

Flagstad, Ø. & Røed, K. H. (2003). Refugial origins of reindeer (*Rangifer tarandus* L.) inferred from mitochondrial DNA sequences. *Evolution* 57, 658–670.

Gordon, B. (2003). Rangifer and man: an ancient relationship. *Rangifer.* 14, 15–28.

Groeneveld, L. F., Lenstra, J. A., Eding, H., Toro, M. A., Scharf, B., Piling, D., Negrini, R., Finlay, E. K., Jianlin, H., Groeneveld, E., Weigend, S. & The GLOBALDIV Consortium (2010). Genetic diversity in farm animals – A review. *Animal Genetics.* 41, 6–31.

Hansen, L. I. & Olsen, B. (2013). *Hunters in Transition: An Outline of Early Sámi History*. Leiden: Brill.

Hermanstrand, H., Kolberg, A., Nilssen, T. R. & Sem, L. (2019). *The Indigenous Identity of the South Saami: Historical and Political Perspectives on a Minority Within a Minority*. New York: Springer Nature.

Holand, Ø., Moen, J., Kumpula, J., Löf, A., Rasmus, S. & Røed, K. (2021). Project ReiGN: Reindeer Husbandry in a Globalizing North – Resilience, Adaptations and Pathways for Actions. *In:* Nord, D. C. (ed.) *Nordic Perspectives on the Responsible Development of the Arctic: Pathways to Actions*. Springer Polar Sciences. New York: Springer Nature. 227–248.

Horstkotte, T. (2013). *Contested Landscapes: Social-Ecological Interactions Between Forestry and Reindeer Husbandry*. Ph.D. diss., Umeå University.

Jernsletten, J. L. & Klokov, K. B. (2002). *Sustainable Reindeer Husbandry*. University of Tromsø, Tromsø, Norway.

Kaufman, D. S., Ager, T. A., Anderson, N. J., Anderson, P. M., Andrews, J. T., Bartlein, P. J., Brubaker, L. N., Coats, L. L., Cwynar, L. C., Duvall, M. L., Dyke, A. S., Edwards, M. E., Eisner, W. R., Gajewski, K., Geirsdottir, A., Hu, F. S., Jennings, A. E., Kaplan, M. R., Kerwin, M. W., Lozhkin, A. V., MacDonald, G. M., Miller, G. H., Mock, C. J., Oswald, W. W., Otto-Bliesner, B. L., Porinchu, D. F., Ruhland, K., Smol, J. P., Steig, E. J. & Wolfe, B. B. (2009). Holocene thermal maximum in the western Arctic (0–180 W). *Quaternary Science Reviews*. 23(5–6), 529–560.

Käyhkö, J. & Horstkotte, T. (2017). *Reindeer Husbandry Under Global Change in the Tundra Region of Northern Fennoscandia*. Report number 1. Turku: University of Turku.

Kharzinova, V., Dotsev, A., Solovieva, A., Sergeeva, O., Bryzgalov, G., Reyer, H., Wimmers, K., Brem, G. & Zinovieva, N. (2020). Insight into the current genetic diversity and population structure of domestic reindeer (*Rangifer tarandus*) in Russia. *Animals*. 10, 1309.

Klokov, K. B. (2012). Changes in reindeer population numbers in Russia: an effect of the political context or of climate. *Rangifer*. 32(1), 19–33.

Kofinas, G., Osherenko, G., Klein, D. & Forbes, B. (2000). Research planning in the face of change: the human role in reindeer/caribou systems. *Polar Research*. 19(1), 3–21.

Kortesalmi, J. J. (2008). *Poronhoidon synty ja kehitys Suomessa*. Helsinki: Suomalaisen Kirjallisuuden Seura.

Koz'min, V. A. (2003). *Olenevodcheskaia Kul'tura Narodov Zapadnoi Sibiri*. Saint-Petersburg: Izdatel'stvo Peterburgskogo Universiteta.

Krupnik I. (1993). *Arctic Adaptations. Whalers and Reindeer Herders of Northern Eurasia*. London: University Press of New England.

Kuusela, J. M., Nurmi, R. & Hakamäki, V. (2016). Co-Existence and colonisation: re-assessing the settlement history of the pre-Christian Bothnian Bay coast. *Norwegian Archaeological Review*. 49(2), 177–203.

Kvie, K. S., Heggenes, J. & Røed, K. H. (2016). Merging and comparing three mitochondrial markers for phylogenetic studies of Eurasian reindeer (*Rangifer tarandus*). *Ecology and Evolution*. 6(13), 4347–4358.

Kvie et al. (in prep) *Genetic Variation and Structure in Wild and Domestic Reindeer Across Eurasia*.

Lönneberg, E. (1909). Taxonomic notes about Palearctic reindeer. *Arkiv for Zoologi*. 6(4), 1–18.

Moore, W. S. (1995). Inferring phylogenies from mtDNA variation: mitochondrial-gene trees versus nuclear-gene trees. *Evolution*. 49(4), 718–726.

Murashkin, A. I., Kolpakov, E. M., Shumkin, V. Y., Khartanovich, V. I. & Moiseyev, V. G. (2016). Kola Oleneostrovskiy grave field: a unique burial site in the European Arctic. *ISKOS*. 21, 186–99.

Næss, M. W. (2010). Contradictory evidence as a guide for future research: Investigating the relationship between pastoral labour and production. *Nomadic Peoples*. 14(1), 51–71.

Nieminen, M. (1992). Siperian matkakuvia ja kokemuksia. II. *Porojen maitotalous entisessa Neuvostolutossa. Poromies*. 9, 10–15.

Nieminen, M. (2013). Response distance of wild forest reindeer (*Rangifer tarandus fennicus Lönnb*) and semi-domestic reindeer (*R.t. tarandus L.*) to direct provocation by a human on foot/snowshoes. *Rangifer*. 33(1), 1–15.

Pape, R. & Löffler, J. (2012). Climate change, land use conflicts, predation and ecological degradation as challenges for reindeer husbandry in northern Europe: what do we really know after half a century of research? *Ambio*. 41(5), 421–434.

Riseth, J. Å., Tømmervik, H. & Bjerke, J. W. (2016). 175 years of adaptation: North Scandinavian Sámi reindeer herding between government policies and winter climate variability (1835–2010). *Journal of Forest Economics*. 24, 186–204.

Røed, K. H., Flagstad, Ø., Nieminen, M., Holand, Ø., Dwyer, M. J., Røv, N. & Vilà, C. (2008). Genetic analyses reveal independent domestication origins of Eurasian reindeer. *Proceedings of the Royal Society B-Biological Sciences*. 275(1645), 1849–1855.

Røed, K. H., Flagstad, Ø., Bjørnstad, G. & Hufthammer, A. K. (2011). Elucidating the ancestry of domestic reindeer from ancient DNA approaches. *Quaternary International*. 238(1–2), 83–88.

Røed, K. H., Bjørnstad, G., Flagstad, Ø., Haanes, H., Hufthammer, A. K., Jordhøy, P. & Rosvold, J. (2014). Ancient DNA reveals prehistoric habitat fragmentation and recent domestic introgression into native wild reindeer. *Conservation Genetics*. 15(5), 1137–1149.

Røed, K. H., Bjørklund, I. & Olsen, B. J. (2018). From wild to domestic reindeer – Genetic evidence of a non-native origin of reindeer pastoralism in Northern Fennoscandia. *Journal of Archaeological Science: Reports*. 19, 279–286.

Røed, K. H., Kvie, K. S., Bårdsen, B-J., Laaksonen, S., Lohi, H., Kumpula, J., Aronsson, K-Å., Åhman, B., Våge, J. & Holand, Ø. (2021). Historical and social-cultural processes as drivers for genetic structure in Nordic domestic reindeer. *Ecology and Evolution*. 11, 8910–8922.

Sammalahti, P. (1998). *The Saami Languages. An Introduction*. Davvi Girji, Kárášjohka, Norway.

Sandström, P. (2015). *A toolbox for co-production of knowledge and improved land use dialogues: the perspective of reindeer husbandry*. Ph.D. diss., Swedish University of Agricultural Sciences.

Skarin, A. & Åhman, B. (2014). Do human activity and infrastructure disturb domesticated reindeer? The need for the reindeer's perspective. *Polar Biology*. 37(7), 1041–1054.

Stépanoff, C., Marchina, C., Fossier, C. & Bureau, N. (2017). Animal autonomy and intermittent coexistences: North Asian modes of herding. *Current Anthropology*. 58(1), 57–81.

Storli, I. (1993). Sami Viking age pastoralism-or the fur trade paradigm' reconsidered. *Norwegian Archaeological Review*. 26(1), 1–20.

Tornæus, J. & Wiklund, K. B. (1900). *Johannis Tornæi Berättelse om Lapmarckerna och deres tilstånd*. Uppsala: Wretmans tryckeri.

Vorren, Ø. (1973). Some trends of the transition from hunting to nomadic economy in Finnmark. *In:* Berg, G. (ed.) *Circumpolar Problems: Habitat, Economy and Social Relations in the Arctic*. Oxford: Pergamon Press. 185–194.

Vors, L. S. & Boyce, M. S. (2009). Global declines of caribou and reindeer. *Global Change Biology*. 15(11), 2625–2633.

Weldenegodguad, M., Pokharel, K., Ming, Y., Honkatukia, M., Peippo, J., Reilas, T., Røed, K. H. & Kantanen, J. (2020). Genome sequence and comparative analysis of reindeer (*Rangifer tarandus*) in northern Eurasia. *Scientific Reports*. 10, 8980.

Zabrodin, V. A. & Borozdin, E. K. (1989). Deer. *In:* Dmitriev, N. G. & Ernst, L. K. (eds) *Animal Genetic Resources of the USSR*. Rome: FAO.

Part II

Reindeer in their environment

3 Reindeer behavioural ecology and use of pastures in pastoral livelihoods

Anna Skarin, Jouko Kumpula, Torkild Tveraa and Birgitta Åhman

Seasonality, reproductive phenology and social organization

Reindeer, and in turn reindeer husbandry, are adapted to the arctic/sub-arctic environment where seasonal variation typically provides abundant food in summer, but there is a food shortage in winter (Kerby & Post 2013). Reindeer accumulate body reserves (fat and muscle tissue) in summer and use these reserves during winter (Klein 1986). As a rule of thumb, food abundance in summer determines the growth and size of animals, while food abundance in winter determines density and fecundity (Klein 1965).

Seasonality of reproduction commonly depends on seasonality in the environment, while the synchrony of parturition may be linked to the risk of predation (Kerby and Post 2013). Reindeer parturition starts in early May and may continue until early June (Figure 3.1). The herders' decision about when to start migration is based on the reindeer's willingness to move, snow conditions and grazing conditions in the calving area.

Before calving, the pregnant females separate from the rest of the herd and last year's calf is usually pushed away from its mother (Espmark 1971; Kojola 1993). At parturition, the female also keeps away from other females, and usually spends some days alone with her new calf; this is believed to strengthen the bond between mother and young (Espmark 1971). The calf follows the mother, often during the whole of the first year of life and learns how to find forage in winter (Kojola 1993). Reindeer seem to invest more in female calves as they follow the mother until the next parturition, while male calves are pushed away earlier (Kojola 1993).

During the post-calving period, females with their calves (and male reindeer, separately) merge to form large herds to escape insects as the likelihood of being harassed is lower in groups (Downes et al. 1986; Mörschel & Klein 1997; Fauchald et al. 2007). The gregarious behaviour is most evident when only mosquitoes and species of Simulidae are present, while the reindeer-specific parasites warble and nose bot flies seem to make the reindeer form smaller herds or spread out more (Downes et al. 1986; Mörschel & Klein 1997; Skarin et al. 2004). The tundra-dwelling reindeer ecotypes exhibit more gregarious

DOI: 10.4324/9781003118565-6

behaviour than forest-dwelling ecotypes, which tend to form smaller herds (Helle & Aspi 1984). When the calves are older and stronger, the herders gather the herds for calf marking. This usually takes place from the end of June up to the beginning of August, but in some areas, calf marking takes place in mid-September and some remaining calves are even marked in the following winter.

In late summer, when insect harassment has eased, the herd expands over larger areas. Then in early autumn, the bulls prepare for the rutting season and start to gather harems. The rut starts in late September and usually lasts until mid-October. During the rut males loose bodyweight and afterwards they drop their antlers. Bull slaughter usually takes place just before the rut.

Migration to winter pasture takes place in autumn. Before migration, the herds are gathered for slaughter and separation of the remaining herd into winter grazing groups. The reindeer either move to the winter pastures on foot (by herding or by letting them move freely) or on trucks depending on connectivity between the summer and winter pastures. In southern and central parts of the herding area in Finland, most reindeer are kept in enclosures for supplementary winter feeding (Chapter 12).

Females keep their antlers throughout the winter and defend feeding craters for themselves and their calves (Espmark 1964). Large adult females with big antlers have the most dominant position in the herd. The social interactions

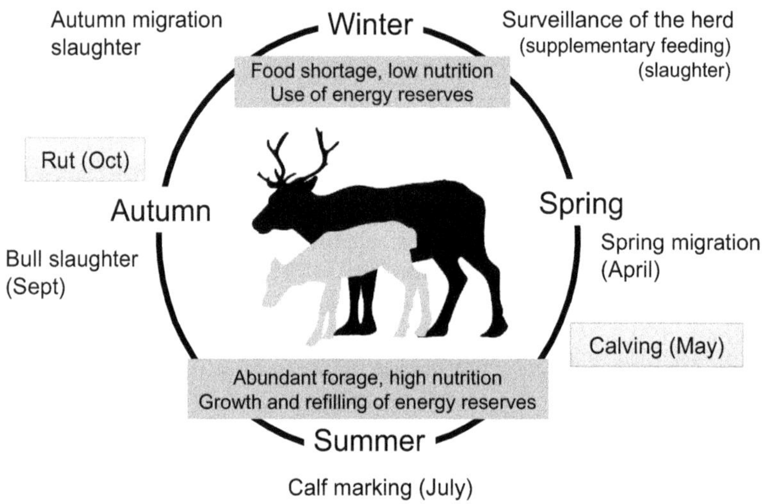

Figure 3.1 The annual cycle of reindeer husbandry reflects the animals' life history, behaviour and spatial utilization of the land and the herders' main operations as practised in most parts of Fennoscandia, with the exception that bull slaughter is not practised in Finland, but is in many herds in Sweden and Norway. Main annual slaughter of calves takes place in connection with autumn migration from mid-October onwards.

in a reindeer herd help winter foraging resources to be allocated primarily to the reproductive females (Kojola 1993; Holand et al. 2012). Antagonistic behaviour among group members is common and dominant animals may, e.g., steal newly dug craters from subordinate animals, and notably from antlerless males. These fine-scale mechanisms may have implications for resource selection at the population level (Torney et al. 2018) and force low-ranked animals (such as reindeer bulls) to forage in less profitable habitats (Holand et al. 2012). Reproductive females are usually those that lead the movements of the herd (Thomson 1975). They also often act as look-outs and defenders and seem to guard the security of the herd. Herders often equip certain adult females with a bell, and nowadays a GPS-collar, as a way to track down the herd.

Habitat selection, foraging behaviour and use of seasonal pastures

Reindeer are intermediate ruminants and generalist feeders moving through the landscape utilizing the most nutritious and digestible forage plants in summer (Hofmann 1989; Trudell & White 1981; Iversen et al. 2014), while in winter lichens and evergreen plants dominate the diet and snow controls where the reindeer can graze (Åhman & White 2018). Reindeer selection of habitat is best understood as a series of behavioural decisions at both large and fine temporal and spatial scales, from selecting seasonal areas to choosing the most nutritious part of a plant (Senft et al. 1987; Mayor et al. 2009; Skarin & Åhman 2014). The reindeer's behavioural response and the herder's actions towards the environment vary with reproductive cycle, season, herd density and availability of land. Reindeer herders often refer to the importance of high-quality pastures free from disturbance, where the reindeer can find 'grazing peace' (Inga 2007). Herders' decisions are integrated with the reindeer behaviour and may be hard to separate. In general, herders have the most impact on the large-scale migrations and selection of regional areas, while shorter movements within designated seasonal pastures and fine-scale selection of grazing patches are a choice made by the reindeer. There is usually more intense herding in winter compared to the snow-free season. In summer, most reindeer herds are freely ranging within the borders of the reindeer herding districts.

Calving and early summer

Reindeer show site fidelity to their calving ground, especially parturient females (Schaefer et al. 2000; Garfelt-Paulsen et al. 2021). Within the calving area, females seek out snow-free patches either in the mountain tundra or, if in the forest, close to open areas to give birth to and nurse the calf (Skogland 1984; Skarin et al. 2008, 2015). Males also seek out snow-free mountains or areas, but well separated from the females. When the new green vegetation starts to emerge, reindeer change from a lichen-rich diet to feed on vascular plants. Newly emerging parts of graminoids (e.g. *Anthoxanthum odoratum, Carex bigelowii, Deschampsia flexuosa, Eriophorum angustifolium, E. vaginatum, Nardus stricta*) and dwarf shrubs (*Vaccinium myrtillus*) are important (Warenberg 1982; Skogland 1984) as they are high in nitrogen (protein) and low in fibre, thus being easily digestible (Klein 1990; Åhman & White 2018). Digestibility of the diet has a major impact on energy intake, and thus on reindeer weight gain (White 1983). As soon as they start to emerge, forbs like *Comarum palustre, Gnaphalium supinum*, and *Potentilla erecta* start to be eaten, and sprouts and leaves of woody taxa, such as willow and dwarf birch (e.g., *Salix lanata* and *Betula nana*), are also preferred (Warenberg 1982; Skogland 1984).

Reindeer are far more sensitive to virtually all sources of anthropogenic disturbance during the calving period than during any other season. This is true for both semi-domesticated and wild reindeer, and it is most likely connected to the fear of predation and the importance of a calm environment for the female and calf (Vistnes & Nellemann 2008; Panzacchi et al. 2013; Skarin & Åhman 2014). Breeding females tend to seek out areas where they can see approaching danger, e.g., predators (Pinard et al. 2012; Sivertsen et al. 2016; Skarin et al. 2018). 'Green-wave surfing' describes how animals are expected to follow waves of resources and select habitats with an optimal balance of forage quality and quantity (Merkle et al. 2016). It has, however, been found that a high abundance of predators (brown bear) may hinder reindeer's optimal use of these resource waves (Rivrud et al. 2018). As the calf grows, the females gradually increase their movements and select ranges in relation to insect harassment and plant phenology (Skarin et al. 2010; Rivrud et al. 2018).

Mid and late summer

Harassment from mosquitoes, blackflies, horseflies and the reindeer-specific parasites, warble and nose bot flies (*Hypoderma tarandi* and *Cephenemiya trompe*), play an important role in habitat selection during summer. In tundra and mountainous regions reindeer typically prefer summits and ridges, and ideally snow patches, to escape the insects (Downes et al. 1986; Hagemoen & Reimers 2002; Skarin et al. 2008, 2010). There is a trade-off between nutritious rich river valleys and wind-exposed summits and ridges (Skarin et al. 2008, 2010). Forest-dwelling reindeer seek out sandy patches, dirt roads, river banks and other open land such as mires and clear cuts (Helle & Aspi 1984). Escaping insects reduces the time available for foraging (Colman et al. 2003), and years with high insect activity have been related to lower slaughter weights in the autumn (Weladji et al. 2003). The negative effects may, however, be reduced in regions with high forage quantity and short distances to insect-free habitats (Skarin et al. 2020). During insect harassment, reindeer may exhibit higher tolerance towards anthropogenic disturbance (Pollard et al. 1996; Skarin et al. 2004): it seems more important to avoid biting insects and warble flies than to avoid other disturbances.

Towards the end of the summer, mushrooms become an important part of the reindeer diet, constituting up to 25 per cent (Boertje 1984; Launchbaugh

& Urness 1993). Northern Sámi herders use the expression *vistit* to explain that 'the reindeer goes after mushrooms' or 'something it likes' (Inga 2007). Herders commonly refer to reindeer spreading out and report that it is hard to gather and herd reindeer during the mushroom season.

Autumn and winter

In autumn, mires are important, because in them reindeer are able to forage green shoots and roots from graminoids and various *Carex* species (Skjenneberg & Slagsvold 1968; Storeheier 2003). As annual plants wither, the role of evergreen perennial plants and lichens increases. Herders usually report that reindeer start eating lichens and dwarf shrubs a while before the snow arrives.

Reindeer have a unique adaptation in being able to digest lichens. Depending on their availability, lichens may comprise up to 70 per cent of the diet of reindeer, although they cannot survive on this alone because of their low nitrogen and macro-mineral content (Storeheier et al. 2003; Åhman & White 2018). The most common terrestrial species eaten by reindeer are the *Cladina* species, although *Cetraria nivalis* have similar digestibility and may also be consumed (Storeheier et al. 2002). Protein sources consist of the green parts of graminoids and some evergreen shrubs (e.g., *Vaccinium myrtillus* and *Empetrum nigrum*) (Boertje 1984; Storeheier et al. 2003; Åhman & White 2018). Mosses may be found in the diet although they are less preferred (White 1983). In forested areas, arboreal lichens (*Alectoria* and *Bryoria*) may constitute an important food source in late winter and when dense snow or ice crust limits the access to ground vegetation.

In winter, the availability of lichens and other ground vegetation is highly dependent on snow conditions (Helle 1984; Inga 2007; Roturier & Roué 2009). In the Sámi languages, there are numerous words for snow and snow conditions related to reindeer grazing (Ryd & Rassa 2001). Good grazing conditions depend on stable temperature and precipitation resulting in soft snow that is easy to dig in throughout the winter. In addition, under the canopy of old trees in mature forests, the snow usually stays softer and thinner (Chapter 4; Inga 2007; Horstkotte et al. 2014). Repeated fluctuations above and below freezing point and extreme snow depth are known to cause problems for reindeer grazing. Herders have different strategies to help the reindeer find food under such circumstances. Reindeer may be split into smaller units with frequent movements according to local variations in grazing conditions or allowed to spread out and seek patches with suitable grazing conditions for themselves. Another possibility is supplementary feeding (see Chapter 12).

Evolving use of pastures

The past wild reindeer populations in Fennoscandia lived in various vegetation and landscapes types using different areas depending on the biogeography and availability of forage. Early reindeer pastoralism (see Chapters 1 and

2) developed with varying degrees of nomadism, longer seasonal migrations between vegetation zones (mountain vs forest areas) in some areas and shorter migrations, primarily between different habitat types within the same vegetation zone (lichen-rich forest vs mires), in others. Use of land has always been adjusted to the reindeer's needs and adapted to the landscape.

Early reindeer herding was not restricted by national borders. In Sweden and Norway, migration typically took place along large river valleys, with winter pasture towards the eastern Baltic coast and summer pasture towards the Atlantic coast in the west. Reindeer further north in Finnmark Norway and east in Finland and Russia migrated from winter ranges in forests in Russia and Finland to summer ranges along the northern Atlantic coast in Norway and Finland. Over time, reindeer herding has however been forced to adapt to the gradual closing of national borders (Chapter 1), leading to sub-optimal utilization of pastures in many areas (Tyler et al. 2021). There were forced relocations of reindeer and Sámi reindeer herders within Sweden (Cramér & Ryd 2012), summer pastures were transformed to winter pastures in Norway (Tveraa et al. 2007) and in a large part of Finland reindeer herding changed to having all seasonal pasture within boreal forest.

Present organization and pasture use

The present use of pastures by reindeer and reindeer husbandry in Norway, Sweden and Finland thus has different ecological and historical backgrounds. Today, three main forms of reindeer herding strategy can be distinguished: (1) seasonal migrations between mountain or Atlantic coast summer pastures to winter pastures in taiga or tundra (Sweden and Norway), (2) seasonal migration between summer pasture on inland mountains and winter pasture by the Atlantic coast (Norway) and (3) year-round grazing in the taiga or mountain area (Sweden and Finland) (Chapter 1). In all countries, reindeer husbandry is divided into reindeer herding districts, the size of the districts and number of reindeer within each district vary considerably, depending on the landscape, historic borders and administrative decisions (Chapter 9).

Norway

The reindeer husbandry is divided into 84 summer reindeer herding districts and, within some districts, herds are further divided into smaller units. The herds in Finnmark (47) move between Atlantic coastal summer pastures and winter pastures in the interior with continental climate and generally shallow snow and good access to forage (Figure 3.2). In Troms, some herds stay year-round on islands, while others make short migrations between summer pastures along the coast and winter pastures further inland. The winter pastures in this region are under the strong influence of the oceanic climate and are frequently inaccessible due to deep and/or crusted snow. In Nordland and Nord-Trøndelag, reindeer use summer pastures in the inland mountains, often close

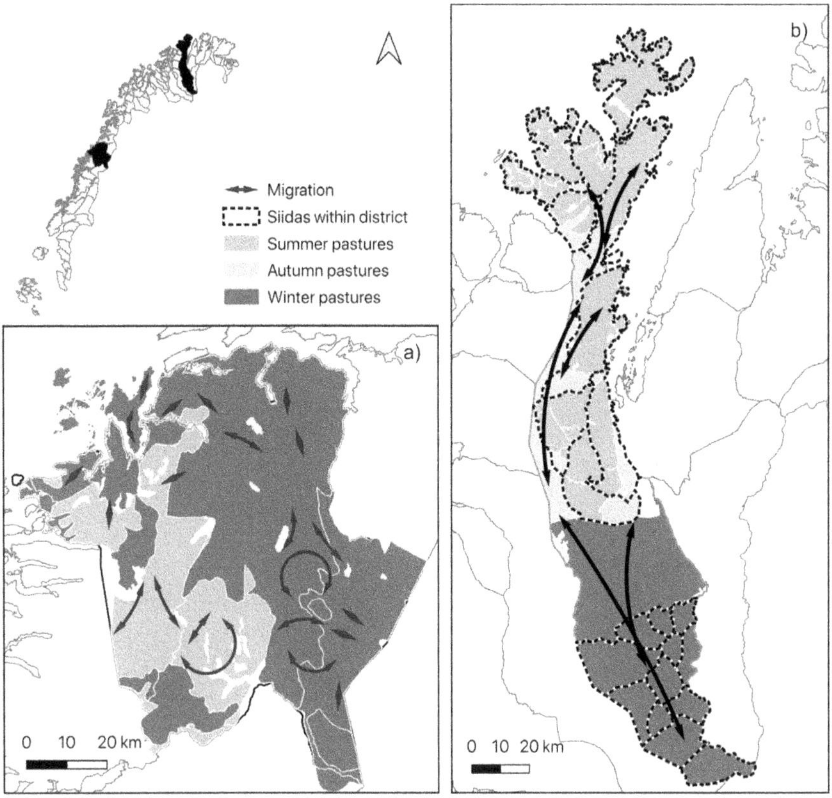

Figure 3.2 Use of and migration routes between summer, autumn and winter pastures in (a) Saltfjellet reindeer herding district in Nordland and (b) Kárašjoga oarjjabealli reindeer herding district in Finnmark.

to the Swedish border, and winter pastures along the Atlantic coast or in the lowlands where precipitation more often falls as rain and the snow is shallower. In Sør-Trøndelag and Hedmark, most reindeer move into continental winter pastures in the Femunden area with generally shallow snow and good access to food. Five reindeer herding districts are operated by non-Sámi people on Concession areas in the mountain range at the southern fringe of the reindeer herding area, adjacent to the wild reindeer herds.

Sweden

In Sweden, the 51 reindeer herding districts are divided into year-round land (used primarily in the snow-free period) and winter pastures where reindeer are only allowed to stay from 1 October to 30 April. There are 33 mountain herding districts. These are mostly long and narrow and use summer pastures

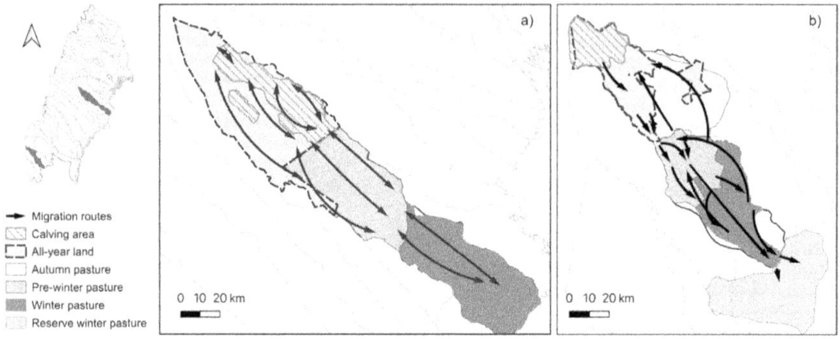

Figure 3.3 Use of and migration between calving area, summer, autumn and winter pastures in (a) Malå forest reindeer herding district in Västerbotten County and (b) Mittådalen mountain reindeer herding district in Jämtland County, Sweden. In April, migration is usually undertaken on foot to the calving and summer (year-round land) pastures solely in the forest or the mountain region, respectively.

in the mountains in the west and winter pastures in the boreal forest in the east towards the Baltic coast (Figure 3.3). There are ten forest herding districts, one in Västerbotten County and nine in Norrbotten County, that use pasture in the boreal forest areas all year. Winter pastures in Sweden are generally influenced by a cold continental climate with proper snow winters; however, pastures close to the Baltic coast may be influenced by a more maritime climate. Migrations between seasonal pastures are undertaken on foot or by truck, depending on the connectivity between the pastures. Eight Concession herding districts rotate the reindeer in the forest region between the Kalix River in the west and the Torne River (Finnish border) in the east.

Finland

In Finland, the reindeer herding area is divided into 54 reindeer herding districts: the 13 northernmost are Sámi reindeer herding districts and the others are Finnish reindeer herding districts. A mosaic of various coniferous forests and mires dominates the landscape in the southern and central parts of the herding area, while in the north tundra and mountain birch forests are more common. The natural scattering and short distance between winter and summer pastures in Finland have made the seasonal pasture rotation system more sedentary than in Sweden and Norway (Figure 3.4). In the small districts in the south and central parts of the reindeer herding area, reindeer move freely between summer and winter pastures. In the larger districts in the middle and northern parts, there is a distinct migration between seasonal pastures. Several of these districts have also separated the summer and winter pastures by fences, to avoid

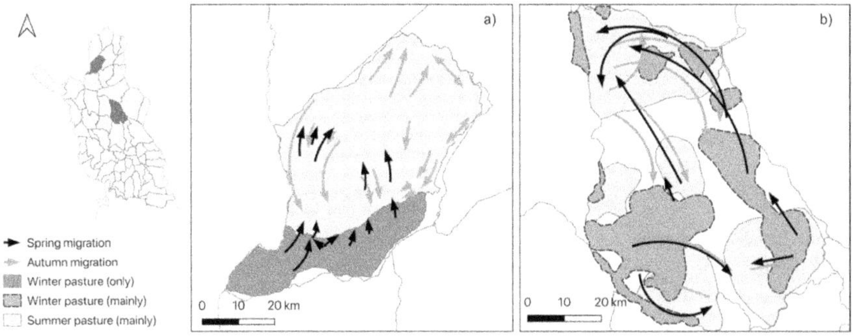

Figure 3.4 Use of and rotation between summer and winter pasture in (a) Muotkatunturi Sámi Finnish reindeer herding district and (b) Oraniemi Finnish reindeer herding district in the northern and central parts of the reindeer husbandry area in Finland, respectively.

trampling of winter lichen pastures by reindeer in summer. In the north, most Sámi districts have either a clear separation of seasonal pastures by means of fences or they herd the reindeer to the different seasonal pastures.

Concluding remarks

The fundamental resource for reindeer herding is reindeer access to pastures in all seasons. Over time, the closing of national borders, loss of land to industry and infrastructure and growing disturbance from various human activities have led to loss of suitable land for reindeer grazing (Chapters 4 and 5).

There has been an actual loss of land, resulting in areas that the reindeer cannot use anymore, barriers in the terrain limiting the access to an area and disturbances making reindeer avoid otherwise suitable grazing areas. Although, reindeer as a species seem to exhibit behavioural plasticity in relation to disturbances, they often move away from disturbance (Helle et al. 2012; Skarin & Åhman 2014), indicating the importance of disturbance-free pastures where they can find grazing peace.

The quality of pastures for reindeer is partly favoured by their own grazing, e.g., stimulating the creation of graminoid meadows on summer ranges and favouring biodiversity in general (Bråthen et al. 2007; Sundqvist et al. 2019). Indeed, closing the border between Finland and Norway and thus preventing reindeer in Finland from leaving the winter grounds and migrating to the summer pastures in Norway caused degradation of lichen-rich heaths and turned them into graminoid heaths with dwarf shrubs and mosses, leaving the herds with little lichen pasture on the Finnish side of the border (Kumpula 2006; Tyler et al. 2021). Biodiversity and availability of high-quality pasture for reindeer are also challenged by the increased greening and shrubification of the tundra caused by climate change (Macias-Fauria et al. 2012). However, recent research suggests that reindeer suppress the growth and regrowth of woody taxa (Bråthen et al. 2017; Skarin et al. 2020), and grazing may thus counteract albedo feedbacks and mitigate climate warming (teBeest et al. 2016; Meredith et al. 2019).

References

Åhman, B. & White, R. G. (2018). Rangifer Diet and Nutritional Needs. In: Tryland, M. & Kutz, S. J. (eds) *Reindeer and Caribou: Health and Disease*. 1st. ed, Boca Raton, Florida: CRC Press, 107–134.

Boertje, R. D. (1984). Seasonal diets of the Denali Caribou Herd, Alaska. *Arctic*. 37(2), 161–165.

Bråthen, K. A., Ims, R. A., Yoccoz, N. G., Fauchald, P., Tveraa, T. & Hausner, V. H. (2007). Induced shift in ecosystem productivity? Extensive scale effects of abundant large herbivores. *Ecosystems*. 10(5), 773–789.

Bråthen, K. A., Ravolainen, V. T., Stien, A., Tveraa, T. & Ims, R. A. (2017). Rangifer management controls a climate-sensitive tundra state transition. *Ecological Applications*. 27(8), 2416–2427.

Colman, J. E., Pedersen, C., Hjermann, D. Ø., Holand, Ø., Moe, S. R. & Reimers, E. (2003). Do wild reindeer exhibit grazing compensation during insect harassment? *Journal of Wildlife Management*. 67(1), 11–19.

Cramér, T. & Ryd, L. (2012). *Tusen år i Lappmarken: juridik, skatter, handel och storpolitik*. Skellefteå: Ord & visor.

Downes, C. M., Theberge, J. B. & Smith, S. M. (1986). The influence of insects on the distribution, microhabitat choice, and behaviour of the Burwash caribou herd. *Canadian Journal of Zoology*. 64(3), 622–629.

Espmark, Y. (1964). Studies in dominance-subordination relationship in a group of semi-domestic reindeer (Rangifer tarandus L.). *Animal Behaviour*. 12(4), 420–426.

Espmark, Y. (1971). Mother-Young relationship and ontogeny of behviour in reindeer (Rangifer tarandus L.). *Zeitschrift für Tierpsychologie*. 29(1), 42–81.

Fauchald, P., Rødven, R., Bårdsen, B.-J., Langeland, K., Tveraa, T., Yoccoz, N. G. & Ims, R. A. (2007). Escaping parasitism in the selfish herd: age, size and density-dependent warble fly infestation in reindeer. *Oikos*. 116(3), 491–499.

Garfelt-Paulsen, I. M., Soininen, E., Ravolainen, V., Loe, L. E., Hansen, B. B., Irvine, R. J., Stien, A., Ropstad, E., Veiberg, V., Fuglei, E. & Pedersen, Å. Ø. (2021). Don't go chasing the ghosts of the past: habitat selection and site fidelity during calving in an Arctic ungulate. *Wildlife Biology*. 2021(2), wlb.00740.

Hagemoen, R. I. M. & Reimers, E. (2002). Reindeer summer activity pattern in relation to weather and insect harassment. *Journal of Animal Ecology*. 71(5), 883–892.

Helle, T. (1984). *Foraging behaviour of the semi-domestic reindeer (Rangifer tarandus L.) in relation to snow in Finnish Lapland*. Report from Kevo Subarctic Research Station. 19, 35–47.

Helle, T. & Aspi, J. (1984). *Do sandy patches help reindeer against insects?* Report from Kevo Subarctic Research Station. 19, 57–62.

Helle, T., Hallikainen, V., Sarkela, M., Haapalehto, M., Niva, A. & Puoskari, J. (2012). Effects of a holiday resort on the distribution of semidomesticated reindeer. *Annales Zoologici Fennici*. 49(1–2), 23–35.

Hofmann, R. R. (1989). Evolutionary steps of ecophysiological adaptation and diversification of ruminants: a comparative view of their digestive system. *Oecologia*. 78, 443–457.

Holand, Ø., Weladji, R. B., Mysterud, A., Røed, K., Reimers, E. & Nieminen, M. (2012). Induced orphaning reveals post-weaning maternal care in reindeer. *European Journal of Wildlife Research*. 58(3), 589–596.

Horstkotte, T., Sandström, C. & Moen, J. (2014). Exploring the Multiple Use of Boreal Landscapes in Northern Sweden: The Importance of Social-Ecological Diversity for Mobility and Flexibility. *Human Ecology*. 42(5), 671–682.

Inga, B. (2007). Reindeer (Rangifer tarandus tarandus) feeding on lichens and mushrooms: traditional ecological knowledge among reindeer-herding Sámi in northern Sweden. *Rangifer*. 27(2), 93–106.

Iversen, M., Fauchald, P., Langeland, K., Ims, R. A., Yoccoz, N. G., Bråthen, & K. A. (2014). Phenology and cover of plant growth forms predict herbivore habitat selection in a high latitude ecosystem. *PLoS ONE*. 9(6), e100780.

Kerby, J. & Post, E. (2013). Reproductive Phenology of Large Mammals. In: Schwartz, M. D. (ed.) *Phenology: An Integrative Environmental Science*. Dordrecht: Springer. 467–479.

Klein, D. R. (1965). Ecology of Deer Range in Alaska. *Ecological Monographs*. 35(3), 259–284.

Klein, D. R. (1986). Latitudinal variation in foraging strategies. In: Gudmundsson, O. (ed.) *Grazing Research at Northern Latitudes*. New York: Plenum Press. 237–246

Klein, D. R. (1990). Variation in quality of caribou and reindeer forage plants associated with season, plant part, and phenology. *Rangifer*. Special Issue No. 3, 123–130.

Kojola, I. (1993). Early maternal investment and growth in reindeer. *Canadian Journal of Zoology*. 71(4), 753–758.

Kumpula, T. (2006). Very High Resolution Remote Sensing Data in Reindeer Pasture Inventory in Northern Fennoscandia. In: Forbes, B. C., Bölter, M., Müller-Wille, L., Hukkinen, J., Müller, F., Gunslay, N., & Konstantinov, Y. (eds) *Reindeer Management in Northernmost Europe: Linking Practical and Scientific Knowledge in Social-Ecological Systems*. Berlin: Springer. 167–185.

Launchbaugh, K. L. & Urness, P. J. (1993). Mushroom consumption (mycophagy) by North America cervids. *Great Basin Naturalist*. 52(4), 321–327.

Macias-Fauria, M., Forbes, B. C., Zetterberg, P. & Kumpula, T. (2012). Eurasian Arctic greening reveals teleconnections and the potential for structurally novel ecosystems. *Nature Climate Change*. 2(8), 613–618.

Mayor, S. J., Schaefer, J. A., Schneider, D. C. & Mahoney, S. P. (2009). The spatial structure of habitat selection: A caribou's-eye-view. *Acta Oecologica*. 35(2), 253–260.

Meredith, M., Sommerkorn, M., Cassotta, S., Derksen, C., Ekaykin, A., Hollowed, A., Kofinas, G., Mackintosh, A., Melbourne-Thomas, J., Muelbert, M., Ottersen, G., Pritchard, H. & Schuur, E. A. G. (2019). Chapter 3: Polar Regions. *IPCC Special Report on the Ocean and Cryosphere in a Changing Climate*. UNEP/WMO.

Merkle, J. A., Monteith, K. L., Aikens, E. O., Hayes, M. M., Hersey, K. R., Middleton, A. D., Oates, B. A., Sawyer, H., Scurlock, B. M. & Kauffman, M. J. (2016). Large herbivores surf waves of green-up during spring. *Proceedings of the Royal Society of London B: Biological Sciences*. 283, 20160456.

Mörschel, F. M. & Klein, D. R. (1997). Effects of weather and parasitic insects on behaviour and group dynamics of caribou of the Delta Herd, Alaska. *Canadian Journal of Zoology*. 75(10), 1659–1670.

Panzacchi, M., Van Moorter, B. & Strand, O. (2013). A road in the middle of one of the last wild reindeer migration routes in Norway: crossing behaviour and threats to conservation. *Rangifer*. 33(2), 15–26.

Pinard, V., Dussault, C., Ouellet, J.-P., Fortin, D. & Courtois, R. (2012). Calving rate, calf survival rate, and habitat selection of forest-dwelling caribou in a highly managed landscape. *The Journal of Wildlife Management*. 76(1), 189–199.

Pollard, R. H., Ballard, W. B., Noel, L. E. & Cronin, M. A. (1996). Parasitic insect abundance and microclimate of gravel pads and tundra within the Prudhoe Bay oil field, Alaska, in relation to use by Caribou. *Canadian Field-Naturalist*. 110, 649–658.

Rivrud, I. M., Sivertsen, T. R., Mysterud, A., Åhman, B., Støen, O.-G. & Skarin, A. (2018). Reindeer green-wave surfing constrained by predators. *Ecosphere*. 9(5), e02210.

Roturier, S. & Roué, M. (2009). Of forest, snow and lichen: Sámi reindeer herders' knowledge of winter pastures in northern Sweden. *Forest Ecology and Management*. 258(9), 1960–1967.

Ryd, Y. & Rassa, J. (2001). *Snö: en renskötare berättar*. Stockholm: Ordfront.

Schaefer, J. A., Bergman, C. M. & Luttich, S. N. (2000). Site fidelity of female caribou at multiple spatial scales. *Landscape Ecology*. 15(8), 731–739.

Senft, R. L., Coughenour, M. B., Bailey, D. W., Rittenhouse, L. R., Sala, O. E. & Swift, D. M. (1987). Large herbivore foraging and ecological hierarchies. *BioScience*. 37(11), 789–799.

Sivertsen, T. R., Ahman, B., Steyaert, S., Rönnegard, L., Frank, J., Segerstrom, P., Stoen, O. G. & Skarin, A. (2016). Reindeer habitat selection under the risk of brown bear predation during calving season. *Ecosphere*. 7(11), e01583.

Skarin, A. & Åhman, B. (2014). Do human activity and infrastructure disturb domesticated reindeer? The need for the reindeer's perspective. *Polar Biology*. 37(7), 1041–1054.

Skarin, A., Danell, Ö., Bergström, R. & Moen, J. (2004). Insect avoidance may override human disturbances in reindeer habitat selection. *Rangifer*. 24(2), 95–103.

Skarin, A., Danell, Ö., Bergström, R. & Moen, J. (2008). Summer habitat preferences of GPS-collared reindeer Rangifer tarandus tarandus. *Wildlife Biology*. 14(1), 1–15.

Skarin, A., Danell, Ö., Bergström, R. & Moen, J. (2010). Reindeer movement patterns in alpine summer ranges. *Polar Biology*. 33(9), 1263–1275.

Skarin, A., Nellemann, C., Rönnegård, L., Sandström, P. & Lundqvist, H. (2015). Wind farm construction impacts reindeer migration and movement corridors. *Landscape Ecology*. 30(8), 1527–1540.

Skarin, A., Sandström, P. & Alam, M. (2018). Out of sight of wind turbines—Reindeer response to wind farms in operation. *Ecology and Evolution*. 8(19), 9906–9919.

Skarin, A., Verdonen, M., Kumpula, T., Macias-Fauria, M., Alam, M., Kerby, J. & Forbes, B. C. (2020). Reindeer use of low Arctic tundra correlates with landscape structure. *Environmental Research Letters*. 15(11), 115012.

Skjenneberg, S. & Slagsvold, L. (1968). *Reindriften og dens naturgrunnlag*. Oslo/Bergen/Tromsø: Scandinavian University Books – Universitetsforlaget.

Skogland, T. (1984). Wild reindeer foraging-niche organisation. *Ecography*. 7(4), 345–379.

Storeheier, P.V. (2003). *Food intake and forage utilisation in reindeer during winter*. Ph.D. diss., University of Tromsø.

Storeheier, P.V., Mathiesen, S. D., Tyler, N. J. C. & Olsen, M. A. (2002). Nutritive value of terricolous lichens for reindeer in winter. *The Lichenologist*. 34(3), 247–257.

Storeheier, P.V., Van Oort, B. E. H., Sundset, M. A. & Mathiesen, S. D. (2003). Food intake of reindeer in winter. *The Journal of Agricultural Science*. 141(1), 93–101.

Sundqvist, M. K., Moen, J., Björk, R. G., Vowles, T., Kytöviita, M.-M., Parsons, M. A. & Olofsson, J. (2019). Experimental evidence of the long-term effects of reindeer on Arctic vegetation greenness and species richness at a larger landscape scale. *Journal of Ecology*. 107(6), 2724–2736.

te Beest, M., Sitters, J., Ménard, C. B. & Olofsson, J. (2016). Reindeer grazing increases summer albedo by reducing shrub abundance in Arctic tundra. *Environmental Research Letters*. 11(12), 125013.

Thomson, B. R. (1975). Leadership in wild reindeer in Norway. In: Luick, J. R., Lent, P. C., Klein, D. R., & White, R. G. (eds) *Proceedings of First International Reindeer and Caribou Symposium*, 1975. 462–472. University of Alaska.

Torney, C. J., Lamont, M., Debell, L., Angohiatok, R. J., Leclerc, L.-M. & Berdahl, A. M. (2018). Inferring the rules of social interaction in migrating caribou. *Philosophical Transactions of the Royal Society of London B: Biological Sciences*. 373(1746), 20170385.

Trudell, J., & White, R. G. (1981). The effect of forage structure and availability on food intake, biting rate, bite size and daily eating time of reindeer. *Journal of Applied Ecology*. 18, 63–81.

Tveraa, T., Fauchald, P., Yoccoz, N. G., Ims, R. A., Aanes, R., & Hogda, K. A. (2007). What regulate and limit reindeer populations in Norway? *Oikos*. 116(4), 706–715.

Tyler, N. J. C., Hanssen-Bauer, I., Førland, E. J. & Nellemann, C. (2021). The Shrinking Resource Base of Pastoralism: Saami Reindeer Husbandry in a Climate of Change. *Frontiers in Sustainable Food Systems*. 4, 585685.

Vistnes, I. & Nellemann, C. (2008). The matter of spatial and temporal scales: a review of reindeer and caribou response to human activity. *Polar Biology*. 31(4), 399–407.

Warenberg, K. (1982). Reindeer forage plants in the early grazing season: Growth and nutritional content in relation to climatic conditions. *Acta Phytogeografica Suecia*, 70, 71.

Weladji, R. B., Holand, Ø. & Almøy, T. (2003). Use of climatic data to assess the effect of insect harassment on the autumn weight of reindeer (Rangifer tarandus) calves. *Journal of Zoology*. 260(1), 79–85.

White, R. G. (1983). Foraging patterns and their multiplier effects on productivity of northern ungulates. *Oikos*. 40(3), 377–384.

4 Pastures under pressure

Effects of other land users and the environment

Tim Horstkotte, Jouko Kumpula, Per Sandström, Hans Tømmervik, Sonja Kivinen, Anna Skarin, Jon Moen and Stefan Sandström

Introduction

Reindeer habitats cover most vegetation zones of northern Fennoscandia, from boreal forest to coastal and alpine areas (Chapters 1 and 3). These vegetation zones are used during different seasons for reindeer grazing and determine specific seasonal grazing patterns within each herding district. This traditional practice of freely roaming reindeer herds depends on the availability of and accessibility to these season-specific pastures, as well as their ability to move between them. However, it is challenged by almost every other form of parallel land use in the reindeer herding area.

In northern Fennoscandia, large-scale industrial resource developments started to increase from the early to mid-20th century. These included infrastructure, hydropower and the spread of commercial forestry. Such anthropogenic pressures can cause loss, fragmentation or degradation of reindeer pastures, or result in disturbances affecting reindeer behaviour, so that behavioural changes exacerbate biophysical losses (Skarin & Åhman 2014; Tyler et al. 2021). Shrinking pastures therefore threaten not only the ecological basis of reindeer foraging but also the traditional practice of utilizing natural grazing resources intrinsic to reindeer husbandry. To address the complexity of pastures under pressure, a clarification of the term "pasture" is required.

What is a "pasture"?

Biomass and quality of forage resources are the basic factors characterizing a feeding site (Senft et al. 1987, Chapter 3). Forage availability at the feeding site, i.e., the quantity of forage accessible to the animal, depends on environmental conditions, such as snow cover. In order to respond to changes in environmental conditions or to distribute grazing pressure, the ability to move between different feeding sites is essential. Functional reindeer pastures, therefore, are a multi-dimensional concept in time and space. This is also reflected in reindeer herders' classification of vegetation types, grounded in their experience-based

DOI: 10.4324/9781003118565-7

or traditional ecological knowledge. For example, the Northern Sámi term *guohtun* describes pasture quality as the presence of grazing resources, but also their accessibility depending on snow conditions and reindeer behaviour (Roturier & Roué 2009). Furthermore, Sámi reindeer herders consider a good pasture to be an area that may be grazed under a variety of climatic conditions, but also as a place where reindeer can access enough forage without disturbance (*guohtun ráfi*, "grazing peace", Northern Sámi) (Inga 2007). In short, "pasture" is a dynamic and composite term encompassing forage biomass and availability, environmental influences, different landscape functions and the behavioural ecology of reindeer. Thus, a pasture in the sense of Sámi or other reindeer herding people only becomes a pasture when reindeer are able to use it for feeding (Chapter 3).

Other land users who make use of natural resources in the reindeer herding area (RHA) are competing for space and are affecting the condition, availability and accessibility of pastures, calving grounds or migration routes. For that reason, we characterize them here as "extractive land uses", as they reduce the areas that could be used as pastures, irrespective of the resources, renewable or not, that are the focus of these land uses.

This chapter examines how different forms of extractive land use affect reindeer pastures and habitat, and their consequent impact on reindeer and reindeer herders. Following a description of the major forms of extractive land use, we describe (i) their direct effects on grazing resources, with a particular emphasis on winter. Further encroachments appear as (ii) direct loss and fragmentation of pastures, (iii) barrier effects that disrupt landscape connectivity or (iv) objects or activities that reindeer avoid. We further consider the effects of (v) environmental drivers on pasture dynamics and discuss (vi) cumulative effects of these different pressures. Finally, we explore (vii) possibilities for pasture restoration.

Extractive forms of land use in northern Fennoscandia and their historical trajectories

The degree to which reindeer habitats are exposed to extractive forms of land use differs within and between countries (Figure 4.1). For example, summer pastures in northern Norway are more strongly impacted than are winter pastures, while the opposite is true for most of the Swedish and Finnish RHA (Hausner et al. 2020). Therefore, generalizations about impacts need to be regarded with caution, because the specific impact depends on the local conditions in each herding district. The following condensed overview of the major extractive forms of land use – hydropower, wind power, forestry and mining – provides the context for their respective impact on pastures.

Norway

Hydropower delivers 90% of Norway's energy production (NVE 2020a) in 2019, of which approximately 30% is produced in the RHA. The construction

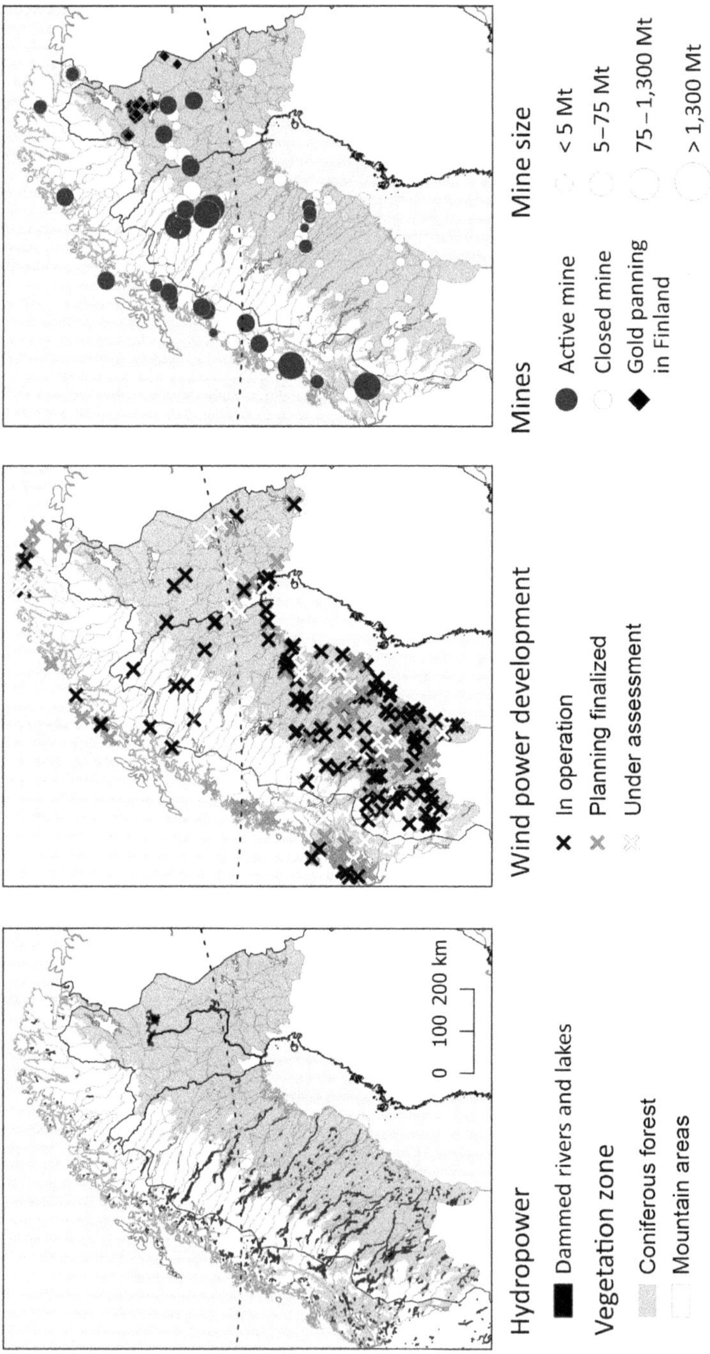

Figure 4.1 Examples of extractive forms of land use in the reindeer herding area of Fennoscandia.

of the Alta hydropower station in Finnmark, completed in 1987, was preceded by environmental and Indigenous rights protests and subsequent downscaling of the original plans (Briggs 2006). In 2019, permission was granted for 17 construction or enlargement projects associated with large developments (>10 mW) in the Norwegian RHA, while permission was granted for a total of 103 smaller hydropower plants (NVE 2020a).

Wind power in Norway provides 7.5% of national energy (NVE 2020a). Within the Norwegian RHA, there were 2 sites for wind power developments with 30 turbines in 2002, increasing to 25 sites with a total of 559 turbines in 2020. Planning has been finalized for a further 43 developments and 13 sites are in the application phase (NVE 2020b).

There are 17 active mines in the Norwegian RHA. Two of these are metal mines and the others produce mainly carbonates and quartz. During the past century, 64 mines in the RHA have been closed. Peat extraction has affected large mire areas, particularly along the treeless coastal zone (Barthelmes et al. 2015).

Sweden

Forestry, i.e., the management of forests to deliver timber and other wood-based products, is the major and dominant land use in the Swedish and Finnish RHA. In northern Sweden, industrial forest exploitation started in the 1860s, primarily by selective cutting of large trees in old-growth forest (Östlund et al. 1997). The increased demand for wood products from the mid-20th century onwards entailed considerable changes in forest management to increase forest productivity, such as fertilization, clear cutting and other methods of intensification (Kivinen et al. 2010). Today, nearly all productive forest land with the growth of at least 1 m^3/ha/yr is used by commercial forestry. The amount of gross harvested timber increased from 42.7 mill. m^3 in 1950 to 91.2 mill. m^3 in 2018, of which approximately 41% is harvested in the RHA (Skogsstyrelsen 2019). In Norrbotten and Västerbotten, Sweden's two northernmost and largest counties, approximately 4.5% of productive forests below the tree line are formally protected, with another 4.4% of productive forests voluntarily set aside for biodiversity protection or due to cultural and social values (SLU 2019).

Hydropower delivers 45% of Sweden's energy (Energimyndigheten 2020), of which ca. 80% is produced in the RHA in 2019. There are 207 hydroelectric dams, leaving only 4 major rivers that are not developed. Most development of hydropower ceased in the 1980s, a consequence of environmental protection, but upgrades of existing dams are continually ongoing (Fischer et al. 2018).

Wind power in Sweden provides 15% of the energy (2020). In the 4 northernmost counties, the number of turbines increased from 48 in 2003 to 1577 operating units in 2019, with permission already given for 2155 turbines and 609 more under evaluation (Vindbrukskollen 2020). Wind power production will thus increase the most in the northern part of the country (Energimyndigheten 2019).

Nine of Sweden's twelve active metal mines are located in the RHA, and six active quarries (SGU 2020a). Though the number of active mines is decreasing, ore production is increasing, amounting to 86.5 million tons (Mt) in 2019 (SGU 2020b). The oldest active mine since the 1860s is Kiirunavaara, one of the largest iron ore mines globally. Boliden Minerals AB established the Aitik mine in 1968, today the largest open pit copper mine in Europe covering an area of approximately 50 km^2. In the Swedish RHA, there are approximately 77 closed mines, varying in type, size and age (2020). Exploration for new mining sites peaked between 2000 and 2013 (SGU 2020b). In 2019, there were 361 valid exploration permits and 41 permits within the RHA (SGU 2020b). Although exploration with a permit often takes place close to existing mines, new sites are also prospected.

Peat is extracted for energy production and agricultural purposes. Approximately 47% of the whole production stems from the four northernmost counties (SGU 2020c). Following a peak in the 1990s, the amount of peat extracted had decreased by 60% in 2019, providing 0.2% of energy production.

Finland

Forestry had a similar historical trajectory in Finland as in Sweden. The amount of harvested timber in Finland increased from 30.3 mill. m^3 in 1955 to 71.3 mill. m^3 in 2018. In contrast to Sweden, only approximately 7% of the gross harvest comes from the Finnish RHA (LUKE 2020). Of the productive forests, 14.4% are protected in the Finnish RHA, with the largest proportion in Sámi Homeland, where 62% of all forests are protected (Turunen et al. 2020).

Hydropower delivers 14% of Finland's energy production (Statistikcentralen 2020). Major hydropower stations in the RHA include those on the river Kemijoki, harnessing for hydropower started in 1948. Altogether eight hydroelectric dams with power plants were built in the main riverbed during the 1950s–1960s. In connection with these power plants, the large Lokka and Porttipahta water reservoirs (located mainly in the Lappi herding district) were established in 1967–1970. Additionally, six hydroelectric dams were built in the upper part of the riverbed. Some smaller dams exist in other smaller rivers in the RHA of Finland. Similar to experiences in Sweden and Norway, the concerns of reindeer herders about the effects of these reservoirs were ignored (Mustonen et al. 2010). The undeveloped river Ounasjoki in northern Lapland remains under environmental protection.

Wind power in Finland provides 9% of the energy (Finnish Wind Power Association 2020). In 2020, there were 17 sites for wind power development with a total of 145 turbines in the RHA. A further 79 turbines have been granted permission and 262 turbines are in the planning or application stages (Ethawind 2020).

Two active mines are located in the Finnish RHA. The Kittilä gold mine is the largest gold mine in Europe with 2 Mt of ore extracted annually (AgnicoEagle Finland 2021). The Kevitsa nickel–copper mine is the largest open pit mine in Finland, producing 7.54 Mt ore in 2019. Two smaller quarries produce carbonates and gem stones. A total of 13 metal mines have been closed within the past 65 years in the RHA, all of them significantly smaller than the metal mines currently operating (Kivinen 2017). Five major projects in the RHA aim to open a mine within the next ten years.

Gold panning, i.e., gold exploitation from the soil, is an activity mainly occurring in Sámi Homeland. In 2020, there were 318 active gold panning sites with 147 pending applications (TUKES 2020). Increasingly, mechanical digging systems are used which require much larger areas.

Peat extraction sites are mainly restricted to the southwestern part of the RHA, with decreasing importance for energy production.

Effects of extractive land use on grazing resources

While the summer provides reindeer with a diversity of forage resources, there is a bottleneck in winter with regard to forage quality and availability. Winter grazing resources are, therefore, of particular concern for sustaining reindeer herds. During winter, reindeer forage mainly on terrestrial lichens ("reindeer lichens", *Cladonia* spp., *Cetraria* spp.), constituting up to 80% of their diet depending on their availability (Heggberget et al. 2002). Being tolerant to drought, terrestrial lichens have competitive advantages over mosses and vascular plants on dry and nutrient-poor sites but are outcompeted on moist and fertile soils. Boreal forests, especially those with limited oceanic influences east of the Scandic mountain chain, are prime habitat for reindeer lichens, in particular on oligotrophic soils dominated by Scots pine (*Pinus sylvestris*), as well as in the interior of northern Norway, where there is a comparatively continental climate (Figure 4.2).

When deep or hard snow prevents reindeer from digging for terrestrial lichens, arboreal lichens (*Bryoria* spp., *Alectoria* spp.), growing primarily in old-growth forests, are crucial. In the mountain birch forest, bark lichens (*Hypogymnia* spp., *Parmelia* spp.) growing on trunks of mountain birch (*Betula pubescens* ssp. *czerepanovii*) are of similar importance. Growing conditions for arboreal lichens primarily depend on microclimatic conditions within the forest canopy, including humidity, light, temperature and wind exposure, but their abundance depends in particular on forest age and continuity of key habitats (Esseen et al. 2016).

In Sweden and Finland, primary winter grazing areas are located in boreal forests (Figure 4.2), which are also used for forestry. Different silvicultural practices for timber production have considerably changed forest age structure and composition, with direct consequences for terrestrial and arboreal lichens. This has led to a degradation of the carrying capacity of winter grazing areas and changed reindeer herding practices (Kumpula et al. 2014; Sandström et al. 2016).

Up to the early 20th century, the dominant harvesting approach involved the selective felling of the largest trees, creating favourable conditions for terrestrial lichens by increasing light availability on the forest floor, which is crucial

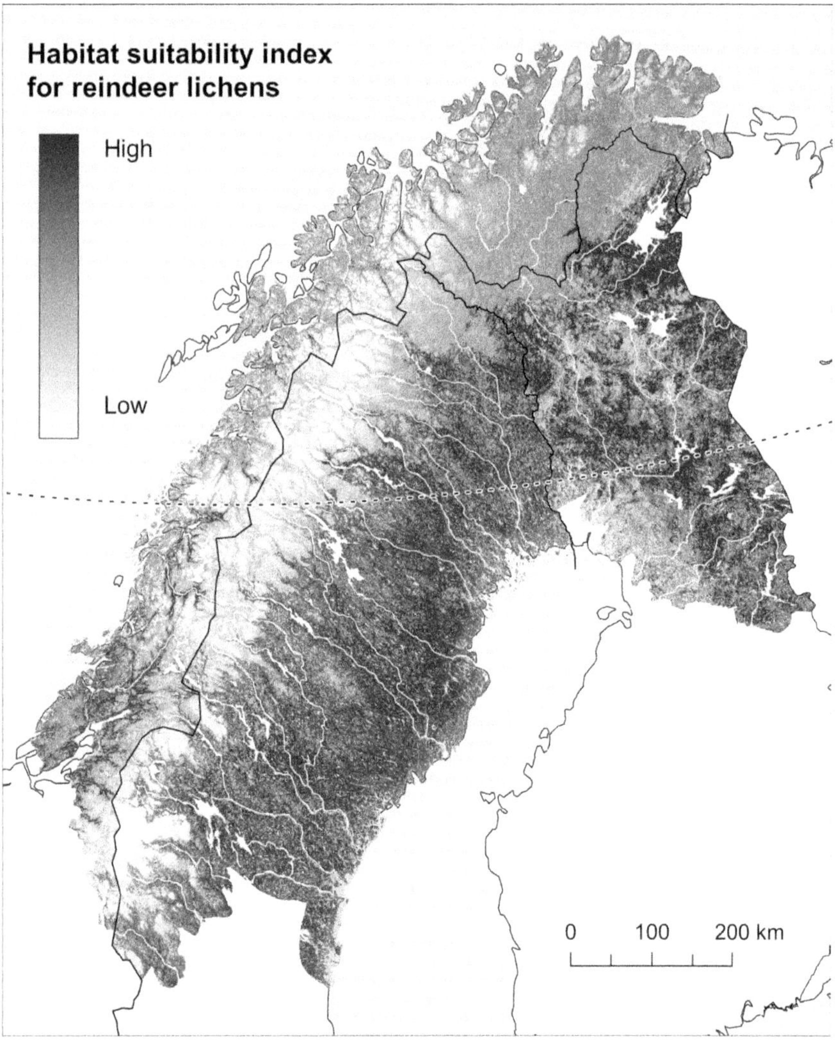

Figure 4.2 Habitat suitability for reindeer lichens. Developed using MaxEnt (Philips et al. 2021), based on data from the Swedish National Forest Inventory, the Natural Resources Institute Finland (Luke) and the Norwegian Institute for Nature Research (NINA). A total of 773 plots with lichen cover > 25% were included. Environmental variables include CORINE land cover, elevation and soil water content (European Soil Data Centre 2016).

for lichen growth (Berg et al. 2008). From the 1950s onwards, clear cutting by removal of trees in a single harvest and subsequent soil preparation had almost completely replaced selective felling (Berg et al. 2008; Kivinen et al. 2012). This form of forest management has led to a significant decrease in the number and size of pristine old-growth forests and increased the proportion of young plantation forests with higher tree density (Sandström et al. 2016).

As this change in forest composition and structure removed the ecological niches occupied by terrestrial and arboreal lichens, their distribution and abundance decreased throughout boreal forests in Fennoscandia (Figure 4.3). In the Swedish RHA, the extent of forests with abundant lichen cover (>50% cover) decreased by 71% between 1953 and 2013 (Sandström et al. 2016). In Sámi Homeland, northern Finland, a decrease of 44% was observed between 1995 and 2018 (data from LUKE). In both countries, there is spatial variation in these changes, and lichen cover in Finland has been able to increase in some areas despite the overall decline.

Clear cutting can have varying effects on terrestrial lichens depending on the site. It can offer high light availability and precipitation reaching the ground, both favourable conditions for lichen growth and (re-)establishment. However, these exposed areas can also suffer desiccation during summer, which may limit lichen's growth. Mechanical disturbance of the soil during forestry operations or exposure of mineral soil to enhance tree growth is particularly damaging to terrestrial lichens. In addition, logging residues left on the ground hinder lichen regeneration and can prevent reindeer from reaching existing lichens (Roturier & Bergsten 2006; Turunen et al. 2020).

The ability of terrestrial lichens to re-establish after tree harvest also depends on the structure of the regenerating forests (Kumpula et al. 2014; Sandström et al. 2016; Horstkotte & Moen 2019). Increased tree density limits light availability at the forest floor, so that mosses and vascular plants may outcompete lichens. Once mosses have gained dominance at the expense of lichens, the latter are unlikely to re-establish on these sites (Sandström et al. 2016; Horstkotte & Moen 2019).

Other forestry operations detrimental to terrestrial lichens include fertilization, promoting growth of vascular species (Strengbom et al. 2008) and in Sweden planting the exotic lodgepole pine (*Pinus contorta*), native to North America. Introduced in the Swedish RHA in the 1960s due to its productivity being 30% higher than native pine, *P. contorta* adversely affects terrestrial lichens due to increased litter and dense canopies (Bäcklund et al. 2018).

Growing conditions for terrestrial lichens can also be affected by mining. Chen et al. (2017) found that lichen cover in shrub tundra decreased from approx. 15% cover at 1000 m from a mining road to absent cover close to roads, depending on increased soil pH as a result of dust deposition. Reindeer herders in Sweden report airborne dust more than 10 km away from an open pit mine (Lawrence & Larsen 2017), but effects on lichens have not been quantified. However, reindeer herders have raised concerns about the harmful impacts of toxic substances on reindeer in the vicinity of both active and closed mining areas (Johnsen et al. 2016; Kivinen et al. 2018).

Habitat for arboreal lichens disappears immediately when trees are harvested. It takes at least 60 years for arboreal lichens to begin accumulating in regenerating forests (Horstkotte et al. 2011), but it takes 140–200 years before they

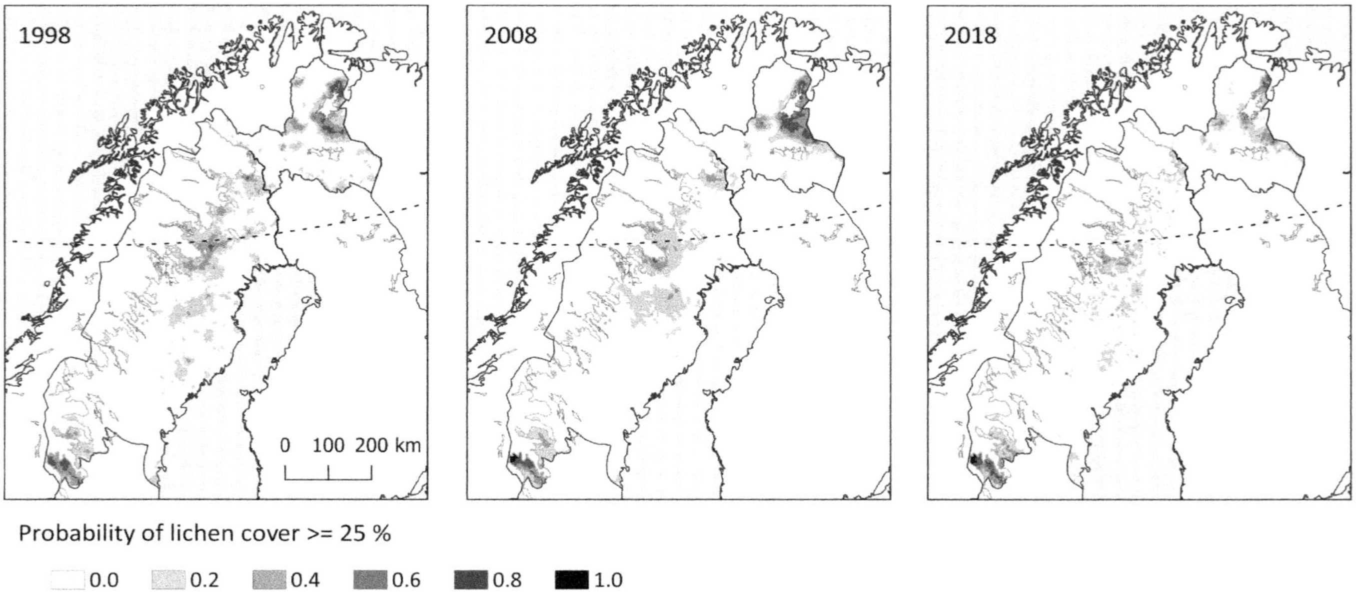

Figure 4.3 Development of lichen cover recorded in national inventories in Sweden and northern Finland between 1998 and 2018. Darker shades indicate a higher probability that lichen cover at the site is at least 25%. Data are aggregated for several years to account for one full survey of all inventory sites (3–5 years), and matched most closely to each other in time. Therefore, maps should be understood as an approximation. Produced by co-kriging with CORINE land cover.

reach sufficient biomass to be considered a reliable forage resource. Current rotation times (the interval between harvests) in Northern Sweden and Finland between 90 and 120 years do not allow such development. Climate change may improve the growth of trees but also increases the risk of damage, e.g., by more frequent storms. In response, rotation times may be shortened in the future (Subramanian et al. 2019), thus increasing the negative effects on arboreal lichens.

Due to the limited dispersal abilities of arboreal lichens, perhaps only about 200 m, the fragmentation of old-growth forests rich in arboreal lichens restricts their ability to colonize new habitats (Dettki & Esseen 2003). The availability of arboreal lichens at the landscape level therefore depends on the composition of the landscape in terms of forest age, size of different aged forest stands and the distances separating them (Horstkotte et al. 2011; Kivinen et al. 2012).

Where reindeer graze in the boreal forest during summer, other factors besides the presence of lichens are important for functional pastures. For example, clear cuts may be a preferred habitat by reindeer during calving and early summer, probably due to the abundance of high-quality vascular plants and as an anti-predator strategy (Skarin et al. 2015). Dense and cool old spruce forests are important for relief from heat and insects but have also been greatly reduced by forestry (Sandström et al. 2006).

Direct loss and fragmentation of pastures by extractive forms of land use

Impacts of extractive forms of land use can go beyond the direct effects on grazing resources. Distribution, abundance and size of suitable grazing patches, as well as connectivity between them, determine whether a landscape can be used for reindeer grazing under different grazing conditions and over an extended time.

On winter grazing areas in the boreal forest, the option to select and access forest stands with different structural characteristics, such as canopy closure, age composition or tree sizes can buffer against variation in grazing conditions. For example, large canopies in old-growth forests intercept more snow and reduce snow accumulation at the forest floor compared to younger forests. However, intercepted snow drops out of the canopy during a warm spell and may result in a harder snow pack compared to younger forests. Today, reindeer herders in Sweden and Finland emphasize the lack of old-growth forests, and thus a decreased functionality of pastures at the landscape level available to respond to such variation in grazing conditions (Horstkotte et al. 2014). Physical loss of pastures or loss of their functionality therefore increases the grazing pressure on the remaining pastures (Axelsson Linkowski et al. 2020).

Mining activities cause direct loss of available pastures over the area of the actual mine or quarry itself. Furthermore, associated facilities and infrastructure fragment the landscape and extend the impact area of the actual mine. For instance, the largest open pit mine in northern Sweden, Aitik copper mine, covers approx. 3 km^2, but its overall physical footprint is approximately 50 km^2, even when excluding infrastructure for access and transport. Peat extraction by ditching and drainage destroys potential pastures, primarily on summer grazing grounds (Blind et al. 2015; Reindeer Herders' Association 2014).

Hydropower development submerged important habitats along rivers and lakes. These were often important calving grounds, migration routes or pastures early in the growing season. The proportional loss of important reindeer habitats can far exceed the total area that was submerged. Herders from Lappi herding district, Sámi Homeland in northern Finland, report that the construction of the Lokka and Porttipahta reservoirs, which commenced in 1967 and 1970, flooded 11% of the district's area. However, the actual loss of important pastures was as high as 25% (herder in Lappi herding district, pers. comm. 2015). Reindeer had to learn new migration routes, a process that can take up to a decade, and collaboration with neighbouring districts was interrupted by a fence (herder in Lappi herding district, pers. comm. 2015). Likewise in Sweden and Norway, herding groups had to change their practices and interaction with each other in response to the construction of reservoirs (Paine 1982; Össbo & Lantto 2011).

Barrier effects caused by extractive forms of land use

Connectivity and the option to choose and move between different grazing patches, both for reindeer on their own and during herding operations, are particularly important to allow responses to changes in grazing conditions, e.g., due to adverse weather effects on snow or other disturbances (St John et al. 2016). However, landscape connectivity can become disrupted by barriers that hinder free movement between different areas, requiring more work by herders. For instance, infrastructure such as busy roads, dense plantations of *P. contorta* or unstable and unsafe ice conditions on regulated rivers with fluctuating water levels makes movements across landscapes difficult or dangerous for both animals and people. Ditches from peat extraction hinder movement and can turn into deadly traps for reindeer calves, which can drown in such ditches (Blind et al. 2015; Reindeer Herders' Association 2014).

Other barriers, such as mines, roads or railroads fenced against wildlife, can completely or partly cut off areas and require herders to reorganize their land use patterns. Hydropower development has, in several cases, completely eliminated migration routes. Passages constructed to bypass infrastructure or to compensate for losses, such as alternative migration routes or bridges/weirs, often remain non-functional. Lost connectivity therefore necessitates transport of reindeer by truck between seasonal grazing areas (herder in Ohredahke herding district, Sweden, pers. comm. 2018).

Behavioural changes caused by extractive forms of land use

Reindeer can be sensitive to anthropogenic activity or related visual or noise-related disturbance. Altered or decreased pasture use by reindeer defines the effective zone of influence (ZOI) of these disturbances. Levels of disturbance that determine the strength of avoidance, and thus size of the ZOI, can vary over time and are often connected to several environmental and anthropogenic effects on pastures (Boulanger et al. 2021).

Wind power development and mining sites produce noise and visual disturbance. Depending on local conditions and season, the effects of wind power development on reindeer behaviour and pasture use differ. During the construction of wind turbine sites, reindeer migration routes may be cut off and reindeer avoid these sites most probably as a result of the frequent human activity (Colman et al. 2013; Skarin et al. 2015). However, during the operation phase, reindeer have been found to shift their habitat use to areas where operating wind turbines are hidden from sight (Skarin et al 2018). Compared to the situation prior to wind power development, reindeer were found to decrease their use of sites by 13% at 5 km where wind turbines were visible, while they increased the use of sites where the turbines were obscured by topography by 79% at 5 km distance (Skarin et al. 2018). In open areas in Norway, the ZOI can extend up to 13 km from the turbines (Eftestøl et al. 2021). Avoidance behaviour is particularly strong during calving and early summer and in the autumn and also occurs in winter (Skarin et al. 2018; Eftestøl et al. 2021; Skarin et al. 2021). During winter, wind power developments increase the workload of the herders, e.g., gathering and migrating with their animals or keeping them away from public roads. In summer and during periods of insect harassment or when wind turbines are placed in less important grazing areas, avoidance behaviour seems to be less pronounced. Studies on peninsulas and islands on the Atlantic coast, where reindeer have limited options to evade disturbance, found less pronounced or no effects of the wind turbines on habitat use (Colman et al. 2013; Tsegaye et al. 2017).

Mining sites require more intense, regular use of infrastructure and human presence compared to sites for wind power development during operation, e.g., for transport of mining products or blasting of rocks. Reindeer avoid these areas, particularly during periods of high anthropogenic activity. In Ivalo herding district, Sámi Homeland in northern Finland, reindeer avoided gold panning sites for a distance of 1.5 km during the summer season when human activity peaks (Anttonen et al. 2011). In Finnmark, reindeer on a peninsula reduced habitat use by ca. 35% within 1.4 km of open pit mines during high activity periods compared to 0.9 km during weeks of low activity (Eftestøl et al. 2019). However, studies on reindeer with the option to evade such disturbance over larger areas are lacking from Fennoscandia. Caribou in Eastern Canada were found to avoid underground mines, maintaining a distance of 3 km to 21 km in open terrain (Plante et al. 2018), and approximately 14 km around open pit mines (Boulanger et al. 2012). The latter finding was in part attributed to dust deposition and its negative effects on vegetation.

While some studies report a higher tolerance of technical installations than human disturbance (Anttonen et al. 2011; Colman et al. 2013), long-term studies providing a full understanding of the effect of these and similar disturbances on pasture availability mediated by behavioural responses are lacking.

Environmental drivers of pasture dynamics

Reindeer grazing

Reindeer themselves influence their grazing resources, either directly by consuming them or indirectly by trampling and fertilization. While the net effect of reindeer on abundance and diversity of vascular plants, from graminoids to shrubs, is highly context-dependent, e.g., on seasonal timing of grazing, history, animal densities or site productivity, they generally reduce cover and biomass of terrestrial lichens (Bernes et al. 2015). However, reindeer grazing can positively affect lichen growth by reducing competition from vascular plants (Olofsson et al. 2010).

The effect of reindeer on terrestrial lichens needs to be understood in the context of herding practices and the presence of other forms of land use. In particular, the size of the herding district and the pattern of seasonal grazing rotation affect the relationship between herd size and lichen biomass (Kumpula et al. 2014; Sandström et al. 2016). Larger herd sizes often result in lower lichen biomass, but these relationships vary over time as they depend on environmental conditions, such as availability of lichen-type pastures per reindeer and snow cover (Tømmervik et al. 2009; Kumpula et al. 2014). However, reindeer trampling of lichen-rich areas especially during summer is particularly harmful, because dry lichens are easily damaged. This is a particular challenge where a separation between summer and winter grazing areas is not possible, such as in many herding districts in the southern parts of the Finnish RHA. Where reindeer remain in the forest for the whole year, lichen cover is lower compared to areas only used during winter (Kumpula et al. 2014; Horstkotte & Moen 2019; Uboni et al. 2019).

As grazing areas are shrinking and become increasingly fragmented, it becomes difficult or even impossible for reindeer herders to access alternative grazing sites in the winter ranges or to allow pastures to rest to facilitate the recovery of lichens. Consequently, remaining areas need to be used more intensely, which may contribute to the observed trend in lichen decline (Axelsson Linkowski et al. 2020; Uboni et al. 2020).

Climate and climate change

Climate and weather have direct consequences on pasture resources but also affect competition between different vegetation communities or shape grazing pressure in space and time.

The growth of terrestrial lichens increases with precipitation, while warmer temperatures can limit the time in a wet state during which lichens are able to grow (Kumpula et al. 2014; Tømmervik et al. 2012). Warmer summer temperatures can improve growing conditions for some arboreal lichens in the lower canopy (*Alectoria sarmentosa, Usnea* spp.) but have a negative effect on those that grow in exposed parts of the canopy (*Bryoria fuscescens*) (Esseen et al. 2016).

Climate change will affect forage resources differently depending on the season. Earlier springs and longer and warmer growing seasons may increase the abundance of vascular plants, but decrease their nutritional value (N/C-ratio) (Turunen et al. 2009). However, these responses are species-specific and depend on other environmental factors, such as soil nutrient availability. Above the tree line, experimental warming has been found to reduce lichen abundance due to increased competition with vascular plants (Alatalo et al. 2015). Similarly, warmer temperatures can adversely affect lichen abundance in boreal forests if increased density of the tree layer and vascular plants on the forest floor decrease light availability for lichen growth (Hedwall et al. 2016).

A warmer climate and more extreme weather events result in deep and /or hard snow cover or ice crusts, making foraging through the snow pack energy demanding or impossible (Chapter 5). While deep snow negatively affects the growth of lichens (Kumpula et al. 2014), the relationship between snow depth and hardness, grazing pressure and lichen growth is dynamic. As reindeer select areas with shallower and softer snow, deep or hard snow can locally reduce grazing pressure and allow the recovery of grazed lichens at these sites in the following growing season (Tømmervik et al. 2012; Axelsson Linkowski et al. 2020). In contrast, shallower snow cover enables reindeer to graze in places that, under normal conditions, are inaccessible, resulting in a wider distribution of grazing pressure (Tømmervik et al. 2009).

Cumulative impacts on pastures

The effects of extractive land use on reindeer husbandry by pasture loss and fragmentation, barriers in the landscape and avoidance behaviour cannot be fully understood in isolation from each other, nor from environmental pressures. Rather, these different pressures interact with each other and with legacies of pressures from the past. For example, in Sirges herding district in Sweden, the Lule River was previously used for migration, and crossings were possible at several places (Larsen et al. 2020). Today, due to the effects of damming and fluctuating water levels on ice conditions, herders are rarely able to migrate on the river and only one narrow and dangerous crossing remains. During migration, herders have to make sure that all the gathered herd crosses to the desired riverbank at this particular site. However, intense forestry has made it difficult to keep the reindeer gathered during the winter. Thus, the combined effects of forestry and water power development decrease the possibility of using what is left of the historic migration routes.

Cumulative impacts are case-specific, hard to predict and often difficult to communicate. This can occur particularly when the cumulative impacts have not only ecological causes and consequences but also social and cultural consequences. For example, Marin et al. (2020) report that herds in Finnmark with a high proportion of females, following management recommendations by the state to increase productivity, result in a loss of "bull pastures", as females – in particular those with calves – are more sensitive to anthropogenic disturbance than are males. Grazing areas can, therefore, be "*available on the map, but not in reality*" thus increasing the tension regarding "ideal" herd composition (Herder from Finnmark, pers. comm. 2015).

Cumulative impacts also depend on fluctuating weather or snow conditions that govern access to grazing resources within and between years; what constitutes a good pasture can differ considerably between years. Such a fluctuating access to grazing resources due to climatic drivers is an integral part of reindeer husbandry (Sara 2009). However, the continuing decline and fragmentation of pastures, or options to access them, strongly limit the adaptive capacity of reindeer herders to respond to these climatic drivers and find alternative pastures (Chapter 5).

Gradual pasture loss, as a long-term process, can affect the expectations of reindeer herders; what young herders currently regard as good or average pastures are considered low quality by older generations (Axelsson Linkowski et al. 2020). Intergenerational sharing of local knowledge becomes particularly important to prevent this shift towards lowering of the accepted thresholds with respect to pasture conditions.

Restoration of pastures

Alternative forest management and increased landscape connectivity

The contribution of forest management to the establishment and persistence of terrestrial and arboreal lichens could be improved by several strategies. However, all these involve several trade-offs, including the economic dimension, as some aspects of high productivity of both trees and lichens can be difficult to combine (Horstkotte et al. 2016).

More severe thinning, especially of young forests, can benefit both lichen growth, due to increased availability of light and humidity (Jonsson Čabrajič et al. 2010), and timber production. Likewise, continuous cover forestry avoiding the clear cutting stage, also known as uneven-aged forestry, keeps the age and layer structure of trees diverse over time, and the canopy relatively open (Peura et al. 2018). Lichen re-establishment can also be accelerated by spreading lichen fragments (Roturier et al. 2007), as well as by avoiding soil preparation in dry lichen-type pastures after harvest (Roturier & Bergsten 2006).

At the landscape level, connectivity between different grazing areas and intact migration routes are important to facilitate movements within and between seasons. However, it remains challenging to upscale forest management, often conducted at the stand level, to the landscape level, which involves different forest owners, in order to deliver the spatial requirements of pastures as described by herders (Horstkotte et al. 2014). Landscape connectivity can also be increased through the construction of ecoducts, i.e., at strategically important sites to assist movement across barriers such as roads or across unsafe lakes.

Restoration of post-mining sites

As pioneer species, lichens are able to colonize disturbed soils such as post-mining environments, if their propagules or fragments reach these sites either naturally or artificially. At post-mining sites in arctic Canada, lichens have been found to establish on till-soil and bare gravel soils (Naeth & Wilkinson 2014), but an initial substrate such as thin moss cover can improve establishment by providing adhesive structures, wind protection and moisture retention (Roturier et al. 2017, Duncan 2015). The timing of recolonization is site-specific, but it can take at least 30–50 years before thick lichen mats have formed (Duncan 2015).

Remaining metal pollutants can be particularly challenging for restoration. In a gradient from a zinc smelter, *C. mitis* was found at less than 5% cover at sites ca. 3.5 km away, indicating low tolerance to metal contamination present in the mining dust (Rola & Osyczka 2014). There is a need to better understand how mining sites can be restored once their mineral resources have been depleted.

Grazing rotation

Rotation between grazing areas allows resources to recover from the previous grazing or protect lichen-rich areas during summer when they are particularly vulnerable to trampling (Kumpula et al. 2014). The latter is of significance in northern Finland, where a mosaic of winter and summer pastures, the limited size of herding districts and fences between them often do not permit high spatial flexibility in the season-specific use of pastures. A decline in grazing resources forces herders to use any available pastures, with no or little option to let them rest to allow some years for recovery (Axelsson Linkowski et al. 2020). While there is debate about "the optimal herd size" in different areas or districts between herders, researchers and state authorities (Chapter 9), the need for grazing rotation is emphasized by all parties.

The practice of such rotation periods is originally grounded in traditional ecological knowledge amongst both Sámi and Nenet herders in Western Siberia. Nenets use the traditional and skilful herd-navigation rule "*ya puna hayoda*" ("land after us remains") for a pasture that is set aside to rest, sustaining even large herds of reindeer (Golovnev 2017). To meet this challenge, it is necessary to foster decision-making based on knowledge co-production between different actors within processes that embrace herders' knowledge to provide a holistic understanding of pasture dynamics. However, the integration of traditional ecological knowledge into policies to increase resilience of reindeer husbandry, including for land use management and climate change adaptation, has so far been insufficient (Eira et al. 2018).

Concluding remarks

Extractive forms of land use in the reindeer husbandry area are adversely affecting reindeer pastures by decreasing size, number, accessibility and quality of pastures. A holistic understanding of the different pressures and their cumulative impacts, spatially and temporally, is required to prevent tipping points being passed, beyond which reindeer husbandry practices may change in a way that will no longer allow a return to practices that reflect the cultural preferences of reindeer herders (Chapter 14).

Pressures on pastures, and their cumulative impacts, accumulate not only over time but also in space and reduce the diversity of pastures. Sources of diversity have been identified as the most important factor for adaptation to act upon (Chapin et al. 2009). This becomes evident in diverse landscape elements that can buffer, for instance, the effects of extreme weather events during winter and provide options for grazing under different winter conditions. Diverse landscapes are also necessary to reflect the different grazing resource niches, particularly on winter grazing grounds. For example, pristine old-growth forests are of high value for arboreal and terrestrial lichens but do not necessarily guarantee high continuous abundance of terrestrial lichens if they develop dense understory vegetation (Kivinen et al. 2010). However, planning at the landscape level that takes into account the variation in grazing conditions in time and space may be hampered by misalignment in spatial planning between reindeer husbandry and extractive forms of land use and an imbalance with respect to herders' capacity to influence decision-making in land use planning.

The responses of reindeer herders to these different encroachments and losses of pastures include providing reindeer with supplementary feeding to buffer a lack or inaccessibility of grazing resources (Chapter 12), transport by truck between seasonal pastures and a forced need to utilize the remaining pastures more intensely (Uboni et al. 2020). This is particularly the case where the spatial extent of herding districts is comparatively small, as limited spatial flexibility alone can lead to competition with other land use, and all the available areas of the district may be affected.

Supplementary feeding has been implemented since the 1960s in the southern part of the Finnish RHA (Chapter 12). However, the need to provide reindeer with supplementary feeding involves several ecological, social, cultural and economic challenges. While this strategy in the short term can compensate for the lack of natural food and alleviate urgent crises with regard to availability or accessibility of grazing resources, the underlying problem of pasture encroachments and diminishing grazing resources remains unresolved.

Restoration of pastures by (re-)establishment and recovery of grazing resources and reconnecting grazing areas, as well as considerate use by other land users, is, therefore, fundamental for preserving the social-ecological identity of reindeer husbandry as a livelihood based on free-ranging animals feeding on natural forage. In particular in the Finnish RHA, as well as in Finnmark, northern Norway, the impact of reindeer on their winter grazing resources is a topic with high social conflict potential (Chapter 9). While grazing rotation to protect lichen-rich areas from reindeer trampling during the snow-free season is crucial, it has also been argued that decreased herd sizes are required to allow depleted winter grazing resources to recover. However, herd sizes and slaughter weights as indicators of overgrazing are heavily disputed, because the relationship between forage resources, slaughter weights and herd sizes is complicated by variations in weather and climate, characteristic of ecosystems at high latitudes (Sara 2009, Marin et al. 2020). Increased grazing pressure needs, rather, to be seen in the context of those effects caused by different other land users. To maintain herd sizes, more frequent and intense use of remaining pastures becomes necessary. However, this can lead to a cycle with further decreasing pasture resources, if no action is taken to restore pastures lost by the cumulative pressures they are exposed to and that antagonize the coexistence between reindeer husbandry and other land users in multiple-use landscapes.

To understand the accumulation of diverse pressure on pastures, spatially and temporally, as well as to navigate decision-making processes in land use and resource development, engagement between reindeer herders, representatives of extractive forms of land use and state authorities is needed. While exploitation of natural resources in the early 20th century was wrought with the colonial and paternalistic attitude of the Nordic states towards Sámi livelihoods (Chapter 1; Össbo & Lantto 2011), several legal instruments at present require that reindeer herders are involved in land use planning. However, the significance of these consultation procedures for practical application is often contested (Larsen & Raitio 2019). For example, consultations between reindeer herders and forestry representatives have received criticism for coming too late in the planning phase of forestry operations, leaving little room for reindeer herders to negotiate or decreasing the likelihood of reaching consensus (Widmark & Sandström 2012). Furthermore, differences in status and power relations between reindeer herders and other forms of land use representing national interests in resource development can impede agreements on land use planning for coexistence being reached. The same power dynamics can affect the acceptance of different knowledge systems as an equal evidence base for evaluating the ways in which pressures affect pastures. This imbalance in power in the discourse on natural resource management in Fennoscandia and elsewhere on Indigenous lands restricts options to identify alternative ways of managing natural resources and other ways to identify a desirable future based on diverse ontologies (Jääskelinen 2020).

References

AgnicoEagle Finland (2021). http://agnicoeagle.fi/ [14.05.2021]

Alatalo, J. M., Jägerbrand, A. K. & Molau, U. (2015). Testing reliability of short-term responses to predict longer-term responses of bryophytes and lichens to environmental change. *Ecological Indicators*. 58, 77–85.

Anttonen, M., Kumpula, J. & Colpaert, A. (2011). Range selection by semi-domesticated reindeer (Rangifer tarandus tarandus) in relation to infrastructure and human activity in the boreal forest environment, northern Finland. *Arctic*. 64(1), 1–14.

Axelsson Linkowski, W., Fjellström, A. M., Sandström, C., Westin, A., Östlund, L. & Moen, J. (2020). Shifting strategies between generations in Sami reindeer husbandry: the challenges of maintaining traditions while adapting to a changing context. *Human Ecology*. 48(4), 481–490.

Bäcklund, S., Jönsson, M., Strengbom, J. & Thor, G. (2018). Tree and stand structure of the non-native Pinus contorta in relation to native Pinus sylvestris and Picea abies in young managed forests in boreal Sweden. *Scandinavian Journal of Forest Research*. 33(3), 245–254.

Barthelmes, A., Couwenberg, J., Risager, M., Tegetmeyer, C. & Joosten, H. (2015). *Peatlands and Climate in a Ramsar Context: A Nordic-Baltic Perspective* (Vol. 2015:544). Copenhagen: Nordic Council of Ministers.

Berg, A., Östlund, L., Moen, J. & Olofsson, J. (2008). A century of logging and forestry in a reindeer herding area in northern Sweden. *Forest Ecology and Management*. 256(5), 1009–1020.

Bernes, C., Bråthen, K. A., Forbes, B. C., Speed, J. D. & Moen, J. (2015). What are the impacts of reindeer/caribou (Rangifer tarandus L.) on arctic and alpine vegetation? A systematic review. *Environmental Evidence*. 4(1), 1–26.

Blind, A.-C., Kuoljok, K., Axelsson Linkowski, W. & Håkan T. (2015). *Myrens betydelse för renskötseln – biologisk mångfald på myrar i renskötselland*. (Rapport 92). Uppsala: Centrum för biologisk mångfald.

Boulanger, J., Poole, K. G., Gunn, A. & Wierzchowski, J. (2012). Estimating the zone of influence of industrial developments on wildlife: a migratory caribou Rangifer tarandus groenlandicus and diamond mine case study. *Wildlife Biology*. 18(2), 164–179.

Boulanger, J., Poole, K. G., Gunn, A., Adamczewski, J. & Wierzchowski, J. (2021). Estimation of trends in zone of influence of mine sites on barren-ground caribou populations in the Northwest Territories, Canada, using new methods. *Wildlife Biology*. 2021(1), 1–16.

Briggs, C. M. (2006). Science, local knowledge and exclusionary practices: Lessons from the Alta Dam case. *Norsk Geografisk Tidsskrift-Norwegian Journal of Geography*. 60(2), 149–160.

Chapin III, F. S., Kofinas, G. P. & Folke, C. (eds.) (2009). *Principles of ecosystem stewardship: resilience-based natural resource management in a changing world*. New York: Springer.

Chen, W., Leblanc, S., White, H., Prevost, C., Milakovic, B., Rock, C., Sharam, G., O'Keefe, H., Corey, L., Croft, B., Gunn, A., van der Wielen, S., Football, A., Tracz, B., Pellissey, J. & Boulanger, J. (2017). Does Dust from Arctic Mines Affect Caribou Forage? *Journal of Environmental Protection*. 8(03), 258–276.

Colman, J. E., Eftestøl, S., Tsegaye, D., Flydal, K. & Mysterud, A. (2013). Summer distribution of semi-domesticated reindeer relative to a new wind-power plant. *European Journal of Wildlife Research*. 59(3), 359–370.

Dettki, H. & Esseen, P. A. (2003). Modelling long-term effects of forest management on epiphytic lichens in northern Sweden. *Forest Ecology and Management*. 175(1–3), 223–238.

Duncan, S. J. (2015). Woodland caribou alpine range restoration: An application for lichen transplants. *Ecological Restoration*. 33(1), 22–29.

Eftestøl, S., Flydal, K., Tsegaye, D. & Colman, J. E. (2019). Mining activity disturbs habitat use of reindeer in Finnmark, Northern Norway. *Polar Biology*. 42(10), 1849–1858.

Eftestøl S., Tsegaye D., Flydal, K. & Colman J. E. (2021). *VindVal rein og vindkraft. Sluttrapport – Arealkonflikt mellom vindkraftverk og tilhørende infrastruktur og reindriftt*. Institutt for Biovitenskap, Universitetet i Oslo og NMBU.

Eira, I. M. G., Oskal, A., Hanssen-Bauer, I. & Mathiesen, S. D. (2018). Snow cover and the loss of traditional indigenous knowledge. *Nature Climate Change*. 8(11), 928–931.

Energimyndigheten (2019). *Kraftigt ökad vindkraftsproduktion fram till 2021*. www.energimyndigheten.se/nyhetsarkiv/2019/kraftigt-okad-vindkraftsproduktion-fram-till-2021 [27.12.2020].

Energimyndigheten (2020). *Vattenkraft*. www.energimyndigheten.se/fornybart/vattenkraft [27.12.2020]

Esseen, P. A., Ekström, M., Westerlund, B., Palmqvist, K., Jonsson, B. G., Grafström, A. & Ståhl, G. (2016). Broad-scale distribution of epiphytic hair lichens correlates more with climate and nitrogen deposition than with forest structure. *Canadian Journal of Forest Research*. 46(11), 1348–1358.

Etha Wind (2020). Finnish Wind Farms. www.ethawind.com/suomen-tuulivoimapuistot [27.12.2020].

European Soil Data Centre (2016). *Maps of indicators of soil hydraulic properties for Europe*. https://esdac.jrc.ec.europa.eu/resource-type/datasets [20. 03. 2021].

Finnish Wind Power Association (2020). *Wind power projects in Finland*. https://tuulivoimayhdistys.fi/en/wind-power-in-finland [27.12.2020].

Fischer, R., Elfgren, E., & Toffolo, A. (2018). Energy supply potentials in the northern counties of Finland, Norway and Sweden towards sustainable Nordic electricity and heating sectors: A review. *Energies*. 11(4), 751.

Golovnev, A. V. (2017). Challenges to Arctic nomadism: Yamal Nenets facing climate change era calamities. *Arctic Anthropology*. 54(2), 40–51.

Hausner, V. H., Engen, S., Brattland, C. & Fauchald, P. (2020). Sámi knowledge and ecosystem-based adaptation strategies for managing pastures under threat from multiple land uses. *Journal of Applied Ecology*, 57(9), 1656–1665.

Hedwall, P. O. & Brunet, J. (2016). Trait variations of ground flora species disentangle the effects of global change and altered land-use in Swedish forests during 20 years. *Global Change Biology*. 22(12), 4038–4047.

Heggberget, T. M., Gaare, E. & Ball, J. P. (2002). Reindeer (Rangifer tarandus) and climate change: importance of winter forage. *Rangifer*. 22(1), 13–31.

Horstkotte, T. & Moen, J. (2019). Successional pathways of terrestrial lichens in changing Swedish boreal forests. *Forest Ecology and Management*. 453, 117572.

Horstkotte, T., Lind, T. & Moen, J. (2016). Quantifying the implications of different land users' priorities in the management of boreal multiple-use forests. *Environmental Management*. 57(4), 770–783.

Horstkotte, T., Moen, J., Lämås, T. & Helle, T. (2011). The legacy of logging – estimating arboreal lichen occurrence in a boreal multiple-use landscape on a two century scale. *PLoS One*. 6(12), e28779.

Horstkotte, T., Sandström, C. & Moen, J. (2014). Exploring the multiple use of boreal landscapes in northern Sweden: The importance of social-ecological diversity for mobility and flexibility. *Human Ecology*. 42(5), 671–682.

Inga, B. (2007). Reindeer (Rangifer tarandus tarandus) feeding on lichens and mushrooms: traditional ecological knowledge among reindeer-herding Sami in northern Sweden. *Rangifer*. 27(2), 93–106.

Jääskeläinen, T. (2020). The Sámi reindeer herders' conceptualizations of sustainability in the permitting of mineral extraction–contradictions related to sustainability criteria. *Current Opinion in Environmental Sustainability*. 43, 49–57.

Johnsen, K. I. (2016). Land-use conflicts between reindeer husbandry and mineral extraction in Finnmark, Norway: contested rationalities and the politics of belonging. *Polar Geography*. 39(1), 58–79.

Jonsson Čabrajič, A. V., Moen, J. & Palmqvist, K. (2010). Predicting growth of mat-forming lichens on a landscape scale–comparing models with different complexities. *Ecography*. 33(5), 949–960.

Kivinen, S. (2017). Sustainable post-mining land use: are closed metal mines abandoned or re-used space? *Sustainability*. 9(10), 1705.

Kivinen, S., Berg, A., Moen, J., Östlund, L. & Olofsson, J. (2012). Forest fragmentation and landscape transformation in a reindeer husbandry area in Sweden. *Environmental Management*. 49(2), 295–304.

Kivinen, S., Moen, J., Berg, A. & Eriksson, Å. (2010). Effects of modern forest management on winter grazing resources for reindeer in Sweden. *Ambio*. 39(4), 269–278.

Kivinen, S., Vartiainen, K. & Kumpula, T. (2018). People and post-mining environments: PPGIS mapping of landscape values, knowledge needs, and future perspectives in northern Finland. *Land*. 7(4), 151.

Kumpula, J., Kurkilahti, M., Helle, T. & Colpaert, A. (2014). Both reindeer management and several other land use factors explain the reduction in ground lichens (Cladonia spp.) in pastures grazed by semi-domesticated reindeer in Finland. *Regional Environmental Change*. 14(2), 541–559.

Larsen, R. & Raitio, K. (2019). Implementing the state duty to consult in land and resource decisions: Perspectives from Sami communities and Swedish state officials. *Arctic Review*. 10, 4–23.

Larsen, R., Skarin, A., Stinnerbom, M., Vannar, J., Alam, M., Kuhmunen, M., Lawrence, R., Nygård, J., Raitio, K., Sandström, P., Sandström, S., Stinnerbom, J., Wik-Karlsson, J. & Österlin, C. (2020). *Omtvistade landskap – Navigering mellan konkurrerande markanvändning och kumulativa effekter*. (Rapport 6908). Stockholm: Naturvårdsverket. www.naturvardsverket.se/Documents/publ-filer/978-91-620-6908-7.pdf?pid=25984

Lawrence, R. & Larsen, R. (2017). The politics of planning: assessing the impacts of mining on Sami lands. *Third World Quarterly*. 38(5), 1164–1180.

Luke (2020). *Finnish forest statistics*. Helsinki: Luonnonvarakeskus. https://stat.luke.fi/sites/default/files/suomen_metsatilastot_2020_verkko.pdf

Marin, A., Sjaastad, E., Benjaminsen, T. A., Sara, M. N. M. & Borgenvik, E. J. L. (2020). Productivity beyond density: A critique of management models for reindeer pastoralism in Norway. *Pastoralism*. 10, 1–18.

Mustonen, K., Mustonen, T., Aikio, A. & Aikio, P. (2010). *Drowning reindeer, drowning Homes: Indigenous Sámi and hydroelectricity development in Sompio, Finland*. Vaasa: Snowchange.

Naeth, M. A. & Wilkinson, S. R. (2014). Establishment of restoration trajectories for upland tundra communities on diamond mine wastes in the Canadian Arctic. *Restoration Ecology*. 22(4), 534–543.

NVE (2020a). *Kraftproduksjon*. www.nve.no/energiforsyning/kraftproduksjon [27.12.2020].

NVE (2020b). *Vindkraftdata*. www.nve.no/energiforsyning/kraftproduksjon/vindkraft/vindkraftdata [27.12.2020].

Olofsson, J., Moen, J. & Östlund, L. (2010). Effects of reindeer on boreal forest floor vegetation: Does grazing cause vegetation state transitions? *Basic and Applied Ecology*. 11(6), 550–557.

Össbo, Å. & Lantto, P. (2011). Colonial tutelage and industrial colonialism: reindeer husbandry and early 20th-century hydroelectric development in Sweden. *Scandinavian Journal of History*. 36(3), 324–348.

Östlund, L., Zackrisson, O. & Axelsson, A. L. (1997). The history and transformation of a Scandinavian boreal forest landscape since the 19th century. *Canadian Journal of Forest Research*. 27(8), 1198–1206.

Paine, R (1982). *Dam a river, Damn a people? Sámi (Lapp) livelihood and the Alta/Kautokeino hydro-electric project and the Norwegian Parliament*. Copenhagen: IWIGA.

Peura, M., Burgas, D., Eyvindson, K., Repo, A. & Mönkkönen, M. (2018). Continuous cover forestry is a cost-efficient tool to increase multifunctionality of boreal production forests in Fennoscandia. *Biological Conservation*. 217, 104–112.

Philips, S.J., Dudík, M. & Schapire, R.E. (2020). Maxent software for modeling species niches and distributions (Version 3.4.1). http://biodiversityinformatics.amnh.org/open_source/maxent/ [03.05.2021].

Plante, S., Dussault, C., Richard, J. H. & Côté, S. D. (2018). Human disturbance effects and cumulative habitat loss in endangered migratory caribou. *Biological Conservation*. 224, 129–143.

Reindeer Herders' Association (2014). *Guide to examining reindeer husbandry in land use projects*. Rovaniemi: Ministry of the Environment. https://paliskunnat.fi/poroyva/PoroYVA_2014_EN_web.pdf

Rola, K. & Osyczka, P. (2014). Cryptogamic community structure as a bioindicator of soil condition along a pollution gradient. *Environmental Monitoring and Assessment*. 186(9), 5897–5910.

Roturier, S. & Bergsten, U. (2006). Influence of soil scarification on reindeer foraging and damage to planted Pinus sylvestris seedlings. *Scandinavian Journal of Forest Research*. 21(3), 209–220.

Roturier, S. & Roué, M. (2009). Of forest, snow and lichen: Sámi reindeer herders' knowledge of winter pastures in northern Sweden. *Forest Ecology and Management*. 258(9), 1960–1967.

Roturier, S., Bäcklund, S., Sundén, M. & Bergsten, U. (2007). Influence of ground substrate on establishment of reindeer lichen after artificial dispersal. *Silva Fennica*. 41(2), 269–280.

Roturier, S., Ollier, S., Nutti, L. E., Bergsten, U. & Winsa, H. (2017). Restoration of reindeer lichen pastures after forest fire in northern Sweden: Seven years of results. *Ecological Engineering*. 108, 143–151.

Sandström, P., Cory, N., Svensson, J., Hedenås, H., Jougda, L. & Borchert, N. (2016). On the decline of ground lichen forests in the Swedish boreal landscape: Implications for reindeer husbandry and sustainable forest management. *Ambio*. 45(4), 415–429.

Sara, M. N. (2009). Siida and traditional Sami reindeer herding knowledge. *Northern Review*. 30, 153–178.

Senft, R. L., Coughenour, M. B., Bailey, D.W., Rittenhouse, L. R., Sala, O. E. & Swift, D. M. (1987). Large herbivore foraging and ecological hierarchies. *BioScience*. 37(11), 789–799.

SGU (2020a) *Mineralstatistik*. www.sgu.se/mineralnaring/mineralstatistik/ [27.12.2020]

SGU (2020b). *Statistics of the Swedish Mining Industry 2019*. (2020:1). http://resource.sgu.se/produkter/pp/pp2020-1-rapport.pdf

SGU (2020c). *Torvbruk*. www.sgu.se/samhallsplanering/planering-och-markanvandning/markanvandning/torvbruk [27.12.2020].

Skarin, A. & Åhman, B. (2014). Do human activity and infrastructure disturb domesticated reindeer? The need for the reindeer's perspective. *Polar Biology*. 37(7), 1041–1054.

Skarin, A., Nellemann, C., Rönnegård, L., Sandström, P. & Lundqvist, H. (2015). Wind farm construction impacts reindeer migration and movement corridors. *Landscape Ecology*. 30(8), 1527–1540.

Skarin, A., Sandström, P. & Alam, M. (2018). Out of sight of wind turbines – Reindeer response to wind farms in operation. *Ecology and Evolution*. 8(19), 9906–9919.

Skarin, A., Niebuhr, B. B., Sandström, P., Alam, M., & Adler S. (2021). *Renar och vindkraft III – studier av vindkraft i drift på ren och renskötsel i vinterbeteslandet och året-runt-markerna*. Rapport 7011. Stockholm: Naturvårdsverket.

Skogsst yrelsen (2019). *Skogsstyrelsens statistikdatabas.* http://pxweb.skogsstyrelsen.se/pxweb/sv/Skogsstyrelsens%20statistikdatabas/Skogsstyrelsens%20statistikdatabas__Bruttoavverkning [27.12.2020].

SLU (2019). *Forest statistics 2019. Official Statistics of Sweden.* Umeå: Swedish University of Agricultural Sciences.

St John, R., Öhman, K., Tóth, S. F., Sandström, P., Korosuo, A. & Eriksson, L. O. (2016). Combining spatiotemporal corridor design for reindeer migration with harvest scheduling in Northern Sweden. *Scandinavian Journal of Forest Research.* 31(7), 655–663.

Statistikcentalen (2020). El- och värmeproduktion. www.stat.fi/til/salatuo/2019/salatuo_2019_2020-11-03_tie_001_sv.html [27.12.2020].

Strengbom, J. & Nordin, A. (2008). Commercial forest fertilization causes long-term residual effects in ground vegetation of boreal forests. *Forest Ecology and Management.* 256(12), 2175–2181.

Subramanian, N., Nilsson, U., Mossberg, M. & Bergh, J. (2019). Impacts of climate change, weather extremes and alternative strategies in managed forests. *Écoscience.* 26(1), 53–70.

Tømmervik, H., Johansen, B., Riseth, J. Å., Karlsen, S. R., Solberg, B. & Høgda, K. A. (2009). Above ground biomass changes in the mountain birch forests and mountain heaths of Finnmarksvidda, northern Norway, in the period 1957–2006. *Forest Ecology and Management.* 257(1), 244–257.

Tømmervik, H., Bjerke, J. W., Gaare, E., Johansen, B. & Thannheiser, D. (2012). Rapid recovery of recently overexploited winter grazing pastures for reindeer in northern Norway. *Fungal Ecology.* 5(1), 3–15.

Tsegaye, D., Colman, J. E., Eftestøl, S., Flydal, K., Røthe, G. & Rapp, K. (2017). Reindeer spatial use before, during and after construction of a wind farm. *Applied Animal Behaviour Science.* 195, 103–111.

Tukes (2020). *Finnish Safety and Chemicals Agency.* https://tukes.fi/en/map-files [01.06. 2021].

Turunen, M., Soppela, P., Kinnunen, H., Sutinen, M. L. & Martz, F. (2009). Does climate change influence the availability and quality of reindeer forage plants? *Polar Biology.* 32(6), 813–832.

Turunen, M. T., Rasmus, S., Järvenpää, J. & Kivinen, S. (2020). Relations between forestry and reindeer husbandry in northern Finland–Perspectives of science and practice. *Forest Ecology and Management.* 457, 117677.

Tyler, N. J., Hanssen-Bauer, I., Førland, E. J. & Nellemann, C. (2021). The shrinking resource base of pastoralism: Saami reindeer husbandry in a climate of change. *Frontiers in Sustainable Food Systems.* 4, 585685.

Uboni, A., Blochel, A., Kodnik, D. & Moen, J. (2019). Modelling occurrence and status of mat-forming lichens in boreal forests to assess the past and current quality of reindeer winter pastures. *Ecological indicators.* 96, 99–106.

Uboni, A., Åhman, B. & Moen, J. (2020). Can management buffer pasture loss and fragmentation for Sami reindeer herding in Sweden? *Pastoralism.* 10(1), 1–13.

Vindbrukskollen (2020). *Vindlov.* www.vindbrukskollen.se [27.12.2020]

Widmark, C. & Sandstrom, C. (2012). Transaction costs of institutional change in multiple-use commons: the case of consultations between forestry and reindeer husbandry in Northern Sweden. *Journal of Environmental Policy & Planning.* 14(4), 428–449.

5 Reindeer husbandry and climate change

Challenges for adaptation

Sirpa Rasmus, Tim Horstkotte, Minna Turunen, Mia Landauer, Annette Löf, Ilari Lehtonen, Gunhild Rosqvist and Øystein Holand

Introduction

Observed climate change in the Arctic

In recent decades, the polar regions have warmed faster than the global average. Several processes contribute to this (see, e.g., Serreze & Barry 2011), the most obvious being the ice–albedo feedback, where warming leads to reduced sea ice, and the emerging open water surface contributes to further warming due to increased absorption of solar radiation (Koenigk et al. 2020). Warming has been rapid in northern Fennoscandia (Figure 5.1) during all seasons over the past three decades, but particularly between December and February (e.g., Berglöv et al. 2015).

Changes in precipitation patterns show strong spatial variation. An increase in winter precipitation has been recorded in northern Sweden and Norway during the past 30 years compared to the reference period 1961–1990 (Vikhamar-Schuler et al. 2016). For example, winter precipitation increased by 30% and snowpack thickness varied up to 50% between years in several herding districts in Sweden (Rosqvist et al., unpublished). In contrast, trends in precipitation are not clear within the RHA of Finland (Maliniemi et al. 2018), and no significant changes in snow cover thickness have been observed (Rasmus et al. 2014; Lépy & Pasanen 2017). However, more frequent and more intense rainy periods in winter have been reported (Vikhamar-Schuler et al. 2016; Luomaranta et al. 2019).

The timing of spring snowmelt in reindeer calving areas differs significantly between years. Generally, later formation and earlier melt of the snow cover have been observed in Finland (Lépy & Pasanen 2017; Luomaranta et al. 2019). The growing season degree day (GDD) sum has significantly increased throughout the RHA (Park et al. 2016; Rasmus et al. 2020).

Expected future climate change

Warming is expected to continue at a faster rate in the Arctic and elsewhere in high northern latitudes relative to the rest of the world. In addition, precipitation

DOI: 10.4324/9781003118565-8

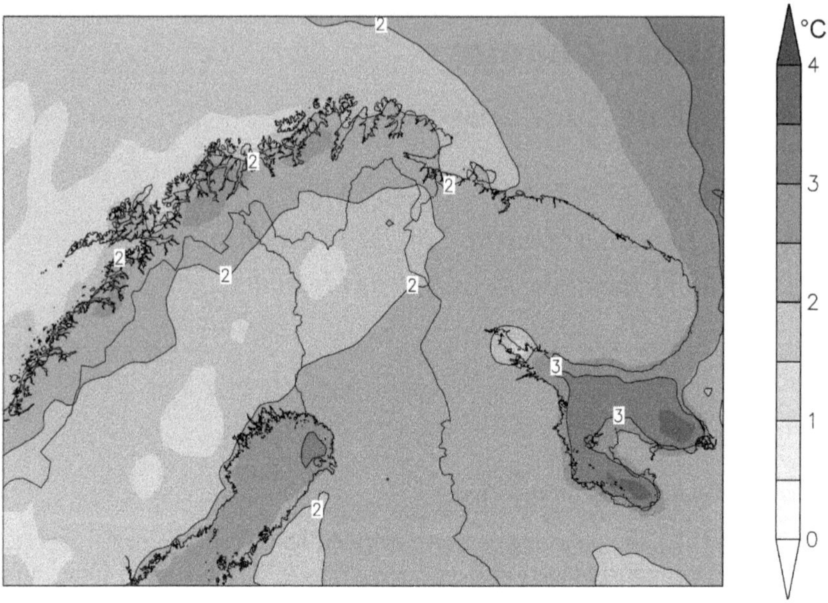

Figure 5.1 Linear trend in annual two-meter air temperature calculated from the ERA5 reanalysis data (Hersbach et al. 2020) between 1979 and 2019. The increasing trend is statistically significant at the 1% risk level across the whole area.

levels are projected to continue to increase in northern Fennoscandia. However, inter-annual variability in seasonal temperature and precipitation will probably remain high, or even increase. During the coming decades, warming is likely to continue at approximately the current rate (Figure 5.2), whereas the rate of warming after the mid-21st century will depend strongly on the evolution of greenhouse gas emissions. Several scenarios for greenhouse gas trajectories, so-called representative concentration pathways (RCP), are adopted by the Intergovernmental Panel on Climate Change (IPCC). The pathways describe different climate futures, all of which are considered possible depending on the volume of greenhouse gases emitted in the coming years. According to RCP4.5, the global emissions peak around 2040, and the warming is thereafter modest compared to the high-emission scenario RCP8.5. Thus, while winter temperatures are projected to increase approximately 5 °C by the year 2100 according to RCP4.5, the increase may be as much as approximately 8 °C under RCP8.5 (Kjellström et al. 2016). Similarly, mean annual precipitation is expected to increase by 20% or 40% by the year 2100, depending on the forcing scenario.

Warmer winters with more precipitation will further increase the number of heavy snow accumulation events, at least in the medium term. In the long term, however, warming will result in a prolonged snow-free season. For example, the

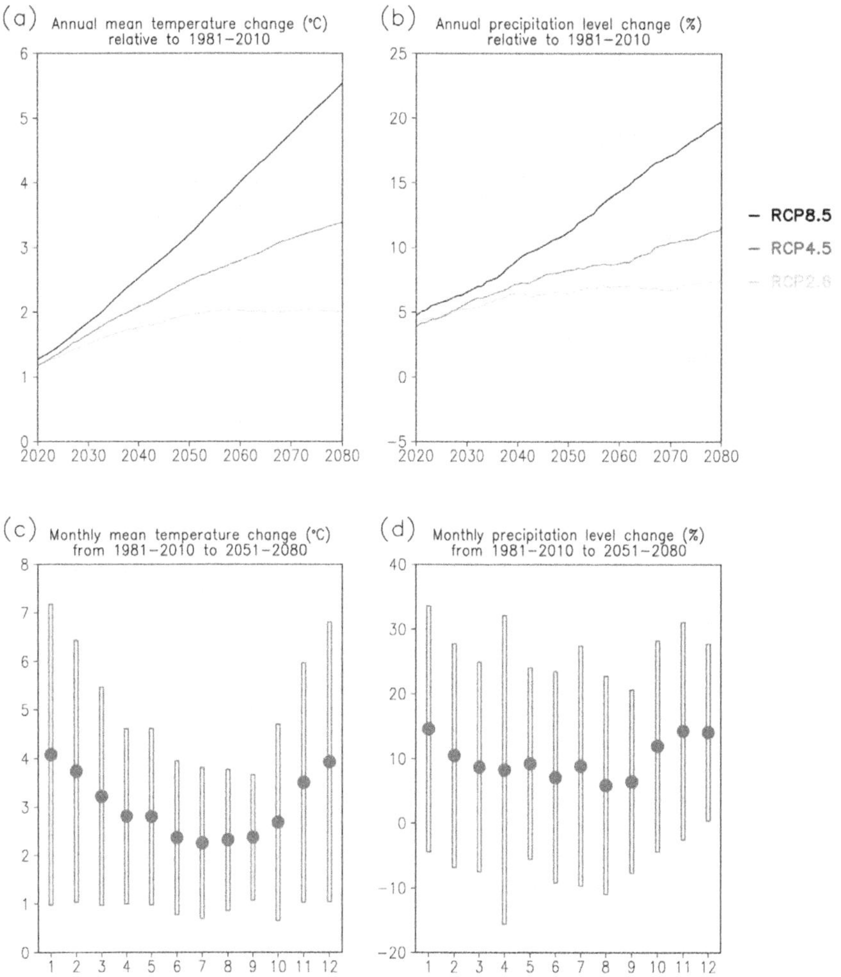

Figure 5.2 Projected multi-model mean change in annual mean temperature (a) and precipitation (b) in Central Finnish Lapland from 2020 to 2080, relative to 1981–2010 values, under three different representative concentration pathways (RCPs). For RCP4.5, the 90% confidence interval is indicated by dotted lines. Projected multi-model mean change with 90% confidence intervals under RCP4.5 is also shown for monthly changes in mean temperature (c) from 1981–2010 to 2051–2080 and precipitation level (d). Projections are based on 28 global climate models participating in the Coupled Model Intercomparison Project Phase 5.

Source: Flato et al. 2013.

number of snow cover days will decrease by between 40 and 60 days in certain mountain catchments in northern Sweden according to IPCC scenarios (Berglöv et al. 2015). The shorter snow season and shrubs protruding through the snow will result in a prolonged period of absorption of solar radiation by tundra vegetation, contributing to regional warming and furthering the growth of shrubs (Te Beest et al. 2016).

Ice crusts on the ground or in the snow are likely to become more frequent due to the increased frequency of thaw-freeze cycles. Increases in winter warming and rain-on-snow (ROS) events have already been reported in the Nordic Arctic regions (Vikhammar-Schuler et al. 2016; Kivinen et al. 2017; Rosqvist et al. 2020), as well as increases in extremely warm events in spring and autumn. Indeed, within this century, "extremes are becoming routine in an emerging new Arctic" (Landrum & Holland 2020).

Adverse weather events and climate change – seasonal impacts and herders' coping strategies

The annual cycle of reindeer ecology and natural pasture use determine seasonal herding activities (Table 5.1). Each season is characterized by particular weather-related risks that are critical in reindeer husbandry and require strategic responses by herders. In a changing climate, some climatic features become more frequent, and others more rare, with both positive and negative impacts on reindeer husbandry (Moen 2008; Turunen et al. 2016). Direct, primary impacts of climate change are related, for example, to weather events; secondary

Table 5.1 Conventional seasons, eight seasons of reindeer husbandry; seasonal features typical for reindeer husbandry and weather and weather-related conditions during each season adversely affecting the practice

Conventional seasons	*Seasons of reindeer husbandry*	*Typical seasonal features, relevant for reindeer husbandry*	*Adverse conditions for reindeer husbandry*
Autumn	Autumn	First frosts and first snow Autumn pastures: mixed forests, bogs Migration to winter pastures Rut Round-ups start	Highly variable weather Temperature varies above and below 0 °C Snow falls on unfrozen ground
	Early winter	Start of the polar night Snow cover formation River/lake ice formation Ground frost formation Migration to winter pastures and round-ups continue Herding Supplementary feeding	Highly variable weather High temperatures Rain-on-snow events Temperature varies above and below 0 °C

Table 5.1 Cont.

Conventional seasons	*Seasons of reindeer husbandry*	*Typical seasonal features, relevant for reindeer husbandry*	*Adverse conditions for reindeer husbandry*
Winter	Winter	Polar night Low temperatures Winter pastures: lichen pastures, old-growth forests Foraging lichen below the snow or arboreal lichen Round-ups continue Herding Supplementary feeding	Highly variable weather Long periods of very low temperatures Deep or hard snow; formation of thick ice crust Temperature varies above and below 0 °C
Spring	Late winter	Foraging lichen below the snow or arboreal lichen Herding Supplementary feeding	Deep soft snow cover with no ice crusts (in forest region)
	Spring	Snowmelt Spring floods Migration to spring/summer pastures: bogs, slopes, deciduous forests	Deep soft snow cover with no ice crusts Low temperatures Late snowmelt
Summer	Early summer	Start of the polar day Start of the growing season Summer pastures: forests, bogs, fjells Calving	Low temperatures, sleet Late snowmelt Late start to the growing season
	Summer	Polar day High temperatures Insect harassment Calf-marking Hay-making for supplementary feeding	Long heat and drought Plenty of insects Forest fires Lack of snow patches on the mountains Rainy summers (especially in hay-making districts)
	Late summer	Light decreases End of the growing season Mushrooms	Poor mushroom yields due to dry weather

impacts include ecosystem responses to changing climate, e.g., spreading new zoonotic diseases or vegetation shifts; tertiary impacts are political or cultural consequences such as changes in traditional knowledge and skills (Butler & Harvey 2010; Magga et al. 2011; Turi 2016). While these dynamics are challenging in themselves, they may be amplified by multiple pressures originating from factors such as competing forms of land use, constraining herders' adaptation options (e.g. Hovelsrud et al. 2021; Landauer et al. 2021; see Table 5.2 for definitions of key concepts used in this chapter).

Table 5.2 Definitions of key concepts used in this chapter

Concept	*Definition*
Climate change risk	The potential for adverse consequences for human or ecological systems or communities, arising from impacts of climate change and human responses to climate change.
Vulnerability	Set of conditions and processes resulting from physical, social, economic and environmental factors, which increase the likelihood that climate change will have negative impacts on a system or community. Combination of exposure, sensitivity of the system to climatic conditions and the adaptive capacity.
Adaptive capacity	Potential of a system or a community to adapt to climate change, to moderate potential damage, to utilize the opportunities or to manage the consequences. Increased adaptive capacity results in reduced vulnerability.
Climate change adaptation	Adjustments in response to actual or expected climatic stimuli (including variability and extremes) and their impacts; changes in processes, practices and structures to moderate potential damage or to benefit from opportunities associated with climate change. Successful adaptation results in reduced vulnerability.
Coping/reactive adaptation	Adjustments implemented in order to maintain basic functioning of the system, often in the short or medium term.
Planned/proactive adaptation	Adjustments resulting from deliberate policy/livelihood decisions, implemented in order to return to, maintain or achieve a desired state of the system also in the long term, often before certain impacts are observed.
Institutional adaptation	Adaptive actions taken and/or implemented at the (formal) institutional level, e.g. in terms of policy or rules.
Maladaptation	Adaptive actions resulting in increased vulnerability of the system, or other actors or sectors, even in the future.

Sources: Smit & Wandel, 2006; Parry et al. 2007; Pelling et al. 2010; IPCC 2012.

Impacts and strategies in autumn

Warm, wet early autumns may favour mushrooms, a vital forage to increase the body condition of reindeer by the onset of winter. However, warm late autumns with unfrozen soils may result in the growth of mycotoxin-producing microfungi (moulds) below the snow, adversely affecting forage resources (Kumpula et al. 2000).

Early slaughter is optimal, since calves start losing weight after the snow cover forms, especially if access to forage is limited. However, to obtain sufficient meat to sell, slaughter cannot happen too early – not before calves have gained enough weight (Näkkäläjärvi et al. 2020). In addition, reindeer need enough time for rutting and to recover from it before round-ups for slaughtering. Timing of the rut is affected by weather and grazing conditions during the previous spring and summer (Paoli et al. 2020), but herders have also noted that warm autumns can cause late or unsynchronized rutting (Rasmus et al. 2020).

Precipitation and temperature shifts during late autumn and early winter determine snow conditions at the base of the snowpack and thus access to ground vegetation (e.g. Löf et al. 2012; Rosqvist et al. 2021). A late rut, the absence of snow, formation of ground ice due to variable weather – combined with low lichen biomasses (Chapter 4) – can cause herds to disperse over a wide area while searching for food. This makes gathering and moving them to round-up sites difficult, and delays the slaughter (Turunen et al. 2016; Rasmus et al. 2020). Controlling the herds becomes difficult when the snow cover is thin. A herder from a fell district in Finland explains that "*if our herds have already been separated according to herding groups/siidas, but snow is scarce, it is difficult to keep the herds apart*" (all direct quotes in this chapter are from Rasmus et al. 2020).

Prolonged autumns, but also earlier springs, require herders to change the timing of other strategic events, such as migration between seasonal pastures. For example, in Finnmark, Norway, the prolonged growing season may allow longer time spent on coastal summer pastures and calf slaughter before migration, sparing the winter grazing areas (Riseth & Tømmervik 2017). At the same time, prolonged grazing in these summer pastures can prevent increased growth of shrubs and trees, perceived as detrimental to both migration and valuable grazing resources (Horstkotte et al. 2017; Riseth & Tømmervik 2017). In Sweden, herders have also chosen to stay longer on the summer pastures in the mountains or on the transitional spring/autumn pastures (Löf et al. 2012).

Due to late formation of permanent snow cover, increasingly, reindeer need to be gathered and moved to the round-up sites using all-terrain vehicles or helicopters instead of snowmobiles (Löf et al. 2012, Turunen et al. 2016). Late and weak ice formation on waterbodies and late freezing of bogs can make gathering even more difficult and hinder migration between seasonal pastures. As the bearing capacity of ice is decreased, there are risks to both reindeer and herders (Näkkäläjärvi et al. 2020). On the other hand, late ice formation can facilitate herding, because open water bodies can provide effective barriers. In Sweden, trucks may be necessary in some herding districts to transport reindeer between different seasonal pastures because of lost migration routes or unsafe ice conditions (Löf et al. 2012).

Impacts and strategies in winter

Herders note that weather has become more variable in all seasons (Vuojala-Magga et al. 2011; Löf et al. 2012; Risvoll & Hovelsrud 2016; Horstkotte et al. 2017). Higher temperatures, increased windiness, more frequent rainfall and increased snow-loads on trees in winter are observed by herders in all three countries (Horstkotte et al. 2017; Rasmus et al. 2020). Some herders also report increased snow depth, but later snow cover formation and earlier snowmelt (Näkkälärvi et al. 2020). Younger herders, therefore, conclude that they have hardly lived through a "normal winter" compared to those experienced by the previous generation (Löf 2013; Axelsson Linkowski et al. 2020).

Winters with long snowless periods or thin snow cover can provide better opportunities for grazing, and warmer weather can help reindeer maintain good body condition (Helle & Kojola 2008). However, warm winters have more frequent and longer-lasting thawing events (air temperatures above zero). More frequent freeze-thaw cycles or ROS events cause the formation of very dense snow or ice layers on the ground or within the snowpack (Rasmus et al. 2018; Nilsen et al. 2020). Formation of ground ice makes the ground vegetation layer harder to access ("locked pastures"). Even a single intensive snowfall on unfrozen ground or ROS event can severely affect grazing conditions for the rest of the winter (Rasmus et al. 2018; Rosqvist et al. 2021). Therefore, deep and/or icy snow may increase reindeer mortality and reduce calving success.

The responses of herders to such events vary and depend on the local context, including pasture environment, herding system and culture. Difficult grazing conditions can be avoided by making use of pasture diversity and mobility, e.g. migration to the coast, usually used during summer in Northern Norway (Eira et al. 2018). Herders exposed to oceanic climate in their winter areas in Norway may need to reverse their former grazing rotation. Previously, coastal pastures were more often locked by ice crusts than inland ones. Nowadays, coastal areas are often snow-free, while inland pastures are more likely to become locked (Risvoll & Hovelsrud 2016). However, coastal pastures are fragmented and are shared with many other forms of land use. Some herders in Norway, therefore, use winter pastures further inland across the Swedish border, based on informal agreements with the herders in Sweden (Risvoll & Hovelsrud 2016). Tyler et al. (2007) report "trading snow"; neighbouring herding partnerships (*siidas*) may allow one another to exploit an area of undisturbed snow. In Sweden, herders might migrate in early winter with their herds to particularly lichen-rich grazing grounds, to avoid the risk that these become inaccessible later (Axelsson Linkowski et al. 2020).

Utilizing local topographical diversity can buffer against adverse weather events; small-scale topographical variation may allow grazing at least in some places (Löf et al. 2012; Horstkotte et al. 2014; Riseth & Tømmervik 2017; Ohredahke herding community 2018). Depending on the snow conditions that need to be avoided, herds can be moved to wind-exposed habitats with less snow, or to forest regions with softer snow and arboreal lichens (where these still remain). However, herders do not consider using spring pastures in the mountains during winter to be a viable long-term strategy as grazing resources on spring pastures are vital during calving.

Under difficult grazing conditions, reindeer tend to disperse in search of grazing resources, particularly arboreal lichens (Eira et al. 2018; Horstkotte et al. 2014). Under such conditions, more active gathering, moving and monitoring of animals by herders are needed to prevent traffic accidents or losses to carnivores (Ohredahke herding community 2018; Axelsson Linkowski et al. 2020). This increases the need for snowmobiles, terrestrial vehicles, drones or helicopters. Herders also report letting the reindeer roam free as a "strategy of last resort" when pastures are locked. However, this choice is associated with increased stress and concern, negatively affecting herders' well-being (Löf et al. 2012).

Deep snow cover affects reindeer even without ice layers therein. A herder from Finland explains problems with deep and soft snow: "*Reindeer get tired of digging for forage … The snow cover does not support their weight, so it is not possible to forage for epiphytic lichens either. Predators catch reindeer easily in deep snow.*" Increasing the herd's percentage of old bulls or castrates, which are better than females at breaking hard snow or digging through deep snow, is a strategy resembling a more traditional herd composition, instead of a focus on maximum productivity with a high percentage of females (Oskal et al. 2009; Riseth & Tømmervik 2017).

Supplementary feed has always been provided for reindeer when needed, e.g. by felling lichen-rich trees (Berg et al. 2011; Turunen & Vuojala-Magga 2014). It also plays an important role in adapting to changing winter conditions. This is particularly seen in the southern and central part of the RHA in Finland, with its long history of forestry. Forestry has reduced winter grazing resources, compounding the negative impacts of climate change. Changing winter conditions and increasing land use pressure enhance the need for supplementary feeding in all three countries (chapter 12). However, in Norway and Sweden many herders are adamant that supplementary feeding is not a preferred adaptation strategy and, indeed, that it increases vulnerability in the long run (Horstkotte et al. 2020).

The majority of the herders in the forest districts in Finland report that changing winter conditions have increased the need for feeding and enclosing reindeer (Rasmus et al. 2020). Enclosure feeding started early in southern districts in Finland. Now this experience is partly seen as an advantage: "*Our use of enclosures, which started at the turn of the 60s into the 70s due to the decreased amount of old forests, has reduced the impact of climate change.*" Nevertheless, keeping enclosures clean and reindeer in them healthy are becoming more important and also problematic: "*It is more challenging than before due to mild weather and rains.*"

Impacts and strategies in spring

Timing of calving is critical for the survival and growth of calves. Calves born early will suffer if there is rain or sleet, amplified by wind, in their first days (Cuyler & Øritsland 2004). Calves born late are weak and vulnerable when reindeer gather into large summer herds. Timing of calving depends on timing of the rut in the previous autumn, and also on the winter and spring weather and grazing conditions. In northern Finland, calving currently happens approximately one week earlier than in the 1970s (Paoli et al. 2018).

Early snowmelt and start of the growing season help reindeer recover from a difficult winter and are particularly favourable for lactating reindeer and their newborn calves (Turunen et al. 2009; Vuojala-Magga et al. 2011; Tveraa et al. 2013). When spring comes early, supplementary feeding can be discontinued early, reducing expenses. In Finland, reindeer fed in enclosures can be released to summer pastures much earlier (Rasmus et al. 2020). Nevertheless, early spring can also cause problems. In Sweden, herders report that early snowmelt can force them to move their animals earlier to spring pastures, because reindeer are more difficult to keep gathered when snow is disappearing (Ran herding community 2018). However, as spring pastures often are located at higher elevations, they do not show the same trend of early spring onset; snowmelt may even be delayed due to increased winter precipitation (Beniston et al. 2018). Consequently, herders may need to provide supplementary feeding, often within enclosures, upon early arrival (Ohredahke herding community 2018). Similarly, early spring requires herders in coastal winter pastures in Norway to move their animals to spring pastures in the mountains to avoid conflicts with agriculture, with the risk of facing difficult snow conditions (Riseth & Tømmervik 2017).

Snow and ice conditions affect moving and migration of herds to spring pastures and calving grounds. Some herders have noted that long periods of hard snow, favourable for migration, have become more rare in recent springs (Näkkäläjärvi et al. 2020). Rapid snowmelt may be problematic for calves, as explained by herders from fell districts in Finland: "*When the warming happens fast, small rivers flood severely. Often the female reindeer have to cross these rivers and the calves may drown.*"

Impacts and strategies in summer

The number of hot summer days is increasing in some regions within the RHA but, more notably, many herders report increased precipitation or heavy rains in summer (Näkkäläjärvi et al. 2020; Rasmus et al. 2020). Heat increases thermal stress in cold-adapted reindeer (Soppela et al. 1986; Klokov et al. 2019). Young calves particularly suffer from long periods of hot weather and insect harassment, but also from heavy rains and cold weather during summer. Insect harassment affects weight and reproduction and increases mortality, because stressed reindeer spend less time grazing, and their energy expenditure increases (Weladji et al. 2003). Heat and insect harassment draw reindeer into large herds, e.g. on snow patches, which facilitates gathering and moving animals for calf-marking in June–July. Snow patch habitats are threatened during the warming climate.

During heatwaves, handling causes extra stress. Therefore the calves are often marked at night. Stressing calves can be avoided by rescheduling the calf-markings or leaving calves unmarked until the autumn round-ups (Turunen et al. 2016; Rasmus et al. 2020). Gathering reindeer for calf-marking has become more difficult in some places because the timing of warm periods and insect harassment has changed. In some districts, this has been the reason for giving up summer calf-marking. A herder from a forest district in Finland also explains that "*heat during the calving period causes reindeer to gather into summer herds earlier. When the large herd moves around, many young female reindeer lose their [weak] calves that were born late.*"

Climate change affects the geographic distribution and epidemiology of climate-sensitive infectious diseases, many of which are zoonotic. This creates new risks for herders. For example, warmer and wetter seasons and increase in shrub and forest vegetation increase tick distribution and abundance (Hovelsrud et al. 2020). Warmer summers with increased precipitation may lead to more frequent parasite epidemics and new invasive alien species (Härkönen et al. 2010; Laaksonen et al. 2010). When reindeer are gathered in enclosures for calf-marking, there is a significant risk of disease outbreaks and parasite transmission, especially in wet, muddy conditions. This can be mitigated by frequently relocating the calf-marking site (Riseth et al. 2020). Wet summers also have negative impacts on hay-making and the quality of winter forage (Rasmus et al. 2020).

As a consequence of the lengthening growing seasons, forests will become more dense and expand northwards and to higher elevations (Karlsen et al. 2017). Climate models predict that pronounced increases in temperature and precipitation could transform more than half of the tundra into shrublands before the next century (Pearson et al. 2013). These vegetation transformations will also have consequences for herding strategies. For example, calf-marking sites may need to be relocated. Grazing can partly mitigate this development and keep landscapes open (Horstkotte et al. 2017). The dry summer of 2018 caused extensive forest fires in Sweden, including the RHA. Climate change may increase the likelihood of fires (Lehtonen et al. 2016), meaning a major adaptation challenge ahead for reindeer husbandry.

Towards holistic adaptation – constraints and ways forward

Knowledge as a key component of adaptive capacity

Knowledge and learning are important components of adaptive capacity (Buchanan et al. 2016; Ford et al. 2016a). Herders have coped with adverse weather conditions for centuries using their knowledge and skills. This knowledge is known as traditional ecological knowledge (TEK), Indigenous knowledge if referring to Indigenous knowledge holders, local knowledge or practitioners' knowledge (Alexander et al. 2011) and has been developed through context-situated learning. New knowledge and practices are accumulated, incorporated, exchanged within the herding community and transmitted from one generation to another (Turunen & Vuojala-Magga 2014).

In response to rapidly changing conditions that affect reindeer herding, herders creatively seek new solutions to add to their traditional knowledge (Axelsson Linkowski et al. 2020). Examples include veterinary education about new diseases (see chapter 13), new infrastructural development and other forms of technical innovation such as mobile slaughterhouses and using drones and GPS to support herding activities. GPS collars provide new information to herders and may facilitate dialogue with other actors (Andersson & Keskitalo 2017). Ideally, these tools demonstrate the pressures on and needs of herding communities and could help to create a platform for more collaborative approaches to land use planning (Sandström et al. 2020). However, besides increasing the financial burdens for herders, this requires balanced arenas for negotiation and interactions between actors – technological innovations alone are not enough (Kuoljok 2019).

Due to the rapid development of technology used in reindeer husbandry and a changing environment, the knowledge and know-how of generations can vary greatly (Axelsson Linkowski et al. 2020; Näkkäläjärvi et al. 2020). Rapid environmental change also creates conditions never experienced before, challenging traditional strategies and knowledge about landscapes and animals and no adaptation strategies may currently exist (Löf 2013; Eira et al. 2018; Ran herding community 2018). This also places new demands on scientific knowledge production for monitoring changes and developing appropriate responses, during all seasons. Winter is considered the most critical season for reindeer husbandry; it is also the most studied season in impact and adaptation research. Nevertheless, survival and productivity of reindeer depend on the cumulative impacts of conditions during the entire herding year (Paoli et al. 2018, 2020).

Adaptation or maladaptation?

Adaptation implies deliberation – to achieve or maintain a desired state of a system in response to change (Smit & Wandel 2006). While many responses described above sustain the essential functioning of reindeer husbandry in the short term, they may result in outcomes that are far from optimal in the long run and from a holistic perspective, including ecological, economic and cultural aspirations. Indeed, strategies can be maladaptive rather than adaptive. Instead of building long-term capacity to adapt to change, they can lay groundwork for future conflict, or increase vulnerability (Noble et al. 2014). For example, changes in seasonal pasture rotation may compromise pasture quality during other seasons, some technical solutions may lead to loss of knowledge and skills, and intensive supplementary feeding may affect reindeer behaviour, increase the risk of disease and be incompatible with herders' views of what constitutes sustainable herding (Horstkotte et al. 2020).

There are also limits to adaptation. Reindeer physiology and behaviour impose certain limits (Chapters 3 and 10). Herders emphasize that they need to work according to the biological rhythm of reindeer, rather than struggling against their reindeer's instincts. Lack of time, workforce or resources can hinder the implementation of adaptative actions. The options are dependent on geographical space available for adjustments and changes to practice as well as variation in topography and vegetation. In many regions, pastures are shrinking (Chapter 4), and growing predator populations further limit the use of certain areas (Chapter 6, Turunen et al. 2020).

Herders often perceive a rift between the demands placed on them to adapt and the lack of power to influence institutional decision-making processes (Löf 2013; Kløcker Larsen et al. 2017; Risvoll & Kaarhus 2020). They describe being stuck in an "adaptation squeeze" (Löf 2013). In analysing adaptation, it becomes evident that herders' options are increasingly restricted and affected by socio-political factors and competing forms of land use (Chapter 8; Hovelsrud et al. 2021; Landauer et al. 2021). These restrict access to or availability of pastures (Chapter 4), which hampers desirable adaptations and exacerbates the impacts of climate change (Risvoll 2016). The overall situation for reindeer husbandry, characterized by unclear rights and lack of balanced institutional arenas, places a heavy adaptation burden on herding communities (Chapter 8; Tyler et al. 2021). Therefore, planned adaptation initiatives at an institutional level are required – initiatives that are able to address and reconsider power structures and relationships between different actors in the landscape.

Towards adaptation policies?

The first step towards institutional adaptation is developing and harmonizing policies. While the governments in Norway, Sweden and Finland all acknowledge the severity of climate change and the challenges facing reindeer husbandry in particular (e.g. SOU 2007:60), they still lack policies to address the structural dimension of adaptation. National policies for climate change adaptation in reindeer husbandry vary, but they typically emphasize technical solutions or compensatory schemes, rather than long-term solutions. These include financial instruments (state-implemented subsidies and compensation) to alleviate the detrimental impacts of natural events. For example, the Finnish Act on compensation for damage inflicted on reindeer husbandry (987/2011 and 655/2016) aims to support herders financially to cope with extensive and unexpected damage resulting from natural events, in practice compensating the supplementary feeding costs. In Norway and Sweden compensation schemes are also available to counter catastrophic grazing conditions, as these are needed as short-term solutions (Näkkäläjärvi 2020).

However important emergency support is for herders to cope financially during these conditions, it is not a long-term solution. It neither addresses nor seeks to govern the multiple goal conflicts apparent between reindeer husbandry and competing forms of land use. The structural needs to balance existing power asymmetries between actors in consultation and planning processes (see Chapter 8) thus remain. As herders, state actors and other land users perceive problems and potential solutions differently, it is profoundly difficult to assess the effectiveness of the support provided. Consequently, herders often find that current adaptation policies implemented by governments only address the symptoms, while the herders' voices and perspectives are left unheard and their knowledge considered less relevant than scientific knowledge (Chapters 7, 8 and 9).

How future land use, including anticipated adaptation actions by other sectors, encroach on adaptation options for reindeer husbandry is of particular concern but is ignored in the current development of institutional adaptation. For instance, in its strategy for the Arctic Region, Sweden expresses an intention to strengthen knowledge about Sámi livelihoods, including reindeer husbandry and the necessary pathways to adaptation to climate change (Regeringskansliet 2020). While the importance of intact ecosystems to act as functional calving grounds, migration routes and provide connectivity between seasonal grazing areas is mentioned, the same strategy also emphasizes the importance of developing extractive land use, such as mining and wind power development on the *same* lands without recognizing the incompatibility. In Finland, adaptation to climate change is guided by The National Climate Change Adaptation Plan 2022 (MAF 2014). Measures to mitigate the adverse effects of climate change on reindeer husbandry are mentioned, including maintaining the connectivity and diversity of pasture areas, environmental protection and considering reindeer husbandry in the legislation steering land use planning. The concrete tools to implement these aims are, however, lacking. Likewise, Norway's Arctic Strategy (Norwegian Ministries 2017) emphasizes the contribution of reindeer husbandry to value-creation and the ambition to maintain the livelihood alongside other forms of land use without specifying how.

Concluding remarks

The current instrumental approaches to climate change adaptation place a considerable burden on herding communities and individual actors. Our analysis emphasizes the need for more proactive and deliberate adaptation actions, backed by political support. This requires just and equitable dialogue about desired adaptation strategies and outcomes for various livelihoods within the RHA, and about the pathways leading there. With such a holistic perspective on adaptation, there is the potential to link climate change adaptation to broader policy goals, such as implementing Indigenous rights and developing just regional land use planning. Including herders' knowledge in these processes increases the likelihood of success – if the needed adjustments in the herding system are rooted in local customs and decision-making processes (Ford et al. 2016 b). In pastoralist systems, adaptation to change requires flexibility. Fixed policies or rigid governing instruments do not work well (Marin et al. 2020). Collaborative processes are essential.

Herders express optimism about continuing with their livelihood, even if, simultaneously, they worry whether adaptation leads to cultural losses. It is, therefore, crucial to co-develop adaptation strategies that are sustainable not only economically and environmentally but also culturally and socially. Success in this respect will shape the future of reindeer husbandry for decades to come.

References

Alexander, C., Bynum, N., Johnson, E., King, U., Mustonen, T., Neofotis, P., Oettlé, N., Rosenzweig, C., Sakakibara, C., Shadrin, V., Vicarelli, M., Waterhouse, J. & Weeks, B. (2011). Linking indigenous and scientific knowledge of climate change. *BioScience*. 61(6), 477–484.

Andersson, E. & Keskitalo, E. C. H. (2017). Technology use in Swedish reindeer husbandry through a social lens. *Polar Geography*. 40(1), 19–34.

Axelsson-Linkowski, W., Fjellström, A. M., Sandström, C., Westin, A., Östlund, L. & Moen, J. (2020). Shifting strategies between generations in Sami reindeer husbandry: the challenges of maintaining traditions while adapting to a changing context. *Human Ecology*. 48(4), 481–490.

Beniston, M., Farinotti, D., Stoffel, M., Andreassen, L. M., Coppola, E., Eckert, N., Fantini, A., Giacona, F., Hauck, C., Huss, M., Huwald, H., Lehning, M., López-Moreno, J.-I., Magnusson, J. Marty, C., Morán-Tejéda, E., Morin, S., Naaim, M., Provenzale, A., Rabatel, A., Six, D., Stötter, J., Strasser, U., Terzago, S. & Vincent, C. (2018). The European mountain cryosphere: a review of its current state, trends, and future challenges. *The Cryosphere*. 12(2), 759–794.

Berglöv, G., Asp, M., Berggren- Clausen, S., Björck, E., Axén Mårtensson, J., Nylén, L., Ohlsson, A., Persson, H. & Sjökvist, E. (2015). *Framtidsklimat i Norrbottens län – enligt RCP-scenarier*. (SMHI Report 32). Norrköping: Swedish Meteorological and Hydrological Institute.

Buchanan, A., Reed, M. G. & Lidestav, G. (2016). What's counted as a reindeer herder? Gender and the adaptive capacity of Sami reindeer herding communities in Sweden. *Ambio*. 45(3), 352–362.

Butler, C. D. & Harley, D (2010). Primary, secondary and tertiary effects of eco-climatic change: the medical response. *Postgraduate Medical Journal*. 84(1014), 230–234.

Cuyler, C. & Øritsland, N. A. (2004). Rain more important than windchill for insulation loss in Svalbard reindeer fur. *Rangifer*. 24(1), 7–14.

Eira, I. M. G., Oskal, A., Hanssen-Bauer, I. & Mathiesen, S. D. (2018). Snow cover and the loss of traditional indigenous knowledge. *Nature Climate Change*. 8(11), 928–931.

Flato, G., Marotzke, J., Abiodun, B., Braconnot, P., Chou, S. C., Collins, W., Cox, P., Driouech, F., Emori, S., Eyring, V., Forest, C., Gleckler, P., Guilyardi, E., Jakob, C., Kattsov, V., Reason, C. & Rummukainen, M. (2013). Evaluation of Climate Models. In: Stocker, T. F. et al. (eds) *Climate Change 2013: The Physical Science Basis. Contribution of Working Group I to the Fifth Assessment Report of the IPCC*. Cambridge: Cambridge University Press. 741–866.

Ford, J. D., Cameron, L., Rubis, J., Maillet, M., Nakashima, D., Willox, A. C. & Pearce, T. (2016a). Including indigenous knowledge and experience in IPCC assessment reports. *Nature Climate Change*. 6(4), 349–353.

Ford, J. D., Stephenson, E., Cunsolo Willox, A., Edge, V., Farahbakhsh, K., Furgal, C., Harper, S., Chatwood, S., Mauro, I., Pearce, T., Austin, S., Bunce, A., Bussalleu, A., Diaz, J., Finner, K., Gordon, A., Huet, C., Kitching, K., Lardeau, M.-P., McDowell, G., McDonald, E., Nakoneczny, L. & Sherman, M. (2016b). Community-based adaptation research in the Canadian Arctic. *WIREs Climate Change*. 7(2), 175–191.

Härkönen, L., Härkönen, S., Kaitala, A., Kaunisto, S., Kortet, R., Laaksonen, S. & Ylönen, H. (2010). Predicting range expansion of an ectoparasite – the effect of spring and summer temperatures on deer ked Lipoptena cervi (Diptera: Hippoboscidae) performance along a latitudinal gradient. *Ecography*. 33(5), 906–912.

Helle, T. & Kojola, I. (1994). Body mass variation in semidomesticated reindeer. *Canadian Journal of Zoology*. 72(4), 681–688.

Hersbach, H., Bell, B., Berrisford, P., Hirahara, S., Horányi, A., Muñoz-Sabater, J., Nicolas, J., Peubey, C., Radu, R., Schepers, D., Simmons, A., Soci, C., Abdalla, S., Abellan, X., Balsamo, G., Bechtold, P., Biavati, G., Bidlot, J., Bonavita, M., De Chiara, G., Dahlgren, P., Dee, D., Diamantakis, M., Dragani, R., Flemming, J., Forbes, R., Fuentes, M., Geer, A., Haimberger, L. Healy, S., Hogan, R.J., Hólm, E., Janisková, M., Keeley, S., Laloyaux, P., Lopez, P., Lupu, C., Radnoti, G., de Rosnay, P., Rozum, I., Vamborg, F., Villaume, S. & Thépaut, J.-N. (2020). The ERA5 global reanalysis. *Quarterly Journal of the Royal Meteorological Society*. 146(730), 1999–2049.

Horstkotte, T., Sandström, C. & Moen, J. (2014). Exploring the multiple use of boreal landscapes in northern Sweden: The importance of social-ecological diversity for mobility and flexibility. *Human Ecology*. 42(5), 671–682.

Horstkotte, T., Utsi, T.A., Larsson-Blind, Å., Burgess, P., Johansen, B., Käyhkö, J., Oksanen, L., Forbes, B. C. (2017). Human–animal agency in reindeer management: Sámi herders' perspectives on vegetation dynamics under climate change. *Ecosphere*. 8(9), e01931.

Horstkotte, T., Lépy, É. & Risvoll, C. (eds.) (2020). *Supplementary Feeding in Reindeer Husbandry: Results from a Workshop with Reindeer Herders and Researchers from Norway, Sweden and Finland*. Umeå: REXSAC.

Hovelsrud G. K., Risvoll C., Riseth J. Å., Tømmervik H., Omazic A. & Albihn A. (2021). Reindeer Herding and Coastal Pastures: Adaptation to Multiple Stressors and Cumulative Effects. In: Nord D. C. (ed.) *Nordic Perspectives on the Responsible Development of the Arctic: Pathways to Action*. Springer Polar Sciences. Cham: Springer. 113–134.

IPCC. (2012). Glossary of Terms. In: Field, C. B. et al. (eds) *Managing the Risks of Extreme Events and Disasters to Advance Climate Change Adaptation*. (IPCC special report) Cambridge: Cambridge University Press. 555–564.

Karlsen S.-R., Tømmervik H., Johansen B. & Riseth J. (2017). Future forest distribution on Finnmarksvidda, North Norway. *Climate Research*. 73(1–2), 125–133.

Kivinen, S., Rasmus, S., Jylhä, K. & Laapas, M. (2017). Long-term climate trends and extreme events in Northern Fennoscandia (1914–2013). *Climate*. *5*(1), 16.

Kjellström, E., Bärring, L., Nikulin, G., Nilsson, C., Persson, G. & Strandberg, G. (2016). Production and use of regional climate model projections – A Swedish perspective on building climate services. *Climate Services*. 2–3, 15–29.

Kløcker Larsen, R. K., Raitio, K., Stinnerbom, M. & Wik-Karlsson, J. (2017). Sami-state collaboration in the governance of cumulative effects assessment: A critical action research approach. *Environmental Impact Assessment Review*. 64, 67–76.

Koenigk T., Key J. & Vihma T. (2020). Climate Change in the Arctic. In: Kokhanovsky A. & Tomasi C. (eds) *Physics and Chemistry of the Arctic Atmosphere*. Springer Polar Sciences. Cham: Springer. 673–705.

Kumpula, J., Parikka, P. & Nieminen, M. (2000). Occurrence of certain microfungi on reindeer pastures in northern Finland during winter 1996–97. *Rangifer*. 20(1), 3–8.

Kuoljok, K. (2019). Without land we are lost: traditional knowledge, digital technology and power relations. *AlterNative: An International Journal of Indigenous Peoples*. 15(4), 349–358.

Laaksonen, S., Pusenius, J., Kumpula, J., Venäläinen, A., Kortet, R., Oksanen, A. & Hoberg, E. (2010). Climate change promotes the emergence of serious disease outbreaks of filarioid nematodes. *Eco-Health*. 7(1), 7–13.

Landauer, M., Rasmus, S., & Forbes, B. C. (2021). What drives reindeer management in Finland towards social and ecological tipping points? *Regional Environmental Change*. 21(2), 1–16.

Landrum, L. & Holland, M. M. (2020). Extremes become routine in an emerging new Arctic. *Nature Climate Change*. 10(12), 1108–1115.

Lehtonen, I., Venäläinen, A., Kämäräinen, M., Peltola, H. & Gregow, H. (2016). Risk of large-scale fires in boreal forests of Finland under changing climate. *Natural Hazards and Earth System Sciences*. 16(1), 239–253.

Lépy, E. & Pasanen, L. (2017). Observed regional climate variability during the last 50 years in reindeer herding cooperatives of Finnish fell Lapland. *Climate*. 5(4), 81.

Löf, A. (2013). Examining limits and barriers to climate change adaptation in an Indigenous reindeer herding community. *Climate and Development*. 5(4), 328–339.

Löf, A., Sandström, P., Baer, K., Stinnerbom, M. & Sandström, C. (2012). *Renskötsel och klimatförändring: Risker, sårbarhet och anpassningsmöjligheter i Vilhelmina norra sameby*. (Political Science Department Research Report 4/2012.) Umeå: Umeå University.

Luomaranta, A., Aalto, J. & Jylhä, K. (2019). Snow cover trends in Finland over 1961–2014 based on gridded snow depth observations. *International Journal of Climatology*. 39(7), 3147–3159.

MAF (Ministry of Agriculture and Forestry). (2014). *Kansallinen ilmastonmuutokseen sopeutumissuunnitelma* 2022. (Maa- ja metsätalousministeriön julkaisuja 5/2014.) Helsinki: Maa- ja metsätalousministeriö.

Magga, O. H., Mathiesen, S. D., Corell, R. & Oskal, A. (2011). *Reindeer Herding, Traditional Knowledge and Adaptation to Climate Change and Loss of Grazing Land*. (EALÁT project report). Alta: EALÁT project.

Maliniemi, T., Kapfer J., Saccone P., Skog A. & Virtanen R. (2018). Long-term vegetation changes of treeless heath communities in northern Fennoscandia: Links to climate change trends and reindeer grazing. *Journal of Vegetation Science*. 29(3), 469–479.

Marin, A., Sjaastad, E., Benjaminsen, T., Sara, M. N. & Langfeldt Borgenvik, E. J. (2020). Productivity beyond density: A critique of management models for reindeer pastoralism in Norway. *Pastoralism*. 10(9), 1–18.

Moen, J. (2008). Climate change: effects on the ecological basis for reindeer husbandry in Sweden. *Ambio*. 37(4), 304–311.

Näkkäläjärvi, K., Juntunen, S. & Jaakkola, J. (2020). *SAAMI – Saamelaisten sopeutuminen ilmastonmuutokseen – hankkeen tieteellinen loppuraportti*. (Valtioneuvoston selvitys- ja tutkimustoiminnan julkaisusarja 2020/25). Helsinki: Valtioneuvosto.

Nilsen, I. B., Hanssen-Bauer, I., Tveito, O. E. & Wong, W. K. (2020). Projected changes in days with zero-crossings for Norway. *International Journal of Climatology*. 41(4), 2173–2188.

Noble, I. R., Huq, S., Anokhin, Y. A., Carmin, J., Goudou, D., Lansigan, F. P., Osman-Elasha, B. & Villamizar, A. (2014). Adaptation Needs and Options. In: *Climate Change 2014: Impacts, Adaptation, and Vulnerability. Part A: Global and Sectoral Aspects. Contribution of Working Group II to the Fifth Assessment Report of the IPCC*. Cambridge: Cambridge University Press. 833–868.

Norwegian Ministries. (2017). *Norway's Arctic Strategy – between Geopolitics and Social Development*. (Publication H-2402 E). www.regjeringen.no/contentassets/fad46f0404e14b2a9b551ca7359c1000/arctic-strategy.pdf [10.01.2021]

Ohredahke herding community. (2018). *Klimatstrategiplan för Ohredahke sameby*. www.sametinget.se/120669 [22.10.2020].

Paoli, A., Weladji, R. B., Holand, Ø., & Kumpula, J. (2018). Winter and spring climatic conditions influence timing and synchrony of calving in reindeer. *PloS One. 13*(4), e0195603.

Paoli, A., Weladji, R. B., Holand, Ø., & Kumpula, J. (2020). The onset in spring and the end in autumn of the thermal and vegetative growing season affect calving time and reproductive success in reindeer. *Current Zoology*. 66(2), 123–134.

Park, T., Ganguly, S., Tømmervik, H., Euskirchen, E. S., Høgda, K. A., Karlsen, S. R., Brovkin, V., Nemani, R. & Myneni, R. (2016). Changes in growing season duration and productivity of northern vegetation inferred from long-term remote sensing data. *Environmental Research Letters*. 11(8), 084001.

Parry, M. L., Canzziani, O. F., Palutikof, J. P., van der Linden, P. J. & Hanson, C. E. (eds.) (2007). *Climate Change 2007: Impacts, Adaptation and Vulnerability. Contribution of Working Group 11 to the Fourth Assessment Report of the IPCC.* Cambridge: Cambridge University Press.

Pearson, R. G., Phillips, S. J., Loranty, M. M., Beck, P., Damoulas, T., Knight, S. J. & Goetz, S. J. (2013). Shifts in Arctic vegetation and associated feedbacks under climate change. *Nature Climate Change*. 3(7), 673–677.

Pelling, M. (2010). *Adaptation to Climate Change: From Resilience to Transformation.* London: Routledge.

Ran herding community. (2018). *Klimatpåverkan på renskötseln nu och i framtiden*. www.sametinget.se/120669 accessed [22.10.2020]

Rasmus, S., Kivinen, S. & Irannezhad, M. (2018). Basal ice formation in Northern Finland snow covers during 1948–2016. *Environmental Research Letters.* 13, 114009.

Rasmus, S., Turunen, M., Luomaranta, A., Kivinen, S., Jylhä, K. & Räihä, J. (2020). Climate change and reindeer management in Finland: Co-analysis of practitioner knowledge and meteorological data for better adaption. *Science of the Total Environment*. 710, 136229.

Regeringskansliet. (2020). *Sveriges strategi för den arktiska regionen*. www.regeringen.se/4a8e8c/contentassets/2c099049a492447b81829eb3f2b8033c/arktisstrategi2020.pdf [10.01.2021.

Riseth, J. Å. & Tømmervik, H. (2017). *Klimautfordringer og arealforvaltning for reindrifta i Norge. Kunnskapsstatus og forslag til tiltak – eksempler fra Troms*. (Norut Rapport 6/2017). Tromsø: Norut.

Riseth, J. Å., Tømmervik, H., & Tryland, M. (2020). Spreading or gathering? Can traditional knowledge be a resource to tackle reindeer diseases associated with climate change? *International Journal of Environmental Research and Public Health*. 17(16), 6002.

Risvoll, C. & Hovelsrud, G. K. (2016). Pasture access and adaptive capacity in reindeer herding districts in Nordland, Northern Norway. *Polar Journal*. 6(1), 87–111.

Risvoll, C. & Kaarhus, R. (2020). Struggling with 'Clear Zoning': Dilemmas of Predator-Pastoral Coexistence in Nordland, Northern Norway. In: Breidlid, A. & R. Krøvel, R. (eds) *Indigenous Knowledges and the Sustainable Development Agenda*. London and New York: Routledge. 185–206.

Rosqvist, G., Inga, N. & Eriksson, P. (2021). Impacts of climate warming on reindeer herding require new land use strategies. *Ambio*. 1–16.

Sandström, P., Myntti, E. L., Sandström, S., Jonsson, N., Lidestav, G. & Jonsson, T. (2020). Who Knew Digitizing and Dialogue Could Change the Course of Reindeer Herding Rights? In: McDonald, J. & Tuulentie, S. (eds) *Sharing Knowledge for Land Use Management*. Elgar Online. 130–143.

Serreze, M. C. & Barry, R. G. (2011). Processes and impacts of Arctic amplification: A research synthesis. *Global and Planetary Change*. 77(1–2), 85–96.

Smit, B. & Wandel, J. 2006. Adaptation, adaptive capacity and vulnerability. *Global Environmental Change*. 16(3), 282–292.

Soppela, P., Nieminen, M. & Timisjärvi, J. (1986). Thermoregulation in reindeer. *Rangifer*. 1, 273–278.

SOU. 2007:60. *Sweden Facing Climate Change—Threats and Opportunities. Final Report from the Swedish Commission on Climate and Vulnerability*. Swedish Government Official Reports.

Te Beest, M., Sitters, J., Ménard, C. B. & Olofsson, J. (2016). Reindeer grazing increases summer albedo by reducing shrub abundance in Arctic tundra. *Environmental Research Letters*. 11(12), 125013.

Turi, E. I. (2016). *State Steering and Traditional Ecological Knowledge in Reindeer-Herding Governance: Cases from Western Finnmark, Norway and Yamal, Russia*. PhD diss. Umeå University.

Turunen, M. & Vuojala-Magga, T. (2014). Past and present winter feeding of reindeer in Finland: herders adaptive learning of the practices. *Arctic*. 67(2), 173–188.

Turunen M., Soppela, P., Kinnunen H., Sutinen M.-L. & Martz, F. (2009). Does climate change influence the availability and quality of reindeer forage plants? *Polar Biology*. 32(6), 813–832.

Turunen, M. T., Rasmus, S., Bavay, M., Ruosteenoja, K. & Heiskanen, J. (2016). Coping with difficult weather and snow conditions: Reindeer herders' views on climate change impacts and coping strategies. *Climate Risk Management*. 11, 15–36.

Turunen, M. T., Rasmus, S., Järvenpää, J., & Kivinen, S. (2020). Relations between forestry and reindeer husbandry in northern Finland–Perspectives of science and practice. *Forest Ecology and Management*. 457, 117677.

Tveraa, T., Stien, A., Bårdsen, B.-J. & Fauchald, P. (2013). Population densities, vegetation green-up, and plant productivity: impacts on reproductive success and juvenile body mass in reindeer. *PLoS One*. 8(2), e56450.

Tyler, N. J. C., Hanssen-Bauer I., Førland, E. & Nellemann, C. (2021). The shrinking resource base of pastoralism: Saami reindeer husbandry in a climate of change. *Frontiers in Sustainable Food Systems*. 4, 585685.

Tyler, N. J. C., Turi, J. M., Sundset, M. A., Bull, K. S., Sara, M. N., Reinert, E., Oskal, N., Nellemann, C., McCarthy, J. J., Mathiesen, S. D., Martello, M. L., Magga, O. H., Hovelsrud, G. K., Hanssen.Bauer, I., Eira, N. I., Eira, I. M. G. & Corell, R. W. (2007). Saami reindeer pastoralism under climate change: applying a generalized framework for vulnerability studies to a sub-arctic social-ecological system. *Global Environmental Change*. 17(2), 191–206.

Vikhamar-Schuler, D., Isaksen, K., Haugen, J. E., Tømmervik, H., Luks, B., Schuler, T. V. & Bjerke, J. W. (2016). Changes in winter warming events in the Nordic Arctic Region. *Journal of Climate*. 29(17), 6223–6244.

Vuojala-Magga, T., Turunen, M., Ryyppö, T. & Tennberg, M. (2011). Resonance strategies of Sami reindeer herding during climatically extreme years in northernmost Finland in 1970–2007. *Arctic*. 64(2), 227–241.

Weladji, R. B., Holand, Ø. & Almøy, T. (2003). Use of climatic data to assess the effect of insect harassment on the autumn weight of reindeer (Rangifer tarandus) calves. *Journal of Zoology*. 260(1), 79–85.

6 Large predators and their impact on reindeer husbandry

Birgitta Åhman, Sirpa Rasmus, Camilla Risvoll, Svein Morten Eilertsen and Harri Norberg

Introduction

Several species of large carnivore – wolf, lynx, wolverine, brown bear and golden eagle – are present in the reindeer herding area (RHA) of Fennoscandia. They all prey on reindeer to a varying degree depending on environmental factors like season, density of the carnivore species and overlap with reindeer ranges. The opportunity for reindeer herders to manage the presence of large carnivores depends on national strategies for the management of carnivores and compensation for losses. This chapter describes the situation for Fennoscandian reindeer husbandry in relation to the presence and distribution of large carnivores, estimated losses, other effects, compensation and management systems.

Predator populations

Large carnivore densities in Fennoscandia have fluctuated widely during the time that reindeer husbandry has been conducted. Once almost eradicated by humans, the populations have increased substantially in northern Europe during the last 50 years (Chapron et al. 2014), mainly as an effect of a gradual introduction of legal protection. Increasing numbers of wolf, Eurasian lynx, wolverine, brown bear and golden eagle have resulted in increased predation on reindeer and growing disturbances to reindeer husbandry (described below).

In Finland, the densities of brown bear, wolverine and lynx have increased and populations have grown, although most predators are found outside the RHA. The estimated numbers of lynx within the RHA are currently a little below 100 individuals, while the number of wolverines is slightly higher (Table 6.1). The estimated number of brown bears within the RHA is around 300. A permanent breeding population of wolves exists close to the southern border of the RHA. This, and the proximity of the Russian border, leads to varying numbers of non-resident immigrant wolves annually in the RHA. However, wolf packs and pairs are also frequently observed near the border. Approximately 90% of the golden eagle territories in Finland are located within the RHA, and the numbers are slowly increasing, especially in the south-eastern part. Predators, wolverine and bear, in particular, also cross the borders from Sweden and Norway to Finland.

DOI: 10.4324/9781003118565-9

Table 6.1 Approximate numbers of large carnivores (individual animals, except for golden eagle) within the reindeer herding area (RHA) of Finland, Sweden and Norway during recent years (2016–2020)

	Finland	*Sweden*	*Norway*	*Notes*
Wolf	10–20	10–50	sporadic	Large variations between years
Lynx	100	700	200	
Wolverine	100–150	700	250	
Brown bear	300	2,000	at least 100	Norway: minimum number
Golden eagle	400	350	500	Nesting pairs

Sources: Estimates based on the following sources: Heikkinen et al. 2021, Norberg 2021, pers. comm. (wolf, Finland), Holmala et al. 2020 (lynx, Finland), Kojola et al. 2020 (wolverine, Finland), Heikkinen et al. 2020 (bear, Finland), Metsähallitus/National Board of Forestry 2021 (golden eagle, Finland, www.metsa.fi/maakotka, August 2021), www.sametinget.se/statistik/rovdjur (wolf, lynx and wolverine, Sweden), Mattisson & Frank 2020 (lynx, Scandinavia); Mattisson et al. 2020a (wolverine, Scandinavia); Kindberg & Swenson 2018 (bear, Sweden); Fløystad et al. 2020 (bear Norway); Wallén et al. 2019 (eagle Sweden); Mattisson et al. 2020b (eagle Norway).

Notes: Estimates of lynx and wolverine are based on annual inventories of family groups and dens, respectively (each representing about six individuals). In Finland, the estimate of wolverine population is based on a combination of different methods (wildlife triangles, areal counts and DNA). Number of bears is based on analysis of DNA (yearly inventories in Norway, single years in Sweden), in Sweden in combination with reports of direct observations of bear, and in Finland based on observations only. The number of nesting pairs of golden eagles is based on observations of occupied nests.

Populations of large carnivores are partly shared between Norway and Sweden and have shown stable or increasing trends during recent decades, except for lynx, which has declined somewhat from a maximum in around 2010 (Mattisson & Frank 2020). The total wolf population has reached about 450 individuals, although most are present outside the RHA (Figure 6.1). Lynx were already present in relatively high numbers in Sweden and Norway when systematic inventories started in the late 1990s, and the total population within the RHA is currently around 900 individuals (Table 6.1). Wolverines are almost exclusively present within the RHA (Figure 6.1), and numbers have almost doubled during the last two decades (Mattisson et al. 2020a). The latest estimates show nearly 700 animals in Sweden and about 250 in Norway (Table 6.1). The brown bear population in Norway seems to be rather stable, and around 150 individuals were identified by DNA analysis in 2020, when the highest number of female bears since 2009 was also recorded (rovdata.no 2021). In Sweden, there are at least 2000 bears within the RHA (Table 6.1). The population of golden eagles seems to be fairly stable, with around 350 and 500 nesting pairs, respectively, within the RHAs of Sweden and Norway (Table 6.1).

Hunting behaviour of large carnivores

Wolves are regarded as the most efficient predator on reindeer, both with regard to how many reindeer they are able to kill and their disturbance of

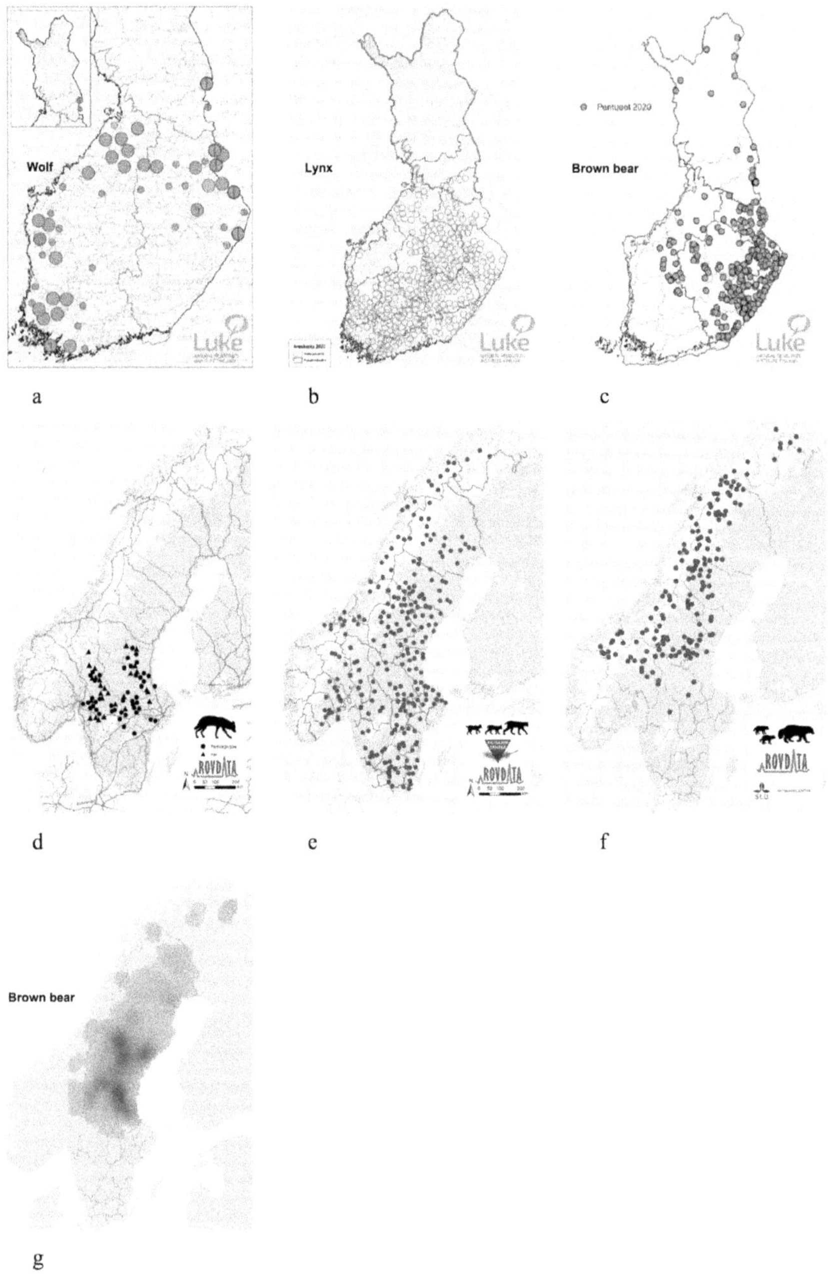

Figure 6.1 Maps for Finland (Luke/Natural Resources Institute Finland, 2021) showing distributions 2019/2020 of (a) wolf (family groups/pairs), (b) lynx (family groups) and (c) bear (family groups). Maps for Sweden and Norway from www.rovdata.no showing distribution of (d) wolf family groups (round dots) and pairs (triangles), (e) lynx family groups, (f) wolverine dens in the winter

caption for Figure 6.1 continued

of 2019–2020. Map (g) showing the density of brown bears (according to inventories 2017), from the Scandinavian Brown Bear Research Project (http://bearproject.info). Each family group of wolf corresponds to about ten individuals, while each reproductive unit (family group or den) of lynx or wolverine corresponds to an average of a little over six individual animals. Published with permission of Luke, Rovdata and the Scandinavian Brown Bear Research Project, respectively.

the reindeer herds (Bjärvall et al. 1990; Ryd 2007; Sikku & Torp 2008). The damage by wolves to Fennoscandian reindeer husbandry is currently limited, since their numbers are strongly regulated within the RHA of all three countries. Nevertheless, wolves have the potential to cause substantial damage where they are present, which is well illustrated by studies on wild reindeer in Finland (Kojola et al. 2004) and caribou in North America (Bergerud & Elliot 1986).

Lynx are present in large parts of the RHA (Figure 6.1), where reindeer are usually their main prey (Mattisson et al. 2011b), although to a lesser extent in areas with high abundance of roe deer (Sunde et al. 2000). An investigation in the Sarek area in Northern Sweden revealed that reindeer constituted 90% of the prey killed by females with kittens (Pedersen et al. 1999), corresponding to six reindeer killed per month. Kill rates, however, vary depending on the time of year, abundance of reindeer and age and sex of the lynx. Lynx generally seem to prefer reindeer calves rather than adults (Mattisson et al. 2014).

Wolverines within the RHA usually utilize reindeer as their main food (Mattisson et al. 2016). They are not such efficient hunters as wolf and lynx and often scavenge on leftovers from other predators (Mattisson et al. 2011a). They can, however, hunt effectively when the snow is dense enough to support the wolverine (due to their large feet), but not the reindeer (Haglund 1966), and in these circumstances, they may kill many reindeer on a single occasion. Kill rates range from 0 to 15 (average 1–2) reindeer per month (Mattisson et al. 2016), with lower rates in areas with high presence of lynx, where wolverines can feed on leftovers from lynx (Mattisson et al. 2011a).

Bears cause damage to reindeer husbandry mostly during the calving period. Recent research in two forest reindeer herding districts (RHDs) in Northern Sweden (Sivertsen 2017) showed an annual average kill rate of 11 reindeer calves per bear present within the calving area. Predation ceased shortly after the last calves were born. Similar results have been obtained from Finland (Norberg & Nieminen 2007). Bears may also kill adult reindeer before or during the calving period, as well as later in the autumn before hibernation.

Golden eagles can be the main predator on young calves in areas where other predators are less common (Norberg et al. 2006). Light calves are at higher risk of being killed by eagles than heavier ones (Nybakk et al. 1999; Norberg et al. 2006). The age and condition of female reindeer may play a role, as young and light females tend to give birth to light calves (Rönnegård et al. 2002), and young females are also less experienced in protecting their calves from eagles (*personal observation*, Norberg).

Losses of reindeer due to predation

Losses of reindeer due to predation are estimated differently in the three countries. In Finland and Norway estimates are based on dead reindeer that have been found and assumed to have been killed by predators. In Sweden, with a different compensation system (see below), assessments are largely based on the presence of predators and their estimated kill rates.

The number of reindeer lost to predators for which compensation is paid in each RHD in Finland is published annually in the journal *Poromies* (by the Reindeer Herders' Association) and was relatively low until 1990. Damage has increased significantly since then, with a maximum level being reached in 2020 (in total, 5,965 predator-killed reindeer were found and for which compensation was paid). As damage statistics in Finland are based on the number of found and reported reindeer carcasses for which compensation has been approved, the numbers are to be considered the minimum loss due to predation. Searching for dead reindeer has become more efficient recently, and data from early decades are therefore not entirely comparable with more recent data. Damage varies between regions and is most apparent in the south-eastern corner of the RHA in Finland (Rasmus et al. 2020). Losses are also high along the eastern border and in northern RHDs. In some hotspots, the damage has recently amounted to more than 10% of the reindeer stock.

As there is no systematic documentation of predated reindeer in Sweden, indirect measurements have to be used. A governmental inquiry (SOU 2012), based on predator numbers and kill rates, estimates the total annual predation of reindeer in Sweden to be between 19,500 and 72,500, mostly by lynx (10,000–50,000 reindeer) and wolverine (6,000–15,000 reindeer). Correlations between the number of predators and harvest rates for different RHDs and years suggest that each reproduction (lynx family group or occupied wolverine den, corresponding to about six individuals) of lynx or wolverine, reduce the harvest of reindeer by about 100 animals (Hobbs et al. 2012). According to a model based on the comparison between expected and observed reindeer survival and reproduction rate (Åhman 2017), the annual loss of reindeer due to predation corresponds to 10–20% of the winter stock for many RHDs.

In Norway, compensation was paid for about 19,000 reindeer killed annually by predators from 2015/16 until 2019/20 (www.rovbase.no). The majority of these (76%) were calves. High calf losses have been reported in most of the RHA in Norway during the last two decades. Compensation for around 63,000 reindeer has been applied for annually but paid for only about 30% of these. Of the compensation payments in 2019/2020, lynx were responsible for 27%, wolverine 32%, golden eagle 34%, brown bear 2%, wolf 1% and unspecified predators 4%.

It has been discussed to which extent predation is compensatory to other causes of mortality, that is, when an animal killed by a predator should have died anyway, e.g., due to disease or starvation (Tveraa et al. 2003). Calf mortality due to other reasons than predation may be high after a winter with exceptionally hard grazing conditions or if the summer is late (Tveraa et al. 2013), if females are generally in poor condition (Rönnergård et al. 2002), or in the case of very adverse weather during the calving period. Otherwise, most research conducted previously, when there were fewer predators (when other causes of mortality were not hidden by high predation), revealed low mortality rates for reasons other than predation. Bjärvall et al. (1990) reported calf mortality due to disease, accidents and so on between 2% and 4% from calf marking in July until autumn gathering in November–December. Skjenneberg and Slagsvold (1968) reported 3–10% annual loss of reindeer in parts of Norway, with predation as a minor cause (1.5%), during the period 1948–1956.

Indirect and long-term effects of predators

According to herders' perceptions, the presence of predators has several detrimental effects on reindeer husbandry (Turunen et al. 2017, Rasmus et al. 2020; Risvoll & Kaarhus 2020). Herds break up, reindeer foraging is disturbed, the best pastures may be impossible to use, reindeer condition declines and calving is disturbed. Predators also affect herding, causing difficulties in keeping herds under control and gathering and moving the herds to round-up sites.

Predation has negative effects on herd productivity by changing the age structure of the breeding population, thus reducing calf production, and the opportunities for genetic selection. In the long run, this can compromise the economic sustainability of the livelihood (Pekkarinen et al. 2020). The unpredictability of the work has increased, and the feeling of autonomy has decreased. Recurring finds of reindeer killed by predators cause physical and mental burdens, and the situation affects families and the overall social life of herders (Pohjola & Valkonen 2012).

If losses of reindeer become too large, reindeer husbandry may reach a tipping point (see Chapter 14) when the number of calves that survive to adulthood is too few to replace adult reindeer that die or become too old to reproduce. This will eventually lead to herd collapse, which was the case in one Swedish RHD when a *siida* group lost not only calves but also 18% of adult female reindeer each year, greatly exceeding the threshold for herd collapse (Åhman et al. 2014). This scenario is also increasingly emerging in Norway due to large losses to predators over time (Risvoll et al. 2022).

All through history, herders have aimed to minimize losses to predators. A variety of strategies have been developed based on practitioners' knowledge related to the behaviour of predators and different means of protecting reindeer (Ryd 2007; Sikku & Torp 2008; Sara 2009). The knowledge is active in reindeer herding cultures throughout Fennoscandia. Before the hunting restrictions, there was active hunting of predators, which gave the herders more control over predator–reindeer interactions. Nonetheless, great efforts are still made to protect reindeer from predators.

Present coping strategies include fencing and feeding of reindeer or constantly tending the reindeer by circling the herd with snowmobiles or skis. Areas with many predators are avoided, leaving potential pasture resources unused. Effective protection of reindeer is almost impossible during snow-free seasons and may also be difficult in winter as predation often takes place at night. Herders increasingly use modern techniques, such as GPS collars on reindeer, drones or wildlife cameras, which can provide more control over predator–reindeer interactions and aid in finding reindeer carcasses (Risvoll et al. 2022).

Economic compensation

The compensation systems in all countries aim at minimizing the financial damage caused to reindeer herders, while maintaining viable predator populations (Strand et al. 2016; Zabel & Holm-Müller 2008; Pekkarinen et al. 2020) but differ considerably between countries.

In Finland, the scheme compensating for damage caused by game animals (including large carnivores, but not golden eagle) is regulated in the "Riistavahinkolaki" (game damage law: www.finlex.fi/fi/laki/ajantasa/2009/20090105). According to this scheme, predator-killed reindeer carcasses that have been found should be reported to the municipality official responsible for rural livelihoods. The compensation is provided either to the reindeer owner or to the RHD. From 5% to 10% of reported cases are inspected in the field by the municipality official, often accompanied by representatives of the RHD and local game management association. The sum paid is 1.5 times the defined value of the lost reindeer to compensate for losses that are not documented. In addition, the RHDs are provided with specific calculated compensation for calves lost, but not found, between birth and 30 November. Nevertheless, herders experience that they do not get full and accurate monetary compensation for indirect costs and extra work caused by the predators. Herders also find the uncertainty and slowness of compensation payments, as well as the allocation of payments among herders, problematic.

Since 1998, reindeer damage in Finland caused by golden eagles has been compensated for based on occupied territories and successful reproduction, and herders have been generally satisfied with the scheme. However, they are generally reluctant to switch to a territory or presence-based compensation scheme for damage caused by other large carnivores. The total annual sum paid in compensation has increased during the 2000s and has amounted to over 6 million euros (M€) since 2012. In 2016 and 2017, cuts in compensation per lost animal took place since the maximum allowed compensation level for Finnish reindeer husbandry (10 M€ on annual basis) set by the EU was exceeded, and (in 2017) also due to limitations in the state budget.

In Sweden, compensating for damage is based on the presence of predators within each of the RHDs (Zabel & Holm-Müller 2008). This system was introduced in 1996. Until then, compensation was paid based on predated reindeer carcasses found (similar to Finland). The compensation is regulated by "Viltskadeförordning" ("Wildlife Damage Ordinance", www.riksdagen.se/sv/dokument-lagar/dokument/svensk-forfattningssamling/_sfs-2001-724) and administered by the Sami Parliament. The money is paid to the RHD, who distribute it internally, or use it for collective costs. At present each reproduction of lynx or wolverine is compensated with 200,000 SEK (20,000 Euro), while a wolf reproduction is compensated with 500,000 SEK. Permanent or occasional presence of these three predators results in lower sums being paid in compensation. Compensation for losses due to brown bear or golden eagle is based solely on the area of the RHD. High levels of documented damage (many dead reindeer) on a single occasion may result in additional payment directly to the affected reindeer owner. During recent years, the annual compensation has been around 50 million SEK (≈5 M€) (www.sametinget.se/statistik/rovdjur), of which approximately 90% was for damage caused by lynx or wolverine. The total sum is limited by the government and has not been raised since 2002.

In Norway, compensation for losses due to protected carnivores is regulated by the Government Regulation: "Forskrift om erstatning for tap og følgekostnader når tamrein blir drept eller skadet av rovvilt" (https://lovdata.no/dokument/SF/forskrift/2001-05-04-468). Compensation is paid via the County Governor for reindeer that are found dead and confirmed as having been killed by a predator (wolf, bear, lynx, wolverine or golden eagle) by someone from, or authorized by, the Norwegian Environmental Protection Agency. Compensation is paid directly to the individual herder, who may also get compensation for costs, inconvenience or losses that are related to loss of the animal. In addition, compensation can be paid for lost but not found reindeer, provided that they are lost in an area and at a time with documented finds of predator-killed reindeer and presence of predators. In this case, there is a deduction from the compensation relative to the expected mortality due to causes other than predation. The annual sum for compensation during the last five years has been, on average, 77.6 million NOK (≈8 M€). There is, however, great frustration among herders about the method used for making judgements and what counts as evidence that reindeer are killed by predators (Risvoll & Kaarhus 2020).

Management policy

In historical times, large carnivores were targeted, and populations were kept low using any means available. After almost total eradication of carnivore populations by the early to mid-1900s, national protection laws and restrictions on hunting were gradually developed. The obligations of the CITES Convention (1976), Bern Convention (1979) and later the Convention on Biological Diversity (1994), as well as the EU Habitats Directive (1992) and Birds Directive (1979), had to be taken into account when considering suitable levels of protection.

In Finland, large predators were gradually protected from 1962 until 1984. There are, however, permits for damage prevention or sport hunting, although based on strict criteria. Management plans are important tools in Finnish predator management policy. The policies for the RHA differ from those to the rest of the country. Damage-based hunting permits for wolves, lynx and brown bear in the RHA can be issued without quotas, and for wolverines based on an annual quota, when the conditions set in the Habitats Directive and Finnish hunting law have been thoroughly considered. The brown bear population in the RHA is mainly regulated by quota-based hunting. Some management-based licenses are also issued to hunt lynx in the RHA, although lynx are mostly hunted outside the RHA.

In Sweden, golden eagle, bear and lynx were protected as early as the 1920s. Wolves were protected in 1966 and wolverines in 1969. Like Finland, Sweden has management plans for all the large carnivores. There is quota-based hunting for bear, lynx and recently also wolverine in the RHA, but under strict regulations. Quotas are decided by the Swedish Environmental Agency. In addition, the RHDs can apply for damage-based hunting. Hunting permits are, however, often appealed against by nature conservation organizations, and in many cases revoked. In 2013, the Swedish Parliament decided that 10% should be the maximum loss to predators for any single RHD in Sweden and that actions should be taken if this level was exceeded. So far, this decision has had limited power. A model for estimating loss (Åhman 2017) is used in appeals for protective hunting, but supporting information verifying predation is generally required in order to obtain permission for damage-based hunting.

In Norway, national conservation policies started with protecting bears during the 1960s, followed by wolverine in southern Norway in 1971, and northern Norway in 1982. Norway ratified the Bern Convention in 1986, implying a commitment to safeguard sustainable populations of all large carnivores. In 2011, the Parliament settled on a "Carnivore Agreement" (Stortinget 2011), and management authority was then delegated from the central government to regional large carnivore committees (RLCC). These have a mandate to take decisions regarding hunting as long as the population goals are reached. The RLCCs are responsible for management plans, which should reduce spatial overlap between large carnivores and grazing domestic livestock (so-called "clear zoning"). There are, nevertheless, large overlaps between areas for reindeer and areas prioritized for carnivores (Strand et al. 2016). There has been ongoing controversy about the size of these areas and the instruments in place to document large carnivores in Norway (Risvoll & Kaarhus 2020). Reindeer herders and sheep farmers point out the difficulty of maintaining zones due, e.g., to topography that affects animal movement and behaviour; in addition, basing lynx registration solely on snow tracks is perceived as too rigid, not considering local context or the great variability in snow conditions (ibid).

Concluding remarks

Reindeer husbandry is greatly affected by the presence of large carnivores. At the same time, reindeer are an essential food source for carnivores. Herders are continuously coping with the presence of predators and trying to minimize reindeer losses, for which their traditional and experience-based knowledge is vital. Nevertheless, this knowledge seems to be insufficient in the rapidly changing operational environment, where institutional, societal and climatic constraints are reducing the space for adaptation.

Compensation schemes aim at easing the co-existence of reindeer husbandry and predators. Herders generally acknowledge that predators belong to the northern natural and cultural heritage and accept their presence provided that losses are bearable and damages are fairly compensated (Sippola et al. 2005, Nykänen & Valkeapää 2016). Compensation schemes differ between countries, but none of them is seen as ideal, and each scheme has benefits and drawbacks. A general criticism is that compensation is too low, because not all predator-killed reindeer are acknowledged, the value of a killed reindeer is set too low, indirect costs are not included, or the numbers of predators are underestimated. There is also frustration among herders who find that their voices are not heard, and their knowledge not recognized when it comes to predator management.

There is friction between predator conservation and local livelihoods globally. The Fennoscandian RHA provides an interesting case; predator populations share the landscape with more or less free-ranging semi-domesticated animals (Chapron et al. 2014) and the people who try to make a living from taking care of them. It may well be that compromises made so far have not sufficiently served either the predators or the livelihood of reindeer herders. What is clear is that conservation goals need to be balanced with livelihood needs and human welfare (Groom & Harris 2008; Sjölander-Lindqvist et al. 2015). Striving towards ecological sustainability and biodiversity targets (both reindeer and predators having a significant role in that) requires that economic and social sustainability of local communities are not being overridden (Sjölander et al. 2020).

References

Åhman, B. (2017). *Beräkning av rovdjursförluster i rennäringen baserat på produktion*. www.sametinget.se/61994

Åhman, B., Svensson, K. & Rönnegård, L. (2014). High female mortality resulting in herd collapse in free-ranging domesticated reindeer (*Rangifer tarandus tarandus*) in Sweden. *PLoS ONE*. 9(10), e111509.

Bergerud, A.T. & Elliot, J.P. (1986). Dynamics of caribou and wolves in Northern British Columbia. *Canadian Journal of Zoology*. 64(7), 1515–1529.

Bjärvall, A., Franzén, R., Nordkvist, M. & Åhman, G. (1990). *Renar och rovdjur. Rovdjurens effekter på rennäringen*. Solna: Naturvårdsverket förlag.

Chapron, G., Kaczensky, P., Linnell, J.D.C., von Arx, M., Huber, D., Andrén, H., López-Bao, J.V. et al. (2014). Recovery of large carnivores in Europe's modern human-dominated landscapes. *Science*. 346(6216), 1517–1519.

Fløystad, I., Brøseth, H., Bakke, B.B., Eiken, H.G. & Hagen, S.B. (2020). *Populasjonsovervåking av brunbjørn. DNA-analyse av prøver innsamlet i Norge i 2019*. (NINA Rapport 1808). Norsk institutt for naturforskning.

Groom, R. & Harris, S. (2008). Conservation on community lands: the importance of equitable revenue sharing. *Environmental Conservation*. 35(3), 242–251.

Haglund, B. (1966). Winter habits of the lynx (*Lynx lynx* L.) and wolverine (*Gulo gulo* L.) as revealed by tracking in the snow. De stora rovdjurens vintervanor. *Viltrevy.* 4, 81–229.Heikkinen, S., Kojola, I., Mäntyniemi, S. (2020). Karhukanta Suomessa 2019. *Luonnonvara- ja biotalouden tutkimus* 26/2020. Luonnonvarakeskus. Helsinki.

Heikkinen, S., Valtonen, M., Härkälä, A., Helle, I., Mäntyniemi, S. & Kojola, I. (2021). Susikanta Suomessa maaliskuussa 2021. *Luonnonvara- ja biotalouden tutkimus.* 39/2021 Luonnonvarakeskus. Helsinki.

Hobbs, N.T., Andrén, H., Persson, J., Aronsson, M. & Chapron, G. (2012). Native predators reduce harvest of reindeer by Sámi pastoralists. *Ecological Applications.* 22(5), 1640–1654.

Holmala, K., Heikkinen, S. & Mäntyniemi, S. (2020). Ilveskanta Suomessa 2020. *Luonnonvara- ja biotalouden tutkimus* 48/2020. Luonnonvarakeskus. Helsinki.

Kindberg, J. & Swenson, J.E. (2018). Björnstammens storlek i Sverige 2017. Skandinaviska björnprojektet, Rapport 2018:3.

Kojola, I., Huitu, O., Toppinen, K., Heikura, K., Heikkinen, S. & Ronkainen, S. (2004). Predation on European wild forest reindeer (*Rangifer tarandus*) by wolves (Canis lupus) in Finland. *Journal of Zoology.* 263(3), 229–235.

Kojola, I., Heikkinen, S., Mäntyniemi, S. & Ollila, T. (2020). Ahmakannan kehitys ja ahmakanta Suomessa 2020. *Luonnonvara- ja biotalouden tutkimus* 68/2020. Luonnonvarakeskus. Helsinki.

Luke/Natural Resources Institute Finland. (2021). *Reports for the population estimates of large carnivores.* http://riistahavainnot.fi/suurpedot/kannanarviointi/lausunnot

Mattisson. J., Andrén, H., Persson, J. & Segerström, P. (2011a). Influence of intraguild interactions on resource use by wolverines and Eurasian lynx. *Journal of Mammalogy.* 92(6), 1321–1330.

Mattisson, J., Odden, J., Nilsen, E.B., Linnell, J.D.C., Persson, J. & Andrén, H. (2011b). Factors affecting Eurasian lynx kill rates on semi-domestic reindeer in northern Scandinavia: Can ecological research contribute to the development of a fair compensation system? *Biological Conservation.* 144(12), 3009–3017.

Mattisson, J., Arntsen, G.B., Nilsen, E.B., Loe, L.E., Linnell, J.D.C., Odden, J., Persson, J. & Andrén, H. (2014). Lynx predation on semi-domestic reindeer: do age and sex matter? *Journal of Zoology.* 292(1), 56–63.

Mattisson, J. & Frank, J. (2020). Bestandsovervåking av gaupe i 2020. *Bestandsstatus for store rovdyr i Skandinavia.* 2–2020.

Mattisson, J., Höglund, L. & Brøseth, H. (2020a). Bestandsovervåking av jerv i 2020. *Bestandsstatus for store rovdyr i Skandinavia.* 3–2020.

Mattisson, J., Nilsen, E.B. & Brøseth, H. (2020b). *Estimering av antall hekkende par kongeørn basert på kjent forekomst i Norge for perioden 2015–2019.* (NINA Rapport 1858). Norsk Institutt for Naturforskning.

Mattisson, J., Rauset, G.R., Odden, J., Andrén, H., Linnell, J.D.C. & Persson, J. (2016). Predation or scavenging? Prey body condition influences decision-making in a facultative predator, the wolverine. *Ecosphere.* 7(8), e01407.

Metsähallitus/National Board of Forestry. (2021). *Reports for the population estimates of Golden eagle.* www.metsa.fi/maakotka

Norberg, H. & Nieminen, M. (2007). Suurpetojen vaikutus poronvasojen kuolleisuuteen Kallioluoman paliskunnassa vuosina 2005–06. *Kala- ja riistaraportteja.* 415.

Norberg, H., Kojola, I., Aikio, P. & Nylund, M. (2006). Predation by golden eagle *Aquila chrysaetos* on semi-domesticated reindeer *Rangifer tarandus* calves in northeastern Finnish Lapland. *Wildlife Biology.* 12(4), 393–402.

Nybakk, K., Kjelvik, O. & Kvam, T. (1999). Golden eagle predation on semidomestic reindeer. *Wildlife Society Bulletin*. 27, 1038–1042.

Nykänen, T. & Valkeapää, L. (eds.) (2016). *Kilpisjärven poliittinen luonto. SKS:n toimituksia 1422*. Helsinki: Suomalaisen Kirjallisuuden Seura

Pedersen, V.A., Linnell, J.D.C., Andersen, R., Andrén, H., Lindén, M. & Segerström, P. (1999). Winter lynx *Lynx lynx* predation on semi-domestic reindeer *Rangifer tarandus* in northern Sweden. *Wildlife Biology*. 5(1), 203–211.

Pekkarinen, A.-J., Kumpula, J. & Tahvonen, O. (2020). Predation costs and compensations in reindeer husbandry. *Wildlife Biology*. 2020(3), 1–14.

Pohjola, A. & Valkonen, J. (2012). *Poronhoitajien hyvinvoinnin uhat ja avun tarpeet*. Rovaniemi: Lapland University Press.

Rasmus, S., Turunen, M., Norberg, H., Kojola, I, Kumpula, J. & Ollila, T. (2020). Mission impossible? Pursuing the co-existence of viable predator populations and sustainable reindeer husbandry in Finland. *Journal of Rural Studies*. 80, 135–148.

Risvoll, C., Hovelsrud, G.K. & Riseth, J.Å. (2022). Falling between the cracks of the governing systems: Risk and uncertainty in pastoralism in northern Norway. *Weather, Climate and Society (WCAS)*. 14, 191–204.

Risvoll, C. & Kaarhus, R. (2020). Struggling with 'clear zoning': Dilemmas of predator-pastoral coexistence in Nordland, northern Norway. In: Breidlid, A. & R. Krøvel, R. (eds.) *Indigenous Knowledges and the Sustainable Development Agenda*. London and New York: Routledge.

Rönnegård, L., Forslund, P. & Danell, Ö. (2002). Lifetime patterns in adult female mass, reproduction and offspring mass in semidomestic reindeer (*Rangifer tarandus tarandus*). *Canadian Journal of Zoology*. 80(12), 2047–2055.

Ryd, Y. (2007). Ren och varg. Samer berättar. Stockholm: Natur & kultur.

Sara, M.N. (2009). Siida and traditional Sámi reindeer herding knowledge. *The Northern Review*. 30, 153–178.

Sikku, O.J. & Torp E. (2008). *Vargen är värst. Traditionell samisk kunskap om rovdjur*. Östersund: Jamtli förlag Jämtlands läns museum.

Sippola, A.-L., Norberg, H., Renko, M., Suopajärvi, K. & Sutinen, T. (2005). *Petovahinkojen sosioekonominen merkitys porotaloudelle Suomessa*. Arktisen keskuksen tiedonantoja 44. Rovaniemi: Gummerus Kirjapaino Oy.

Sivertsen, T.R. (2017). *Risk of brown bear predation on semi-domesticated reindeer calves. Predation patterns, brown bear – reindeer interactions and landscape heterogeneity*. PhD. Diss. Uppsala: Swedish University of Agricultural Sciences.

Sjölander-Lindqvist, A., Johansson, M. & Sandström, C. (2015). Individual and collective responses to large carnivore management: the roles trust, representation, knowledge spheres, communication and leadership. *Wildlife Biology*. 21(3), 175–185.

Sjölander-Lindqvist, A., Risvoll, C., Kaarhus, R., Lundberg, A. K. & Sandström, C. (2020). Knowledge claims and struggles in decentralized large carnivore governance: insights from Norway and Sweden. *Frontiers in Ecology and Evolution*. 8, 120.

Skjenneberg, S. & Slagsvold L. (1968). *Reindriftens og dens naturgrunnlag*. Oslo/Bergen/Tromsø: Universitetsforlaget.

SOU (2012). Mål för rovdjuren. Slutbetänkande av rovdjrusutredningen. *Statens offentliga utredningar*. SOU 2012:22.

Stortinget (2011). Parliament's Document Nr. 8:163 S. (2010–2011) *Representantforslag 163 S*. Available online at: www.stortinget.no/no/Saker-og-publikasjoner/Publikasjoner/Representantforslag/2010-2011/dok8-201011-163/

Strand, G.H., Rekdal, Y., Stornes, O.K., Hansen, I., Rødven, R., Bjørn, T-A., Eilertsen, S.M., Haugen, F-A., Hovstad, K.A., Johansen, L., Mathiesen, H.F., Rustad, L.J., Svalheim, E. & When, S. (2016). Rovviltets betydning for landbruk og matproduksjon basert på norske ressurser. (NIBIO Rapport 2(63)).

Sunde, P., Kvam, T., Bolstad, J.P. & Bronndal, M. (2000). Foraging of lynxes in a managed boreal-alpine environment. *Ecography*. 23(3), 291–298.

Turunen, M., Rasmus, S., Norberg, H. Kumpula, J, Kojola, I. & Ollila, T. (2017). Porot ja pedot – kuinka poronhoidon sopeutuminen petoihin on muuttunut 90 vuodessa? *Suomen Riista*. 63, 19–42.

Tveraa, T., Fauchald, P., Henaug, C. & Yoccoz, N.G. (2003). An examination of a compensatory relationship between food limitation and predation in semi-domestic reindeer. *Oecologia*. 137(3), 370–376.

Tveraa, T., Stien, A., Bårdsen, B.-J. & Fauchald, P. (2013). Population densities, vegetation green-up, and plant productivity: impacts on reproductive success and juvenile body mass in Reindeer. *PLoS ONE*. 8(2), e56450.

Wallén, J., Nilsson, L. & Hellström, P. (2019). *Resultat från inventeringar av kungsörn i Sverige 2019*. Naturhistoriska riksmuseets småskriftserie. 2019:1.

Zabel, A. & Holm-Müller, K. (2008). Conservation performance payments for carnivore conservation in Sweden. *Conservation Biology*. 22(2), 247–251.

Part III

Governance of reindeer pastoralism

7 Implications of norms and knowledge in customary reindeer herding units for resource governance

Tim Horstkotte, Hannu I. Heikkinen, Marius Warg Næss, Mia Landauer, Bruce C. Forbes, Camilla Risvoll and Simo Sarkki

Introduction

Unpredictable environmental conditions are an integral part of many pastoral systems. Fluctuations in abiotic conditions determine resource availability for livestock within and between seasons and from local to regional levels. For pastoralists, intimate knowledge about this heterogeneity is vital to adapt their herding strategies and practices in response (Fernández-Giménez & Le Febre 2006.). Most important among these is the spatial and temporal flexibility of herders' movements across the landscape. To govern these movements and allocate access to fluctuating resources between different herding units, internal rules may be necessary. Responses to an unpredictable environment, therefore, hinge not only on the bio-physical and ecological characteristics of the landscape but also on how people compete or collaborate in herding. Culturally embedded norms, values and customary laws, as part of the cultural and social capital in pastoral societies, are fundamental for building trust and facilitating collaboration between individuals and groups (Forbes 2013; Bodin 2017).

At high latitudes, forage availability for herbivores differs profoundly between seasons. This predictable pattern varies with unpredictable stochasticity in weather events. Furthermore, the directional trend in changing climate increases weather events that are less predictable and more extreme and can have unfavourable impacts on forage accessibility and on pastoralism as a whole (Forbes et al. 2016). These events make the planning of particular herding activities challenging (Chapter 5), while anthropogenic impacts of conflicting forms of land use reduce the availability of pastures for reindeer (Chapter 4). Moreover, colonial influences by the nation states, such as marginalization of reindeer herding communities and invalidation of their customary rights and institutions by superimposing conflicting norms out of the local context, have tarnished the history of reindeer husbandry in Norway, Sweden and Finland up to the present. These cumulative factors reduce the herders' capacity to fully exploit their cultural and social capital in shaping adaptive responses to environmental change.

DOI: 10.4324/9781003118565-11

In light of these historical legacies and present-day challenges of unpredictable availability of and access to grazing resources, this chapter explores how reindeer herders' internal governance systems, including social networks, norms, customary laws and traditional knowledge, shape internal cooperation, as well as their relation to state policies.

Analytical framework

People and nature are interlinked as social-ecological systems (SES), with mutual influences upon each other. In her seminal work on the subject, Ostrom (2007) identifies, among others, the linkages between people (*users*), the *resource system* and the *resource units* within that system as key components of an SES. The capacity of people to manage natural resources and adapt to change depends on several characteristics within the social subsystem in an SES, such as the capacity to implement decisions and solutions that are responsive to ecological patterns and processes.

Social networks, norms and customary law

Through ties such as kinship, affinity and collaboration, individuals or groups build and maintain social networks. As such, social networks can enable people to build mutual trust, share knowledge and economic or social support, and thus enable them to address and solve problems or adapt to change together (Armitage et al. 2011).

To facilitate social interaction, networks create and rely on shared values and norms. Norms are culturally embedded, informal rules composed of beliefs, mental models and motivations instead of explicitly stated rules (Fehr & Schurtenberger 2018). Norms influence individual actions, cooperation and expectations, e.g., what behaviours are approved or taboo (Schelling 1980; Henrich & Muthukrishna 2021). In response to environmental and sociopolitical change, norms and practices, e.g., on resource management, are evaluated and revised. Sustainable use of natural resources, therefore, is more likely to succeed if norms and knowledge to promote such use are shared, respected and agreed upon between users (Ostrom 2007).

Similarly, customary laws and rights can promote sustainable use and protect resources if social groups benefit from such use (Schnegg 2018). These laws and rights are documented and passed on orally as traditions and practices, making them so fundamental to the respective culture, shared values and related worldviews that they are treated as laws.

Indigenous and traditional knowledge

Norms and customary laws are deeply connected to the ways Indigenous people, or others with nature-based livelihoods, use and understand their traditional lands and waters – in material and spiritual ways. Their knowledge systems include language, skills and practices developed through experiences that are transmitted inter-generationally. Continuously tested against contemporary observations of environmental changes, these knowledge systems are dynamic and adaptive and are often described as a place-specific "way of life" (Berkes 2012).

Indigenous knowledge, in particular, embraces ethical aspects of behaviour towards human and non-human actors and spiritual ties to the bio-physical world (Berkes 2012). It is described as holistic and often practice- and language-based. Western concepts and epistemology may risk misinterpreting Indigenous practices, values and motives when not fully comprehending their knowledge base and epistemology (Berkes 2012). Hukkinen et al. (2006) refer to "ways of knowing" or "practitioners' knowledge", which is based not on ethnicity, instead refers to knowledge originating from engagement with the environment. Here, we use *traditional knowledge* to cover knowledge that originates from cultural continuity, independent of ethnicity, with the awareness that "traditional" is constantly revised against changes within the SES.

Internationally, the significance of Indigenous knowledge for biodiversity conservation and sustainable development was first acknowledged in the UN Convention on Biological Diversity (CBD) in 1992. Within their national context and legislation, countries that signed the Convention are obliged to "*respect, preserve and maintain knowledge, innovations and practices of indigenous and local communities*" (article 8j), as well as to "*protect and encourage customary use of biological resources in accordance with traditional cultural practices*" (article 10c). Norway, Sweden and Finland have ratified the CBD. For land use governance, the voluntary Akwé: Kon Guidelines, developed based on article 8j of the CBD, give recommendations about the incorporation of Indigenous knowledge into impact assessment processes concerning lands and waters owned or used by Indigenous and local communities.

Similarly, the United Nations Declaration on the Rights of Indigenous People (UNDRIP), passed in 2007, seeks to reconcile, restore and protect Indigenous cultures and develop their self-determination. Article 31 stipulates that Indigenous people have the right to "*maintain, control, protect and develop their cultural heritage, traditional knowledge and traditional cultural expressions*". Signatories shall ensure that these rights are protected and recognized. However, neither the CBD nor UNDRIP is legally binding or establishes new rights but aims at placing equal value on Indigenous and local knowledge and other forms of knowledge.

Traditional and Indigenous knowledge are also mentioned in the strategies for the Arctic Region to ensure development towards sustainability, both at the national and EU level. Despite these steps and commitments, the recognition and inclusion of traditional livelihoods, their customary laws and knowledge often remain weakly implemented (Chapter 8) and suffer from earlier suppression or delegitimization by laws instituted by state governments (Åhrén 2004). Likewise, holders of traditional knowledge perceive a persistent lack of trust about their ways of knowing (Wheeler et al. 2020).

Customary institutions in reindeer husbandry: *siida* and *tokkakunta*

The heuristic of an SES, i.e., the relationship between users, particular resources they manage and the resource system these resources are embedded in (Figure 7.1), corresponds to the conceptualization of reindeer herders' customary institution: the Sámi *siida* and the Finnish *tokkakunta*.

The *siida* is a herding unit in which herders seek to balance the relationship between reindeer herd size, available workforce within the herding communities and pasture resources, through social arrangements, often based on kinship or affinity (Figure 7.1; Bjørklund 1990; Sara 2009). Finnish herders have similar local herding units based on neighbourhood rather than kin groups. Finnish customary systems have stronger ties to living in local villages and surrounding pastures, resulting in comparatively lower mobility than most Sámi *siidas* (Heikkinen 2002).

Following the SES framework presented by Ostrom (2007), we describe these customary institutions and herders' social networks based on the history of land use that shapes the dependence on resources even today. Sustaining

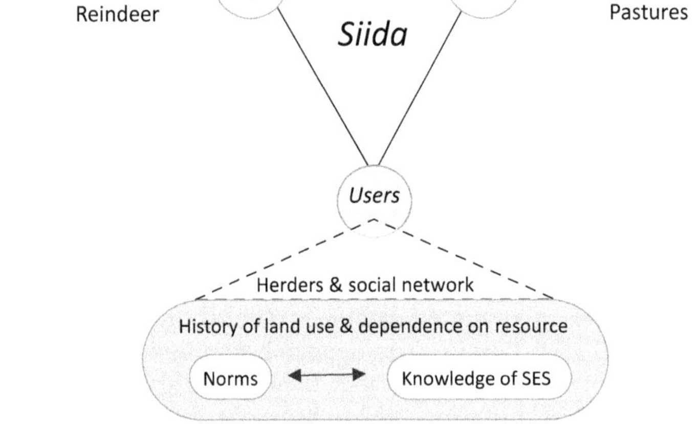

Figure 7.1 Conceptual model of the *siida*, integrated with selected elements of Ostrom's SES framework. The *siida* also interacts with the wider social, economic and political settings, as well as the related ecosystem, including migration routes and other habitats important for reindeer.

these resources involves sharing norms for collective action, as well as knowledge about the SES (Figure 7.1).

The *siida* has been described as the "own and only form of community organization" of Sámi society (Manker 1953:16). The *siida*, therefore, is an institution that existed before any regulations were implemented by the states to govern reindeer husbandry affairs. A *siida* is characterized as a decentralized social network, which establishes a working relationship between households with flexible group composition in space and time based on kinship or affinity (Paine 1994; Bjørklund 2013). Households and families joined as a *siida* live and migrate together, sharing the benefits and costs of the herding work of individually owned reindeer on shared grazing grounds, clearly delineated by borders between *siidas*. However, these borders are permeable due to the customary obligation to grant access to other groups, e.g., during difficult grazing conditions (Marin & Bjørklund 2015).

The land use pattern of a *siida* depends on trust between different *siidas* and interactions between herd structure, reindeer behaviour, weather and topography (Figure 7.1, Sara 2009). The composition and size of a *siida*, comprising people and reindeer, may change with seasons and between years, depending on seasonally changing availability of grazing resources and collective choice arrangements (Bjørklund 1990; Sara 2009). Groups that form a larger *siida* on summer grazing grounds may break up and join as different, smaller groups during migration or on winter pastures to enable a faster response to weather-related deterioration in grazing conditions. Environmental variability and monitoring of the shared grazing resources may, therefore, have social implications by redistributing *siida* members (Sara 2009). This emphasizes the importance of effective relationships within and between *siidas* in the complex meshwork of overlapping rights and territories and high mutual dependence.

Even though the *siida* has undergone changes and transformations in its organization, practices and meaning, its basic principles are still relevant today. However, whether or how the *siida* is understood, implemented and recognized in national law differs between the Nordic countries. The administrative units recognized by the respective state ("reindeer herding districts") encompassing the *siidas* and *tokkakuntas* are the *sameby* in Sweden, *reinbeitesdistrikt* in Norway and *paliskunta* in Finland. In Norway, summer *siidas* are formally recognized as legal units. In Sweden and Finland, there is no legal recognition of the *siida* or *tokkakunta*.

History of siida *land use and their dependence on natural resources*

From the 17th century onwards, reindeer husbandry gradually increased in economic and cultural importance for Sámi livelihoods that combined hunting wild reindeer and fur animals, fishing, gathering and herding for subsistence and taxes. Early *siidas* were primarily organized based on hunting and fishing groups (Tegengren 1952). To maximize the area and minimize distances to be travelled, *siida* areas could have a rounded shape, in particular in the Eastern part of Sápmi, while more elongated borders possibly existed in Western Sápmi (Pennanen & Näkkalajärvi 2002). Natural topography, lakes and rivers, as well as artificial structures, delineated the borders between the land and water bodies, over which families joined in a *siida* had clear ownership (Vorren 1980). From the mid-17th century onwards, paying taxes for these lands [*Lappskatteland*] to the Kingdom of Sweden-Finland secured and confirmed private ownership of them, e.g., in court cases involving conflicts with farmers (Korpijaakko-Labba 1994). However, border conflicts between *siidas* have also been recorded in court cases from each country (Lundmark 1982). As reindeer husbandry increased in economic and cultural importance, herd size also increased, and early pastoralists adapted to the seasonal migration of reindeer. Formerly fixed borders between *siidas* became more fluid as new forms of collaboration developed to share labour between families or households and secure access to grazing grounds (Bjørklund 2013). Ownership structures, land use rights of taxed lands and the *siida* were eroded during the 19th century when the taxed lands fell under the jurisdiction of county administration boards rather than being sovereignly managed by reindeer herding communities (Lundmark 1982).

Geo-political conflicts during the 19th century resulted in border closures between Norway, Sweden and Finland (annexed by Russia from Sweden in 1809) and restricted movement of herders and reindeer across these borders to access season-specific pastures. They also prevented herders, now forced into the confinement of nation states, from maintaining necessary relationships between *siidas* across borders. Border closures resulted in forced relocations of Sámi families from northernmost Sweden, Norway and Finland to the south and east of Northern Fennoscandia. These relocations forced herders into areas that, by Sámi custom, belonged to other *siidas* that originally inhabited the area (Åhrén 2004). Some resultant conflicts persist to this day.

Moving from intensive to extensive herding, where reindeer are spread out over larger areas, e.g., due to the introduction of snowmobiles in the 1960s or to a lack of sufficient grazing resources due to competition with other forms of land use (Helle & Jaakkola 2008), has affected the role of the *siida* today and changed collaboration. For example, if a particular *siida* is more affected by encroachment than another, it may need to access grazing areas that are currently used by other *siidas*, raising the potential for internal competition or conflict (Labba 2015).

The role of the siida *and* siida *autonomy in legislation today*

Partial erosion of *siida* customs arose due to lawmakers' lack of understanding of Sámi traditions, customs and languages. Though the generalized characteristics of a *siida* described above still apply today, the *siida* structures differ between the three Nordic countries. Differences exist between what is meant by a *siida*, and whether and how *siidas* have been or currently are acknowledged by state legislations, including the respective national Reindeer Husbandry Acts. The Reindeer Husbandry Acts construct the right of reindeer herding as fully collective and equal among all members of a herding district, in contrast to individualized and more complex customary rights and practices between *siidas* (Allard 2015).

Norway

The second Reindeer Husbandry Act in Norway (1978) failed to recognize the *siidas*' customary use and complex division of access and usage of the large grazing area of inner Finnmark, and instead defined it as "commons". As a result, state legislation did not take into account *siida* customs, leading to increased internal conflicts between *siidas* and families due to divergent and conflicting perceptions of rights to grazing areas (Turi & Keskitalo 2014). The revised Reindeer Husbandry Act of 2007 attempted to incorporate into law the reindeer herders' view that the traditional *siida* should be granted a greater degree of autonomy to handle internal affairs than previously stated in the Act. While the Act recognized the summer *siidas* as an administrative unit, i.e., herding units grazing their herds within a shared area ("district") during summer, district borders remain as defined by the state, and there were limited practical changes for reindeer herders (Johnsen et al. 2017). As most of the summer areas in Finnmark are managed by only one summer *siida* each, cooperation between different *siida*s on these areas is limited (Hausner et al. 2012). However, much cooperation is needed on migration routes, on both those shared between *siidas* and those which cross other *siidas*' seasonal grazing grounds.

On winter grazing areas, however, *siidas* may re-form into smaller groups of different compositions than summer *siidas*. For that reason, the Act of 2007 abandoned the collective right to undefined broad "commons" on winter grazing areas, and *siidas* need to share and regulate access to overlapping grazing areas between themselves. This creates a network of access rights that can differ between groups and seasons for a given area, based on topography or customary use of these areas. With the devolution of rights to the *siida* level, followed a greater responsibility for the *siidas* to reach policy goals for sustainable reindeer husbandry. This includes determining seasonal grazing patterns, number of reindeer, maintenance of herding infrastructure and division of labour, to be approved by a regional and, in the case of reindeer numbers, national comanagement board (Turi & Keskitalo 2014). However, state governance of reindeer husbandry, based on simplified indicators of sustainability such as carcass weights, does not heed to reindeer herders' complex body of knowledge. As a consequence, disputes between *siidas* can occur over what grazing areas may be used by whom, when, for how long and by how many reindeer (Marin & Bjørklund 2015; Johnsen & Benjaminsen 2017). A specific court has been established (*Indre Finnmark tingrett/Sis-Finnmárkku diggegoddi*) to resolve such conflicts. However, Hausner et al. (2012) found that reindeer herders in Finnmark disagreed about whether the degree of access to customary grazing areas should be regulated by informal agreements or by formalization through the courts or sanctioning of transgressions by an impartial authority.

Sweden

Like Norway, the borders of a *sameby* today do not necessarily correspond to the customary *siida* borders, and the level of recognition of *siida* borders may differ between *samebyar* in Sweden (Labba 2015). In the first Swedish Reindeer Husbandry Act (1886), the right to own and herd reindeer was collectivized for the members of a reindeer herding district (*sameby*). The same is still valid in the current Act (1971), in which the use of the *sameby*'s grazing area is for "the common needs" of its members (Reindeer Husbandry Act 1971:§15). Accordingly, there is no legislative acknowledgement or recognition of the *siida* and *siida* customs of place-specific access rights, which contributes to internal tensions (Allard 2015; Labba 2015). Nonetheless, the *siida* and associated norms still fulfil an important function, in particular in winter grazing areas.

Finland

In the north-eastern part of the then Kingdom of Sweden-Finland, known as Kemi Lappmark, the resident Forest Sámi practised reindeer husbandry with smaller herds in combination with hunting, fishing and gathering (Tegengren 1952). When Finnish settlers entered Sámi lands from the 17th century onwards, they adopted these practices in addition to their farming, fishing and hunting-based livelihood (Heikkinen 2006; Kortesalmi 2008). In 1898, the Senate of Finland legalized this more stationary *paliskunta*-system as the official administrative unit of reindeer husbandry, establishing borders based on the more stationary livelihood rather than on the migratory Sámi *siida*. However, migratory Sámi reindeer herders resisted, and still resist, the system as ill-suited to their way of life and customary rules, as it mainly fitted the needs of settled people, peasants and fishermen of lakes and rivers, including Sámi (Pennanen & Näkkäläjärvi 2002). Even though *siida* or *tokkakunta* arrangements can act as the local units that organize reindeer herding on a daily basis, only the *paliskunta* (herding district) is a legally recognized administrative unit, irrespective of ethnicity, managing all official administration, political power and reindeer herding-related land use planning. The end result is that the majority of Finns are able to dominate reindeer and Sámi-related negotiations (Heikkinen 2002). This complicates the options to defend or uphold reindeer herders' rights and customary rules when Sámi herders and Finnish herders, separately or together, compete with the state and other land users for land and resources.

Different cultures, therefore, (co)exist in the Finnish reindeer husbandry area. While reindeer husbandry is a keystone of Sámi ethnic identity and culture, the cultural dimension is also relevant in the context of reindeer husbandry practised by ethnic Finns, but with no clear relation to ethnic identity (Sarkki et al. 2021). Finnish reindeer herders can be characterized as a minority established through the history of cultivation in northern Finland, passing on their traditional livelihood and way of life, including their own customary rules and reindeer-related dialect.

Norms and customary law

Norms that structure herding practices and social interactions within and between *siidas*, from the nested levels of the individual herder to the household, the family and the *siida* as a whole, are strong in contemporary Sámi reindeer herding communities. Norms also affect the enforcement of borders between neighbouring *siidas*, clearly defined but permeable, in regulating access to grazing resources. Transmitted orally, norms often do not exist in any systemized or written form. However, Finnish court cases from the 18th century documented that herders were already being accused of not following local agreements on pasture use (Kortesalmi 2008).

Norms within *siidas*

Norms for sharing, reciprocity and cooperation between members of a *siida* are fundamental for distributing workloads and risk management, to recover from disasters or to come to shared decisions (Sara 2009). Reciprocity between herders encourages existing or prospective working relationships but can also express affinity or approval of the recipient's capability as a reindeer herder (Thomas et al. 2015). The *siida* can also increase equality among its members, irrespective of differences in wealth, age and domestic status (Paine 1970).

Kinship, e.g., relatedness by blood or marriage, is an important social marker, well-recorded by specific terminology and passed on in Sámi communities (Ruong 1975). In Finnmark, Northern Norway, kinship combined with the capacity to work together in comparatively small groups on summer grazing grounds enabled herders to build up larger herds, compared to non-kin working relationships (Næss et al. 2010). Larger districts or lack of obstacles to reindeer movement such as topography or fences demand more complex cooperation with more groups, so that kinship relations alone may no longer suffice to establish and navigate cooperative behaviour (Næss et al. 2010).

Norms between *siidas*

The *siidas*' social network provides flexibility and stability at the same time. Social fluidity to join groups, based on the spirit of cooperation between individuals (Labba 2015) and to adjust herd sizes provide flexibility to react to, e.g., environmental variability, while clearly defined, but permeable, customary borders between *siidas* and migration routes provide stability by agreed patterns of land use (Sara 2009; Marin & Bjørklund 2015).

Though borders between *siidas* are permeable, no *siida* is supposed to graze their reindeer on another's territory without agreement. Norms between *siidas* strictly regulate this right to access territories of others and the length of stay depending on grazing conditions, so that access to grazing grounds is not free for everyone to exploit (Hausner et al. 2012). Trust between *siidas* is, therefore, important to establish functional relationships. However, trust between *siidas* may erode. For example, in Finnmark, the unclear relationship between customary borders and the legislation of the Reindeer Husbandry Act of 1978, making pastures "commons" for all *siidas* within the same winter grazing area, caused border disputes and loss of trust, as well as having many other implications for Sámi reindeer husbandry (Hausner et al. 2012). Selfish behaviour that disrespects these norms may result in sanctions and retaliation, as it is seen as a conscious act of transgression (Laakso 2008; Marin & Bjørklund 2015). Therefore, *siidas* can be understood as an informal authority with jurisdictive power not enforced by the state, shaping cooperation between and access to different *siidas*. Diversity in strategies and goals exist in reindeer husbandry, such as herd composition and slaughter strategies. The relationship to other herders and other *siidas* can have an important influence on shaping these strategies, often as a response to state regulations (Johnsen & Benjaminsen 2017).

Cooperation also influences one of the most disputed incongruent realities between reindeer herders' customs and state governance: the perception and significance of what constitutes sustainable reindeer herd sizes (Chapter 9). While herders can be concerned about the workload to prevent excessively large herds from different *siidas* from mixing with each other, management authorities are concerned about unsustainable grazing pressure and exceeding "carrying capacity" (Johnsen & Benjaminsen 2017).

From a herder's perspective, herd size is a means to claim the right to grazing grounds, both in interaction within and between *siidas*, or against other forms of land use (Johnsen & Benjaminsen 2017). Kinship ties often imply a high degree of cooperation and shared workload between *siidas*, enabling cooperating herders to increase their herd sizes (Næss et al. 2010). Furthermore, slaughter strategies may depend not only on the number of animals to slaughter within a single herd but also on neighbouring *siidas*' strategies and the cooperation between them regardless of kinship ties (Næss et al. 2012). By not slaughtering more than neighbouring *siidas*, access and claims to winter pastures depending on herd size can be upheld (Næss et al. 2012).

Sámi *siidas* in Finland follow similar norms and strategies in herd management, sometimes contrary to management decisions by the state (Laakso 2008). According to custom, but today also according to the Reindeer Husbandry Act (848/1990), voting rights of individual herders in matters relating to the *paliskunta* depend on the individual's herd size. As the Ministry of Agriculture and Forestry sets the highest permitted reindeer numbers and sanctions at the level of the *paliskunta*, internal struggles and erosion of trust have been evident between Sámi *siidas* and between *tokkakunta* units of Finnish herders. This struggle can create great tensions, e.g., when rebuilding herds following catastrophic winters (Laakso 2008). Where Sámi and Finnish herders compete for access to grazing areas, tensions between them are evident and, in certain places, severe. However, intermarrying, mixed families and local cooperation have been, and still are, common (Kortesalmi 2008).

Indigenous and traditional knowledge in reindeer herding communities

Knowledge about the SES in which reindeer husbandry operates links people to norms and practices (Figure 7.1). It also connects people by knowledge exchange, learning or transmission to subsequent generations. It is, therefore, an irreplaceable resource in order to adapt to changes in local realities brought about by environmental or anthropogenic impacts. Sámi languages are an integral part of Sámi traditional knowledge (*árbediehtu*, "inherited knowledge" in Northern Sámi) and vector for knowledge transmission. However, colonial assimilation practices in the 19th and early 20th centuries in all three countries strongly reduced the degree to which the different Sámi languages are spoken today (Chapter 1). Likewise, traditional knowledge as lived experience is also transmitted in Sámi communities, even where Sámi is not spoken on an everyday basis.

Languages codify knowledge and the Sámi worldview of mutual relationships between people, reindeer and nature (Johnsen et al. 2017). Originating from the need to identify and communicate critical situations and phenomena, a nuanced vocabulary exists about, e.g., reindeer behaviour, morphologies, age classes, as well as weather and snow-related conditions (Magga 2006; Sara 2009; Eira et al. 2013). Complex categories can describe interdependent factors, such as the term *guohtun* (Northern Sámi) describing the relationships between the vegetation community, snow cover and reindeer behaviour that in combination determine the accessibility of grazing resources to reindeer temporally and spatially (Roturier & Roué 2009). Likewise, the vocabulary used by Finnish herders often has its origin in Sámi languages (Heikkinen 2002).

Recognition of traditional knowledge

Reindeer herders' knowledge can still be challenged, questioned or marginalized, and power imbalances between different types of knowledge limit the capacity to find common solutions to shared concerns in multiple-use landscapes or nature conservation (Sjölander-Lindqvist et al. 2020).

Complementarity between different forms of knowledge has gained increased recognition in the scientific community. However, the willingness to incorporate such knowledge as an evidence base into decision-making processes often hinges on whether it fits within current resource management models and paradigms and on power asymmetries between government approaches and local communities (Turi & Keskitalo 2014). A key challenge remains to ensure that Indigenous and local knowledge is not taken out of context, misinterpreted or misused when included in research or environmental management decisions. Accordingly, traditional knowledge of reindeer herders has been recognized or implemented to varying degrees, effect and satisfaction of involved parties in decision-making processes, as the selected cases below illustrate.

Norway

Losses to apex predators are a major concern for reindeer herders in all countries (Chapter 6). Sámi traditional knowledge about these predators has been documented, but much knowledge was also lost when the number of predators declined to near extinction. Due to the recovery of predator populations by successful conservation efforts in the mid-20th century, this knowledge is newly revived (Gaup Eira & Sara 2017).

In Norway, the Nature Diversity Act §8 (2009) refers to the CBD in emphasizing that Sámi traditional knowledge, as well as the Sámi Parliament, needs to be considered in decision-making processes regarding biodiversity conservation. However, reindeer herders call for a more holistic outlook, where interactions between reindeer, predators and the surrounding landscape are seen as interrelated. Predator management in Norway relies on a science-based system that leaves little room for local herders to present their knowledge as legitimate and valid (Risvoll & Kaarhus 2020). Herders have recently expressed concern that neither their traditional knowledge of predators nor their daily realities of living with them is reflected in the national management strategy (Sjölander-Lindqvist et al. 2020). For instance, the methods to document and verify predator abundances and kills are difficult to align with reindeer herders' observations. A mismatch between Western scientific methods and reindeer herders' observations, therefore, threatens to erode mutual trust and may impede finding solutions. One example is the diverging view on reindeer losses if caused primarily by a combination of density-dependence and environmental stochasticity, increasing their vulnerability to predation, or predators as the main source of mortality (Tveraa et al. 2014).

Sweden

Mapping of reindeer herders' Indigenous knowledge about their vital grazing grounds, migration routes, GPS-location of reindeer and other relevant environmental information has been realized in a participatory GIS (*renbruksplaner*, Reindeer Husbandry Plans, Sandström et al. 2012). The resultant maps can bridge Western academic knowledge and the herders' Indigenous knowledge, interpreting their animals' movement based on this knowledge.

Aimed primarily as a tool for conflict resolution with forestry, these digitized maps can visualize cumulative effects and have shown their potential to facilitate both knowledge-based dialogue about mutual influences and collaborative learning processes with representatives of other forms of land use (Sandström et al. 2012) and within the reindeer herding community. However, the tool is time consuming to keep updated. Furthermore, the representation of some of the herders' knowledge in spatial terms can force them to "prove" all their knowledge, as partners in consultation can be unfamiliar with or sceptical about knowledge that is not represented on maps. These plans, therefore, are not a substitute for reindeer herders' knowledge but rather depend on it for continuous adaptation as a living document.

Finland

Disputes and disagreements over local land uses are prevalent, long-term and continuing in Finland. For instance, in the municipality of Eanodat (Enontekiö) in north-western Finland, the state-owned Finnish Forest Enterprise Metsähallitus has developed management plans for so-called "wilderness areas", including the preservation of Sámi livelihoods together with nature conservation, tourism and potential prospecting for minerals (Markkula et al. 2019). These plans follow guidelines articulated in the CBD for co-developing land use plans and to increase knowledge about *siida* customs within the state-owned forest enterprise Metsähallitus. However, as the *siida* system is not officially acknowledged in the Finnish Reindeer Husbandry Act, Sámi reindeer herders saw this gap as a serious concern with respect to customary rights, comanagement of land use and acknowledgement of their Indigenous knowledge. Herders perceived a difference between Indigenous and local knowledge being heard versus actually having an impact on decisions for land use planning and development (Landauer & Komendantova 2018). However, land use planners argued that reindeer herders' knowledge needs to be made more spatially explicit, as verbally communicated knowledge is difficult to integrate into planning processes (Markkula et al. 2019). These dilemmas have been ongoing for decades (Raitio & Heikkinen 2003).

Concluding remarks

The examples presented in this chapter demonstrate how customary laws, norms and traditional knowledge structure the social relationships between reindeer herders, as well as their relevance in responding to unpredictable environmental conditions. The present-day challenges of rapid climate change, resource extraction, growing predator populations and competing national law make it difficult for reindeer herding communities to maintain desired relationships between each other, as well as within the wider social, economic and political settings and the related ecosystems (Figure 7.1). Where the herding community has to adopt undesired responses to such external pressures, these responses may reinforce unsustainable outcomes – culturally, socially and/or environmentally. To escape these traps of reinforcing feedbacks, a revitalization of customary laws could increase the fit to the dynamics of the SES. The engagement of customary laws in broader social and political structures for meaningful and effective participation in environmental governance would offer increased empowerment (Grey & Kuokkanen 2020). One example includes the re-institutionalization of customary rights to distribute access to grazing areas between different *siidas*.

Recognition of customary rights, as well as traditional knowledge, as an evidence base in national laws and international agreements and a reversal of colonial influences of knowledge invalidation, has been identified as pathway to escape social-ecological traps (Eckert et al. 2018). While traditional knowledge and non-Western epistemologies are increasingly recognized by international laws, rules and guidelines, challenges due to power imbalances persist in practice. National governments can dominate the discourse and decision-making processes with practical implications for reindeer husbandry (Johnsen et al. 2015). Power imbalances thus threaten the viability of reindeer herders' customary institutions and thereby impose different norms that may run contrary to the customary ones. This may result in conflicts within the herding society that intensify the severe and increasing pressures from parallel land use or other impacts by the majority of society. As long as reindeer herders' customs and knowledge do not receive legal recognition or contradict national legislation, they are vulnerable, may collapse or lead to internal conflict. Loss of community cohesion and erosion of social ties may thus threaten the internal capacity of the livelihood to escape social-ecological traps (Boonstra et al. 2016).

While reconciliation of the colonial past, in particular Sámi–state relationships including weakening or disempowering customary institutions and traditional knowledge, is to some degree going ahead, relevant resources and genuine opportunities for self-determination and effective participation in environmental governance are still lacking (Kuokkanen 2020). For instance, engagement of reindeer herders in planning and decision-making processes during the early stages is necessary to value their knowledge and foster coproduction with other knowledge systems (Tengö et al. 2014; Landauer & Komendantova 2018). Respectful inclusion of herders' knowledge through collaborative processes that respect the integrity and complementarity of each knowledge system can increase the validity and relevance of decision-making processes; it is a step towards shared power and responsibility in resource governance.

As the unprecedented pace of environmental change challenges herders' traditional knowledge and Western science, integration of knowledge systems may thus become an impactful resource to address the challenge of climate change and to adapt to increasingly unpredictable environmental conditions.

References

Åhrén, M. (2004). Indigenous peoples' culture, customs, and traditions and customary law-the Saami people's perspective. *Arizona Journal of International and Comparative Law.* 21, 63–112.

Allard, C. (2015). Some characteristic features of Scandinavian laws and their influence on Sami matters. In: Allard, C. & Skogvang, S.F (eds.) *Indigenous Rights in Scandinavia–Autonomous Sami Law*. 49–64. Farnham: Ashgate Publishing, Ltd.

Armitage, D., Berkes, F., Dale, A., Kocho-Schellenberg, E. & Patton, E. (2011). Co-management and the co-production of knowledge: Learning to adapt in Canada's Arctic. *Global Environmental Change*. 21(3), 995–1004.

Berkes, F. (2012). *Sacred Ecology*. New York: Routledge.

Bjørklund, I. (1990). Sami reindeer pastoralism as an indigenous resource management system in northern Norway: a contribution to the common property debate. *Development and Change*. 21(1), 75–86.

Bjørklund, I. (2013). Domestication, reindeer husbandry and the development of Sámi pastoralism. *Acta Borealia*. 30(2), 174–189.

Bodin, Ö. (2017). Collaborative environmental governance: achieving collective action in social-ecological systems. *Science*. 357(6352) eaan1114.

Boonstra, W.J., Björkvik, E., Haider, L.J. & Masterson, V. (2016). Human responses to social-ecological traps. *Sustainability Science*. 11(6), 877–889.

Eckert, L.E., Ban, N.C., Tallio, S.C. & Turner, N. (2018). Linking marine conservation and Indigenous cultural revitalization. *Ecology and Society*. 23(4), 23.

Eira, I.M.G. & Sara, M.N. (2017). Reindriftsnæringens erfaringsbaserte kunnskap om reindrift og rovvilt. www.statsforvalteren.no/siteassets/fm-nordland/dokument-fmno/landbruk-og-mat-dokumenter/reindrift-dokumenter/forskningsrapporter/eira-og-sara-2017_reindriftsnaringens-erfaringsbaserte-kunnskap-om-reindrift-og-rovvilt.pdf [01.02.2021]

Eira, I.M.G., Jaedicke, C., Magga, O.H., Maynard, N.G., Vikhamar-Schuler, D. & Mathiesen, S.D. (2013). Traditional Sámi snow terminology and physical snow classification—Two ways of knowing. *Cold Regions Science and Technology*. 85, 117–130.

Fehr, E. & Schurtenberger, I. (2018). Normative foundations of human cooperation. *Nature Human Behaviour*. 2(7), 458–468.

Fernandez-Gimenez, M.E. & Le Febre, S. (2006). Mobility in pastoral systems: Dynamic flux or downward trend? *The International Journal of Sustainable Development and World Ecology*. 13(5), 341–362.

Forbes, B.C. (2013). Cultural resilience of social-ecological systems in the Nenets and Yamal-Nenets Autonomous Okrugs, Russia: a focus on reindeer nomads of the tundra. *Ecology and Society*, 18(4).

Forbes, B.C., Kumpula, T., Meschtyb, N., Laptander, R., Macias-Fauria, M., Zetterberg, P., Verdonen, M., Skarin, A., Kim, K.Y., Boisvert, L.N. & Stroeve, J.C. (2016). Sea ice, rain-on-snow and tundra reindeer nomadism in Arctic Russia. *Biology Letters*. 12(11), 20160466.

Grey, S. & Kuokkanen, R. (2020). Indigenous governance of cultural heritage: searching for alternatives to co-management. *International Journal of Heritage Studies*. 26(10), 919–941.

Hausner, V.H., Fauchald, P. & Jernsletten, J.L. (2012). Community-based management: under what conditions do Sámi pastoralists manage pastures sustainably? *PloS one*. 7(12), e51187.

Heikkinen, H. (2002). *Sopeutumisen mallit–Poronhoidon adaptaatio jälkiteolliseen toimintaympäristöön Suomen läntisellä poronhoitoalueella 1980–2000*. Oulu: University of Oulu.

Heikkinen, H. (2006). Neo-entrepreneurship as an adaptation model of reindeer herding in Finland. *Nomadic Peoples*, 10(2), 187–208.

Helle, T.P. & Jaakkola, L.M. (2008). Transitions in herd management of semi-domesticated reindeer in northern Finland. *Annales Zoologici Fennici*. 45(2), 81–101.

Henrich, J. & Muthukrishna, M., 2021. The origins and psychology of human cooperation. *Annual Review of Psychology*. 72, 207–240.

Hukkinen, J., Müller-Wille, L., Aikio, P., Heikkinen, H. I., Jääskö, O., Laakso, A., Magga, H., Nevalainen, S., Pokuri, O., Raitio, K., & West, N. (2006). Development of participatory institutions for reindeer management in Finland: A diagnosis of deliberation, knowledge integration and sustainability. In: Forbes, B.C., Bölter, M., Müller-Wille, L., Hukkinen, J., Müller, F., Gunslay, N. & Konstantinov, Y. (eds.) *Reindeer management in Northernmost Europe. Ecological Studies*. 184, 47–71. Berlin: Springer.

Johnsen, K.I. & Benjaminsen, T.A. (2017). The art of governing and everyday resistance: "rationalization" of Sámi reindeer husbandry in Norway since the 1970s. *Acta Borealia*. 34(1), 1–25.

Johnsen, K.I., Mathiesen, S.D. & Eira, I.M.G. (2017). Sámi reindeer governance in Norway as competing knowledge systems. *Ecology and Society*. 22(4): 33.

Korpijaakko-Labba, K. (1994). *Om samernas rättsliga ställning i Sverige-Finland: en rättshistorisk utredning av markanvändningsförhållanden och-rättigheter i Västerbottens lappmark före mitten av 1700-talet*. Helsinki: Juristförbundets förlag.

Kortesalmi, J. (2008). *Poronhoidon synty ja kehitys Suomessa. Suomalaisen Kirjallisuuden Seuran toimituksia* 1149. Helsinki: Tiede.

Kuokkanen, R. (2020). Reconciliation as a threat or structural change? The truth and reconciliation process and settler colonial policy making in Finland. *Human Rights Review*. 21, 293–312.

Laakso, A.M. (2008). The shadow field of reindeer management: A case study from Finland. *Acta Borealia*. 25(2), 138–159.

Labba K. (2015). The legal organization of Sami Reindeer Herding and the role of the siida. In: Allard, C. & Skogvang, S.F. (eds) *Indigenous Rights in Scandinavia–Autonomous Sami Law*. 141–154. Surrey: Ashgate.

Landauer, M. & Komendantova, N. (2018). Participatory environmental governance of infrastructure projects affecting reindeer husbandry in the Arctic. *Journal of Environmental management*. 223, 385–395.

Lundmark, L. (1982). *Uppbörd, utarmning, utveckling: det samiska fångstsamhällets övergång till rennomadism i Lule lappmark*. Lund: Arkiv för studier i arbetarrörelsens historia.

Magga, O.H. (2006). Diversity in Saami terminology for reindeer, snow, and ice. *International Social Science Journal*. 58(187), 25–34.

Manker, E.M. (1953). *The Nomadism of the Swedish Mountain Lapps: The Siidas and their Migratory Routes in 1945*. Stockholm: ActaLapponica/Nordiska museet.

Marin, A. & Bjørklund, I. (2015). A tragedy of errors? Institutional dynamics and land tenure in Finnmark, Norway. *International Journal of the Commons*. 9(1), 19–40.

Markkula, I., Turunen, M. & Kantola, S. (2019). Traditional and local knowledge in land use planning: insights into the use of the Akwé: Kon Guidelines in Eanodat, Finnish Sápmi. *Ecology and Society*. 24(1), 20.

Næss, M.W., Bårdsen, B.J., Fauchald, P. & Tveraa, T. (2010). Cooperative pastoral production—the importance of kinship. *Evolution and Human Behavior*. 31(4), 246–258.

Næss, M.W., Bårdsen, B.J. & Tveraa, T. (2012). Wealth-dependent and interdependent strategies in the Saami reindeer husbandry, Norway. *Evolution and Human Behavior*. 33(6), 696–707.

Ostrom, E. (2007). A diagnostic approach for going beyond panaceas. *PNAS*. 104(39), 15181–15187.

Paine, R. (1970). Lappish decisions, partnerships, information management, and sanctions: a nomadic pastoral adaptation. *Ethnology*. 9(1), 52–67.

Paine, R. (1994). *Herds of the Tundra: A Portrait of Saami Reindeer Pastoralism*. Washington DC: Smithsonian Institution Press.

Pennanen, J. & Näkkäläjärvi, K. (2003). *Siiddastallan: From Lapp communities to modern Sámi life*. Inari: Siida Sámi Museum.

Raitio, K. & Heikkinen, H. (2003). Enemmän oma-aloitteisuutta, vähemmän valitusta. Hallinnon näkemyksiä poronhoidon osallistumiseen perustuvien instituutioiden kehittämisestä. Helsinki: Helsinki University of Technology.

Risvoll, C. & Kaarhus, R. (2020). Struggling with 'clear zoning'-Dilemmas of carnivore-pastoral coexistence in Nordland, northern Norway. In: Breidlid, A & Krøvel, R. (eds) *Indigenous Knowledges and the Sustainable Development agenda*. 185–206. New York: Routledge.

Roturier, S. & Roué, M. (2009). Of forest, snow and lichen: Sámi reindeer herders' knowledge of winter pastures in northern Sweden. *Forest Ecology and Management*. 258(9), 1960–1967.

Ruong, I. (1975). *Samerna*. Stockholm: Aldus Bonnier.

Sandström, P., Sandström, C., Svensson, J., Jougda, L. & Baer, K. (2012). Participatory GIS to mitigate conflicts between reindeer husbandry and forestry in Vilhelmina Model Forest, Sweden. *The Forestry Chronicle*. 88(3), 254–260.

Sara, M.N. (2009). Siida and Traditional Sámi Reindeer Herding Knowledge. *Northern Review*. 30, 153–178

Sarkki, S., Heikkinen, H.I. & Löf, A. (2021). Reindeer Herders as Stakeholders or Rights-Holders? Introducing a Social Equity-Based Conceptualization Relevant for Indigenous and Local Communities. In: Nord, D. (ed.) *Nordic Perspectives on the Responsible Development of the Arctic: Pathways to Action*. Cambridge: Springer. 271–292.

Schelling, T.C. 1980. *The Strategy of Conflict: With a New Preface by The Author*. Cambridge: Harvard University Press.

Schnegg, M. (2018). Institutional multiplexity: social networks and community-based natural resource management. *Sustainability Science*. 13(4), 1017–1030.

Sjölander-Lindqvist, A., Risvoll, C., Kaarhus, R., Lundberg, A.K. & Sandström, C. (2020). Knowledge claims and struggles in decentralized large carnivore governance: insights from Norway and Sweden. *Frontiers in Ecology and Evolution*. 8, 120.

Tegengren, H. (1952). *En utdöd lappkultur i Kemi lappmark*. Turku: Åbo Akademi.

Tengö, M., Brondizio, E.S., Elmqvist, T., Malmer, P. & Spierenburg, M. (2014). Connecting diverse knowledge systems for enhanced ecosystem governance: the multiple evidence base approach. *Ambio*. 43(5), 579–591.

Thomas, M.G., Næss, M.W., Bårdsen, B.J. & Mace, R. (2015). Saami reindeer herders cooperate with social group members and genetic kin. *Behavioral Ecology*. 26(6), 1495–1501.

Turi, E.I. & Keskitalo, E.C.H. (2014). Governing reindeer husbandry in western Finnmark: barriers for incorporating traditional knowledge in local-level policy implementation. *Polar Geography*. 37(3), 234–251.

Tveraa, T., Stien, A., Brøseth, H. & Yoccoz, N.G. (2014). The role of predation and food limitation on claims for compensation, reindeer demography and population dynamics. *Journal of Applied Ecology*. 51(5), 1264–1272.

Vorren, Ø. (1980). Samisk bosetning på Nordkalotten. In: Badou, E. & Dahlstedt, K.-H. (eds) *Nord-Skandi-naviens historia i tvärvetenskaplig belysning*. Umeå. 235–261.

Wheeler, H., Danielsen, F., Fidel, M., Hausner, V.H., Horstkotte, T., Johnson, N., Lee, O., Mukherjee, N., Amos, A., Ashtorn, H. & Ballari, Ø. (2020). The need for transformative changes in the use of Indigenous knowledge along with science for environmental decision-making in the Arctic. *People and Nature*. 2, 544–556.

8 Unpacking reindeer husbandry governance in Sweden, Norway and Finland

A political discursive perspective

Annette Löf, Kaisa Raitio, Bruce C. Forbes, Kristina Labba, Mia Landauer, Camilla Risvoll and Simo Sarkki

Introduction

Sociopolitical governance entails processes of formulating and addressing societal issues – or negotiating which "problems" need solving and how (Kooiman 2003; Torfing et al. 2011). The ability to navigate pluralistic problem representations and develop public acceptance for different governance models and interventions (either through procedural mechanisms [so-called input legitimacy] or desired effects [so-called output legitimacy]) (Scharpf 1999) is an essential governance quality. However, as this volume – as well as previous research – demonstrates, reindeer husbandry governance suffers from deficits in relation to both dimensions – procedurally and in delivering results relevant for herders (e.g., Heikkinen 2011; Widmark & Sandström 2012; Löf 2014; Kivinen 2015; Brännström 2017; Larsen et al. 2017; Raitio et al. 2017; Risvoll & Kaarhus 2020; Turunen et al. 2020; Österlin & Raitio 2020; Pekkarinen et al. 2021; Sarkki et al. 2021). In this chapter, we provide explanation for why these deficits occur and are so persistent. We do so by exploring how problem representations in reindeer husbandry governance are constructed, contested and handled. We thus unpack the discursive and political dimensions of reindeer husbandry governance (cf. Arts & Buizer 2009; Bacchi 2009; Voß & Bornemann 2011) and provide, to our knowledge, the first meta-level comparative analysis of reindeer husbandry governance in Norway, Sweden and Finland.

Governance analyses take many different forms. Ours centres on identifying which societal *issues, solutions and opportunities* are recognized, institutionalized and negotiated in reindeer husbandry governance. Our chapter demonstrates how the governing systems of reindeer husbandry have been created to address – and continue to address – problem representations and solutions defined by state and other land use actors – not the herders. We further show how the failure to address and handle herders' most important issues and needs is directly related to the discursive construction of reindeer husbandry as a

DOI: 10.4324/9781003118565-12

policy area. Despite apparent differences between the governing systems in the three countries, reindeer husbandry governance in Fennoscandia is also similar in how it:

- frames reindeer herding solely as an industry, thereby disregarding herders as rights holders and the broader cultural and livelihood dimensions associated with herding practices
- ignores the needs of reindeer herding and fragments reindeer herding lands by separating reindeer husbandry governance from other forms of land use governance and planning
- enables the states to escape responsibility for ensuring conditions necessary to meet the goal of socially, environmentally and economically sustainable reindeer herding practices
- fails to provide reindeer herding communities with tools and opportunities to regulate herding conditions and exert influence over accumulating pressures
- is based on a minimalist understanding of "co-existence" where reindeer herding is repeatedly forced to adapt to others' needs and demands.

Reindeer herding's key issues remain unresolved

Despite herding organizations' repeated and considerable efforts (e.g., Sametinget 2020; Paliskuntain yhdistys 2021; Chapter 15), the key issues identified by herders typically remain unresolved (see also Chapters 4, 5 and 6). Importantly, the negative, interacting and cumulative impacts of competing land uses, predation by large carnivores and climate change lead to *increasing and continued loss of grazing peace, land and flexibility*. Herders' opportunities to deal with this situation by implementing strategies that are desirable in the long run are circumscribed by interactions with competing land users, institutional limitations and structural asymmetries (Löf 2013; Risvoll & Hovelsrud 2016; Holand et al. 2021; Landauer et al. 2021). In other words, while single disturbances may be manageable, navigating the present complex of multiple and interacting drivers with cumulative and often unpredictable impacts under poor institutional conditions proves to be an insurmountable task for most herding communities (Kaiser et al. 2010; Löf 2013; Benjaminsen et al. 2015; Risvoll & Hovelsrud 2016; Sametinget 2020; Österlin & Raitio 2020; Sörlin 2021; Tyler et al. 2021). Taking herders' issues seriously thus requires governance to address herding as a system, not as fragmented components.

A systems approach essentially entails recognizing the interdependencies and relations between different parts and functions. Herders often describe herding as a way of life (Chapter 15) intimately tied to Sámi culture (including community, family, language, histories and futures) and other traditional practices such as hunting, gathering and fishing. As a semi-nomadic pastoral practice, herding also ultimately depends on land and secure access to diverse natural pastures and grazing peace. As the pressures on these lands increase, it becomes increasingly important to recognize and protect existing land rights. We posit that the governance of reindeer herding requires attention to all three dimensions: livelihood, land and rights.

Key terminologies used

We use somewhat different terminologies than other chapters. In order to recognize the cultural and relational dimension, we use the umbrella concept of reindeer herding communities (RHCs) rather than herding districts. We also make an important distinction between *reindeer herding* and *reindeer husbandry*. Herding refers broadly to the herding system and practice, while we use *reindeer husbandry* to denote the construction of reindeer herding as a policy area in public governance. This is more than mere semantics. Acts of labelling and translating both impose and disguise meaning in more or less visible ways but with tangible consequences (Joks et al. 2020). Naming and classifying should therefore be scrutinized critically as they are neither innocent nor neutral actions but part of negotiating meaning and thus acts constitutive, impacting how sociopolitical governance is constructed and performed (Arts & Buizer 2009).

The discursive and political dimensions of governance

Following interactive governance (Kooiman 2003) and other discursive-institutional governance approaches, we consider governance to be embedded in governing structures while at the same time emerging from the interactions between sociopolitical actors. The relationship is dialectic; governing structures shape interactions and agency, while interactions simultaneously shape those very governing structures - including norms, formal institutional contexts and boundaries of governing objects and systems (e.g., Arts & Buizer 2009). In our governance analysis, we consider this dialectic through one of the most central tasks in sociopolitical governance, namely to formulate and address societal issues (problems) with associated "solutions" and opportunities (visions) (Kooiman 2003; Torfing et al. 2011).

The so-called *discursive dimension of governance* (Arts & Buizer 2009) recognizes that governance is not an external or neutral tool for solving societal issues independently of its own conditions. On the contrary, any governing system is based on particular understandings and representations of "problems" that it is designed to address and "solve" (Kooiman 2003; Arts & Buizer 2009; Bacchi 2009). By structuring representations of societal issues, governing systems effectively limit the scope of solutions available for discussion: some are included, while others are excluded. Governing systems, moreover, construct and reproduce specific understandings, so-called governing images, of the objects governed (so-called systems-to-be-governed) (Kooiman 2003). Governing images entail e.g., specific understandings of what reindeer herding is or should be. The perceived boundaries of these systems (the governing and governed) are a direct function of governing images. Because of these structuring effects, a governing system is a force in its own right and plays a critical role in determining the possible outcomes of governance interactions (Voß & Bornemann 2011). Governing systems cannot, therefore, be considered separate from what is governed. In our case, the system-to-be-governed is *reindeer herding* while *reindeer husbandry governance* is the governing system.

The *political dimension* of governance (Voß & Bornemann 2011) recognizes that societies and sociopolitical actors' views and wills are diverse. Governance is always a form of negotiation and display of power. By acknowledging the so-called politics of governance, we direct attention to negotiation and contestation: how the struggle over meaning and competing problem representations between different sociopolitical actors in governing interactions are addressed, dealt with or ignored (Voß & Bornemann 2011). Present reindeer husbandry governance can be seen as an imprint of power relations between actors struggling to gain hegemony over what and whose problem representations should be prioritized (North 1990).

Methodology: an iterative process of unpacking governance

Three broad questions, derived from the discursive and political dimensions, formulated as "*what*", "*how*" and "*for and by whom*" in reindeer husbandry governance guided our initial empirical search. In operationalizing "what", we primarily used the concept of governing images (encompassing preconceived ideas and norms about the system-to-be-governed), descriptions of problems and issues to be addressed and identification of solutions and visions that provide direction for governing interactions (Kooiman 2003; see also Löf 2014). In terms of "how" we considered governance as policy according to Voß and Borneman (2011), how conceptions and perceived problem representations are translated, negotiated and implemented through specific governing instruments (instrumentalized, see also Kooiman 2003) where we also used classic policy typologies (legal/regulatory, economic and agreement/incentive-based) in order to identify the prevalence of different techniques and types (Bevir 2010). Finally, we looked at how key actors are positioned vis-à-vis each other in governance negotiations and interactions (cf., the use of politics proper in Voß & Borneman 2011). In terms of material, we canvased broadly, including goal formulations, strategies, legislation, preparatory works (for an overview of sources cited in the text, see Table 8A.1) and previous research from respective country context.

After our first comparative assessment, it surprised us how similar the governing systems were. Considering the varying institutional and political contexts, we found this an important finding and have therefore chosen to highlight such similarities, often in the form of examples from one or several governing systems. Attempting to adopt a systems approach to reindeer herding, we additionally consider three overlapping and relationally focused themes: (i) mismatches in boundaries between the governing systems and system-to-be-governed, (ii) contestations and tensions between problem representations, instrumentalized solutions and visions and (iii) interactions with other governing systems and competing land use.

Unpacking reindeer husbandry governance

There are few comparative studies in this field. The study by Allard (2015) – the first comparative assessment of reindeer herding legislation and rights on a Nordic scale – is an important exception. While showing how the legal basis for reindeer herding rights have been established similarly (through undisputed long-term use), she reported large differences (particularly between Sweden and Finland compared to Norway) in how these rights are perceived and treated in the different legal systems (Allard 2015). Importantly, she also noted growing tensions between legislating and governing reindeer herding as an *internal* matter (of national concern) and recognizing its more *universal* aspects resting on human rights and international law (Allard 2015; see also Allard & Brännström 2021). However, there are many other relevant studies that examine both broader and specific issues within each governing system, and we have included these as far as possible. The following account combines empirical results with analysis and each section leads with a brief summary.

Boundaries of governing systems: a history of structured fragmentation

In all three countries, reindeer husbandry governance is structured in a way that separates and fragments the herding system (livelihood, land and rights) into separate silos – discursively, politically and administratively. We find that this structured fragmentation places effective limits on problem representations, solutions and visions and represents a major mismatch compared to herders' own understanding of reindeer herding and its challenges.

In Sweden, reindeer husbandry administratively belongs to the Ministry of Enterprise and Innovation (*näringsdepartementet*), in Norway the Ministry of Food and Agriculture (*landbruksdepartementet*) and in Finland the Ministry of Agriculture and Forestry (*maa- ja metsätalousministeriö*). These ministries also govern many other land uses that impact the conditions for reindeer herding, e.g., agriculture, forestry (which in Sweden and Finland are the competing land use with most widespread impact on reindeer pastures [Sandström et al. 2016; Turunen et al. 2020]), mining (Sweden) and large carnivores (Finland). Yet, both practically (due to different administrative units and personnel within the ministries) and formally (due to different sectoral regulations such as the Minerals Acts, Forestry Acts, etc.) the interactions *between* these governing systems are limited. Institutionalized interactions are moreover poorly regulated, particularly the processes that regulate access to and use of reindeer herding land by competing land users (see e.g., Larsen et al. 2017; Sjölander et al. 2020; Österlin & Raitio 2020).

Separating herding practices from Sámi culture and rights

Sámi culture and Indigenous rights (despite the inclusion herein of reindeer herding rights) are governed as entirely other entities, both administratively and discursively. While reindeer herding is not an exclusive right of the Sámi people in Finland (but requires residence in the reindeer herding area), all three states recognize reindeer herding as an inherent part of Sámi culture and emphasize that reindeer herding enjoys constitutional protection (see e.g., Prop. 2009/10:80 pp. 188–191; Anaya 2004 pp.135–138 on Kitok vs. Kitok). The Finnish Reindeer Husbandry Act (848/1990) moreover recognizes and protects reindeer herding as a *traditional livelihood* in Northern Finland for Sámi and Finns. In Sweden and Norway, reindeer herding is often described as a unique Sámi livelihood, tradition and "bearer" of Sámi cultural heritage (e.g., Swedish Ministry of Culture and Democracy 2015).

However, in Finland, governance of Sámi matters falls under the Ministry of Justice, including the implementation of the Sámi people's right to self-determination. This task has, through the Act on Sámi Parliament (974/1995), been focused on ensuring effective consultations between the state and the Sámi Parliament; whereas, reindeer herding rights and the role of Sámi reindeer herding cooperatives as rights holders have received surprisingly little attention. For example, guidelines for implementing the established principle of Free, Prior, Informed Consent (FPIC, see, e.g., the United Nations Declaration of the Rights of Indigenous Peoples and the International Labour Organization Convention no.169) do not specify reindeer herding communities in the Sámi homeland area as FPIC communities (Ministry of Justice in Finland 2017).

In Sweden, Sámi matters, except reindeer herding, belong (since 2014) to the Ministry of Democracy and Culture where, in recent years, rights-focused rhetoric has become more visible. The former Sámi Minister Bah Kunkhe (not to be confused with the Minister of Reindeer Husbandry) officially stated that the Swedish state had the pressing task to update Swedish Sámi politics by taking responsibility for the present and previous injustices, sharing power and increasing self-determination in issues concerning the Sámi (Swedish Gov. 7 February 2018). This demonstrates a significant step for a colonial state that has never officially apologized for past and present wrongdoings (see e.g., Löf 2016) although truth and reconciliation processes are currently underway. The recognized links between Sámi culture and rights stop short of reindeer husbandry governance and have moreover failed to translate into concrete governing action. Initiatives such as developing a formal order of consultation with the Sámi people (Prop. 2020/21:64) have been met with resistance and early in 2021 forced the Swedish government to withdraw the proposition. However, when the new Swedish government took office in fall 2021, the parliament voted for a new legislation that demands consultation with the Sámi people in matters that concern them. The Sámi Parliament is similarly tasked to monitor questions with relevance for Sámi culture and take initiatives promoting

Sámi culture (Sámi Parliament Act 1992:1433) but the influence over reindeer herding remains administrative and the Sámi Parliament, in its current form, is unable to exercise actual self-determination (Sámediggi 2016). Together, these examples demonstrate both diversity and significant gaps between governing rhetoric and implementation.

We conclude that the formal governing systems divide different dimensions of reindeer herding into different administrative silos where the governance of land takes place through multiple, overlapping and fragmented but poorly coordinated land use planning and authorizing processes.

Governing what? The dominant image of reindeer herding as an industry

Reindeer husbandry is constructed almost exclusively as an industry and the broader conception of reindeer herding is thereby deprived of sensitivity to culture, land and rights. The objective of reindeer husbandry is thus reduced to a primarily economic dimension, even when redressed in the meta-narrative of sustainability. Importantly, this positions and reinforces reindeer herders as stakeholders with interests – on par with other industries – instead of recognizing them as rights holders. This, in turn, has important implications for how interactions with other land uses unfold (see also Löf 2014*;* Sarkki et al. 2021*).*

An industry underpinned by economic rationalization

One of the most striking similarities between reindeer husbandry governance in Sweden, Norway and Finland is how it constructs the dominant image of reindeer herding as an *industry*. It is visible in the organization of governing systems (see above) and in the terminology used. For example, the Swedish concept "*rennäring*" (synonymous with industry) and the Finnish term "*porotalous*" ("reindeer economy") are the institutionalized governing terms (e.g., in legislation). Alternative terminologies (e.g., *renskötsel*, "reindeer management") are sometimes used simultaneously, but the industry image maintains a discursive dominance. For example, the Norwegian government acknowledges that "Reindeer husbandry as an industry, culture and way of life is unique, both nationally and internationally" (Ministry of Food and Agriculture 2020), yet claims husbandry (*reindrift*) as an "extensive landscape-based industry" similar to forestry and agriculture (ibid.).

Underlying this "industrialization" is the associated logic of *rationalization*, particularly visible in the Norwegian and Finnish governing systems. While the Norwegian state has implemented various policies and regulations in different historical phases (Johnsen 2018), a major structural change has been ongoing for the past 40 years. The objective to transform Sámi reindeer husbandry into a corporate, market-oriented, economically efficient and environmentally sustainable industry was driven parallel by internal demands and general modernization and the understanding that too many reindeer and people were engaged in pastoralism, risking overgrazing and poor herding economy. While traditional and experience-based knowledge played a major role in Norwegian reindeer husbandry governance prior to the 1970s, rationalization has been underpinned by a Western scientific perspective to increase meat production and the income and welfare of pastoralists (St. Meld. 32 (2016–2017) p. 7; Johnsen et al. 2017).

In a similar fashion, Finnish reindeer husbandry governance has been built around the logic of fewer herders making a full-time living from reindeer husbandry, as opposed to having multiple smaller income streams, as otherwise typical in Sámi culture. Herders are thus expected to make their living from meat production and not from reindeer herding more broadly, e.g., hunting, engaging in tourism, production of handicrafts and so on. How reindeer herding is perceived as a meat production industry is evident in the way regulations and economic incentives steer towards bigger herds per herder, reinforced by EU policies, and informed by agriculture-like rationalities (Hukkinen et al. 2006; Raitio & Heikkinen 2003).

The idea of what constitutes "good" herding practice thus shifts over time. In Sweden for example, the governing objective in the early 1900s was to keep reindeer herding as "traditional" as possible (Mörkenstam 1999) while from the 1970s onwards "good" herding became synonymous with rationality and profit. This demonstrates how seemingly universal ideas are always bound to specific contexts. This includes the currently dominating goal for reindeer husbandry – the meta-narrative of sustainability (cf. Voß & Bornemann 2011).

The narrative of sustainable reindeer husbandry

The central objective for all three governing systems is framed in terms of sustainable reindeer husbandry, broadly drawing on the three pillars of sustainability. The Norwegian Reindeer Husbandry Act (15 June 2007 nr. 40) states that the state shall arrange for *ecologically, economically, and culturally sustainable reindeer husbandry*. The economic dimension is interpreted in line with the above, as profitable and rational economic practices that can provide a sufficient livelihood for the herding population (Norway) or sustain a "reasonable" number of herders (Sweden, Prop. 1971:51, pp. 29–34; Arctic Strategy 2020).

In Norway, the linkages between dimensions are made explicit, with the ecological dimension seen to provide the basis for economic and cultural sustainability (Riksrevisjonen 2011–2012). The Act's implementation of provisions about the reindeer herding *siida* is part of the cultural sustainability dimension (Ot.prp. Nr. 25 (2006–2007), 31–33; NOU 2001:35). Notably, the focus on reindeer numbers is based on the concept of *carrying capacity* (Benjaminsen et al. 2015), which has played a major role in the state's reindeer administration during the past four decades. Here, research demonstrates how the government has established a set of goals and indicators lacking recognition of the complex reality of reindeer herding (Johnsen 2018; Marin et al. 2020). According to the Norwegian Ministry of Food and Agriculture "*It is an explicit political objective to develop an efficient and market-oriented reindeer husbandry industry based on long-term sustainability*" (2020). The statement emphasizes that while sustainability has taken over as a meta-narrative, rationalization remains a dominant and underlying objective in reindeer husbandry governance.

The Finnish Reindeer Husbandry Act similarly frames ecological sustainability in terms of *carrying capacity of winter pasture areas* (§ 21). In an earlier study, the objective was identified – by the public authorities themselves – as promoting reindeer herding based on natural pastures (as opposed to feeding) (Hukkinen et al. 2006). While this reconnects reindeer herding to landscape conditions, in its present interpretation it fails to include the impacts of other land uses (see below) and ignores that most of the cooperatives are already forced to use supplementary feeding as they no longer have access to natural winter pastures due to habitat loss and fragmentation (Forbes et al. 2020; Pekkarinen et al. 2020).

In Sweden, the goal is for reindeer husbandry to remain an ecologically, economically and culturally long-term sustainable practice (1999/2000:MJU9; SOU 2001:101). The Swedish Reindeer Husbandry Act (1971:437) emphasizes ecological aspects and mandates, e.g., that reindeer husbandry considers the maintenance of biodiversity and the long-term productive capacity of natural pasturelands (65a §). The article was added in the 1990s following public debates on overgrazing (Prop. 1992/9332, pp. 122–125). Initially, overgrazing was considered largely a result of *internal* dynamics (i.e., too many reindeer) but external factors (e.g., other land uses) were soon recognized as critical in this regard. The current appreciation of how to maintain this ecological objective reflects that the productive capacity of pasturelands is *primarily externally determined* and therefore dependent on the actions of many land users, complex interactions and changing environmental conditions (including climate change) (SOU 2001:101). However, as we demonstrate below, there are neither instruments nor arenas in the governing system that sufficiently addresses and divides responsibility between different sociopolitical actors.

Governing issues – instrumental and strategic alignment

If sustainable reindeer husbandry is the overarching governance goal, what are the key issues recognized? Here, there is a partial overlap between herders' accounts (see earlier section) and the formal governing systems. For example, in Norway, the current initiatives of highest priority were recently settled in the annual negotiations between the State and the Norwegian Reindeer husbandry Association (NRL) and include protecting grazing land, crisis preparedness and measures to care for herders who cannot utilize reindeer pastures in Sweden (Prop. 189 S 2020–2021).

Traditionally herders moved freely across borders. Since some years Norway and Sweden lack a Grazing Convention that regulates cross-border mobility, and this has enabled some (Swedish) herders to access their traditional lands (Grönvall & Löf 2020). The State Secretary for Agriculture and Food similarly stated that current major challenges for reindeer husbandry are land encroachment, climate change and large carnivore predation (Ministry of Agriculture and

Food 2021). In Sweden, the currently prioritized policy issues and initiatives include large carnivore predation and climate change (Ministry of Enterprise and Innovation 2020) while in Finland, current policy centres on land use conflicts and planning. These include national goals and National Land Use Guidelines (*VAT*) for different developments with particular concern for reindeer management – including industrial and forestry, EU and Arctic policy as a whole (e.g., Arctic railway plans, development of renewable energy and extractive industries) (du Plessis 2020). However, cumulative impacts resulting from other land uses are paid little or no consideration (Kivinen 2015).

Governing how? Instrumentalization and proposed solutions

Despite the basic alignment around problem representations, our analysis demonstrates that major differences remain regarding how key issues are understood and how to address them. We demonstrate that governing tools able to address problems from herders' viewpoints are still lacking and that the proposed "solutions" typically internalize responsibility, favour incremental change and tend to weigh heavier on reindeer herders compared to other actors. That is, governing instruments are geared towards adapting herding practices, not addressing and recognizing the impacts of competing land use on reindeer herding conditions.

Hierarchical governance and internalized "solutions"

Generally speaking, reindeer husbandry governance is characterized by top-down interactions. We identify the Reindeer Husbandry Acts as the most important governing instruments, with the exception of Norway. Here, the Act alongside the yearly negotiations between the state and herding organization that set operative goals (e.g., to stimulate meat production) is most important (NOU 2001:35; St. Meld. 32 (2016–2017), pp.34–50; Johnsen et. al 2017).

Reindeer husbandry legislation is detailed and governs, among other things, the internal organization, maximum number of reindeer and various forms of boundaries (spatial, temporal and organizational). The Swedish Act interprets and specifies e.g., Sámi rights to land, how herding is organized, what activities RHCs can undertake, where herding can be practised and who decides over internal and external matters.

Our mapping shows how translating problem representation into "solutions" typically targets *incremental* and *internal* responses and/or rests on weak collaborative instruments vis-à-vis other actors. For example, both in response to climate change and predation, supplementary feeding and economic compensation for damage are the only tools in place as agreements on tolerance levels are still to be implemented (Sjölander et al. 2020). Reindeer herding is moreover relatively invisible in public policy at large (Löf 2016). Other than a broad focus on sustainability, reindeer husbandry often lacks clear national-level policy objectives (that can be followed up), especially compared with other sectors (e.g., tourism in Finland Työ- ja elinkeinoministeriö (TEM 2019)).

Dialogue and information-based instruments govern relations with others

Competing sectoral legislation is typically not as detailed as the Reindeer Husbandry Acts and, importantly, do not sufficiently recognize herding objectives or consideration towards reindeer herding (e.g., Brännström 2015). Much effort has therefore been placed on developing soft governing instruments based on dialogue and information. For example, so-called reindeer husbandry plans have been developed to enhance consultations and dialogues with other land users (Löf 2014) and to assist reindeer management (statsforvaltaren.no). While they are meant to provide up-to-date information and description of RHCs' land use strategies, a primary objective is to mitigate conflict between different land users and interests (Sandström et al. 2016). The underlying normative assumptions are, however, based on the continued exploitation of the land, and herders are thus left with tools and processes designed by competing actors for different purposes (in this case industrial forestry) which over time risks eroding the recognition of practice-based knowledge (Löf 2014; Kuoljok 2019; Tyler et al. 2021). This effectively renders RHCs with limited prospects of exercising any real influence over for example forestry-herding interactions, and associated costs have moreover been shown proportionally higher for RHCs than forestry (Widmark & Sandström 2012). So, while such tools are potentially useful in particular interactions, they function poorly under unclear and asymmetrical conditions that characterize much of the land use interaction in Sápmi.

Ironically, both in Norway and Finland, another suggested "solution" that instrumentalizes the image of ecological sustainability is to *reduce* the number of reindeer (see Chapter 9). This begs the question, for whom this is a solution and based on what understanding of the problem? It obscures external pressure on lands and transfers responsibility for mitigating the impacts of multiple land users on pastures solely to reindeer herders. Governing through permissible number of reindeer demonstrates a shared path dependence, beginning in the early 1900s. The underlying rationale was primarily informed by the needs of other forms of land use such as forestry, hydropower and agriculture, coupled with rapid industrialization (Mörkenstam 1999; Allard 2015): demands that have only intensified since then (Forbes et al. 2006). During a recent policy process to control reindeer numbers in Finland (2019–2020), the impacts of forestry and other land uses on winter pastures were taken into account for the first time (!). While the process resulted in the suggested maximum numbers remaining unchanged, cooperatives were tasked with preparing so-called pasture management plans for 2022–2030 to assure "rational" and "sustainable" pasture use. Suggested measures to take into account included changing the timing of slaughter, developing pasture rotation systems or voluntarily combining cooperatives (Valtioneuvosto 2019). However, if these internalized strategies are not deemed sufficient, reducing the size of the reindeer population by 7% to decrease trampling pressure remains an alternative "solution".

Governing interactions – weak collaborative instruments privileging competing land uses

Due to the extensive and parallel land use nature of reindeer herding, governing interactions with competing for land uses should be a central component in reindeer husbandry governance. However, this is not reflected in present governance as responsibility over cumulative impact is lacking, and governing interactions are only addressed through weak collaborative instruments that, ultimately, are designed for the sake of competing for land uses. A minimalistic assumption of co-existence plays a key role in upholding this dynamic.

With shared space, other forms of land use limit access to forage and adversely affect grazing peace and quality and cause loss of the pastures that serve as the natural resource base for reindeer herding (Kumpula et al. 2014; Kivinen 2015; Turunen et al. 2020). Research is unanimous in that cumulative effects related to multiple encroachments, disturbance, increased losses from large carnivores and rapid climate change need to be addressed jointly (Löf 2013; Risvoll & Hovelsrud 2016; Österlin & Raitio 2020; Landauer et al. 2021).

The governing systems recognize, in principle and in somewhat varying degrees, that other land uses may adversely affect pastures and grazing peace, for example, through the general requirement in regulations to protect pastures from *significant adverse impacts.* However, the governing systems do not provide sufficient opportunity to address and manage the consequences thereof. The Finnish Act (2.2 §) mandates, e.g., that other forms of land use must not cause "significant harm" to herding in the area specifically intended for reindeer husbandry (*Erityinen poronhoitoalue*) in general and in the Sámi homeland (*Saamelaisten kotiseutualue*) in particular. The southern area (*Muu poronhoitoalue*) does not enjoy the same protection. In Sweden, other forms of land use may not incur "considerable inconvenience" to reindeer herding (30 §) within the year-round grazing grounds, while the winter pastures are without such protection – despite them being considered the "bottleneck" in herding and are additionally more exposed to encroachment and climate change (SOU 2001:101).

The failure to govern interactions

The fragmented governing systems, compared to the systems-to-be-governed, makes them poorly equipped to regulate accumulating and multiplying pressures. Reindeer husbandry governance thus leaves the regulation of land uses in multiple hands with respect to mining, forestry, large carnivores, infrastructure and so on. Irrespective of the capacity of these individual systems, their disparate nature has resulted in a lack of holistic land use planning attuned to the needs of reindeer herding (Sarkki et al. 2016; Larsen et al. 2017; Larsen & Raitio 2019; Sjölander et al. 2020).

Furthermore, these sectoral systems typically position reindeer herding in a *subordinate* position vis-à-vis other land uses, despite formally recognised as an established right. Collaborative instruments, including dialogues and corporate consultations, typically lack specific or adequate regulations as to appropriate procedures or satisfactory outcomes, leaving them susceptible to being dominated by parties with better resources (Widmark & Sandström 2012). This is particularly the case when the state "delegates" its duties towards reindeer herding to corporate consultants, as is common in Swedish land use regulations (Allard 2008; Brännström 2017; Raitio et al. 2020). In Finland, both the Reindeer Husbandry Act (53 §) and the Act covering *Metsähallitus* – the state enterprise managing public lands – require that state authorities must consult with reindeer herding cooperatives about activities with potentially significant impacts. However, as in Sweden, consultations are a vague instrument with limited impact on outcomes and poor participatory qualities and have been repeatedly criticized by researchers and affected actors (Landauer & Komendantova 2018; Raitio 2016; Sámediggi 2021). In Norway, the annual negotiations between the Sámi Reindeer Herders Association of Norway and the state represented by the Ministry of Food and Agriculture have likewise been criticized for not being conducted on equal terms (Johnsen 2018).

Path-dependent principles – co-existence addresses someone else's needs

Reindeer husbandry governance is characterized by strong path-dependencies (Mörkenstam 1999; Löf 2014; Marin et al. 2020). The key objective of the first Reindeer Grazing Act in Sweden was *not* to protect herders' rights, interests or pasturelands but was primarily geared at controlling and managing herders as to enable settlers' and industries' (forestry in particular) establishment on Sámi customary lands (Mörkenstam 1999; Össbo 2014; Allard 2015). The underlying logic then was that herders must inevitably give way to societal development, an idea underpinned by an industrial colonial discourse (Össbo 2014) and a social-Darwinist ideology (Mörkenstam 1999; Allard 2015). Ideas of *parallel land use* and *co-existence* have guided land use governance ever since (SOU 2001:101). Thus, dialogues and consultations come with the *a priori* assumption that reindeer herding and competing land use will be able to co-exist – instead of assessing whether this is the case in each situation through appropriate impact assessment mechanisms (Brännström 2017; Raitio et al. 2020; Arctic Strategy 2020). Rejecting projects that may undermine the conditions for reindeer herding thereby becomes practically impossible.

Similar assumptions underpin large carnivore management (Rasmus et al. 2020; Risvoll & Kaarhus 2020) and the central participatory planning instrument on state-owned land in Finland, Natural Resource Planning by Metsähallitus (Raitio 2012). According to Metsähallitus (2021) "*it operates on the principle that forestry, tourism and reindeer herding can coexist, once a joint agreement has been reached on reconciling these industries*". When it comes to large carnivores, the governments' agency is limited by the EU's nature conservation policy and the Habitats Directive. EU policy aiming to protect large carnivores has changed the nature of reindeer herding in Sweden and Finland over several decades (Heikkinen et al. 2011; Vuojala-Magga 2012).

Sámiid Riikkasearvi (2021), the Reindeer Herding Association in Sweden, points out the need for structural change in relation to forestry, highlighting that "co-existence" is impossible as long as rules remain unbalanced. Tensions arise in particular when softer instruments (e.g., agreements and goals formulated for reindeer husbandry such as tolerance levels for carnivores) collide with formal policy commitments often translated into national legislation. While softer instruments are important, they only function if balanced against competing and interacting governing systems (see also Löf 2014; Risvoll & Kaarhus 2020). In the current situation, participating in planning and permit granting is draining herders' resources and is unlikely to provide them with meaningful influence (Landauer & Komendantova 2018). At best, stopping a new encroachment means slowing down the negative trend of increasing competition over land, not improving the situation (Österlin & Raitio 2020).

We, therefore, argue that the underlying norm and minimalist assumption of co-existence – turning into a prescribed outcome of governing interactions – provides part of the explanation why land use planning and permit processes for other land uses commonly lack adequate assessment of the cumulative impacts of existing and planned uses on reindeer pastures and herding practices (Raitio et al. 2020). This is striking, considering that impact assessments have long been considered a key instrument in environmental governance at large.

Addressing the legitimacy gap in reindeer husbandry governance

Our assessment of reindeer husbandry governance in the Nordic states corroborates findings that the current governing systems fail to effectively address and mitigate the key issues facing reindeer herding (output) and moreover fail in governing the interaction between different actors and providing meaningful arenas for participation and representation (input). We show, however, that regardless of differences in legislative, institutional and administrative contexts present in the three countries, the challenges facing reindeer herding, and how the governing systems fail to accommodate them, are strikingly similar. Our explanation derives from examining the discursive and political dimensions of reindeer husbandry governance, showing how governance acts constitutively. That is, the governing systems create boundaries in relation to problems, solutions and visions (what is needed, possible and desirable?) and in relation to reindeer herding as the system-to-be-governed (what is reindeer herding and why?). Both of these are at odds with herders' understandings, needs and demands.

We moreover show how negotiation and contestation, the struggle over meaning in governance interactions are structured in favour of competing land use actors and provide little or no opportunity for transformative change (see also Löf 2014). Such interactions are governed in a fragmented system of sectoral silos with weak collaborative instruments that fail to protect both the pastures and reindeer herding rights. Influence over the conditions and interactions structuring the situation for RHCs vis-à-vis other land users largely remains outside their control (see also Löf 2014). We argue that the weakness of collaborative instruments is due to the lack of regulations ensuring an adequate knowledge base in the form of cumulative impact assessment or ensuring the consent of the affected RHCs, while co-existence is a prescribed outcome.

When public governance fails – what remains?

When public governance is unable to address its own legitimacy deficits, there are other venues that provide opportunities to continue the negotiation over meaning and formulating the content in desirable futures. One is using litigation as a tool to gain recognition and protection for reindeer herding rights, lands and livelihood (Löf 2014; Raitio 2016; Keskitalo 2018; Allard & Brännström 2021), another includes protests (Persson et al. 2017) and using international media campaigns to create pressure through the markets (e.g., Lawrence 2007; Sarkki & Heikkinen 2010). Increasingly, Sámi civil society and cultural actors are involved in contesting what is perceived as the manifestation of continued colonialism (Sandström 2020).

Both of these venues place external pressure on the states and the governing systems. International criticism by UN bodies concerning violation of Sámi and reindeer herding rights and particularly the failure of the states to meet the requirements of international law to obtain the FPIC of Indigenous communities when planning land use in their area is frequent and encourages the states to increase Sámi and reindeer herders' influence over land use issues. Recent examples include the CERD (the UN Committee for Elimination of Racial Discrimination) decision in late 2020, which urged Sweden to stop a proposed mining project and revise its Minerals Act (CERD/C/102/D/54/2013, 26 November 2020). Similarly, the UN Humans Rights Committee recently urged Finland to develop better mechanisms for impact assessment to ensure genuine FPIC in its land use planning (CCPR/C/FIN/CO/7, 1 April 2020, paras 42 and 43). Although the policy impacts of these decisions are still uncertain, they call for structural change in reindeer husbandry governance.

The national-level courts also demonstrate potential in driving political change (so-called juridification (Kooiman 2003)). A recent example is the so-called Girjas case, where Girjas RHC took the Swedish state to court over who had the definitive right to decide over hunting and fishing on the RHC's customary area. The Supreme Court ruled unanimously in favour of Girjas RHC and moreover stated that the current Reindeer Husbandry Act fails to sufficiently regulate these rights (T 853–18). While the ruling resulted in the government appointing a parliamentary committee with the task of reviewing the Act, long overdue according to leading researchers in the field (Bengtsson

& Torp 2012; Brännström 2017), it is noteworthy how the responsible reindeer husbandry minister, in public discourse, refrains from using a rights-holder perspective and instead maintains the need to include "all interests" in the process (SvT 2021).

Juridification and mobilizing support internationally have a downside, however. Litigation is in conflict with the principles of good governance in the Nordic political and legal system and has also resulted in an escalation of conflicts between RHCs and other local communities. Attempts to stop development projects on reindeer herding lands are faced with resistance (e.g. Larsen et al. 2017; Sehlin MacNeil 2017) and have led to open questioning of the reindeer herding right, increased hate speech, crimes and other expressions of racism towards the Sámi (Kroik & Hellzen 2011). Disempowerment caused by the cumulative effects of decreasing profitability, enduring conflicts and limited opportunity to improve the situation has also resulted in a situation of reduced psycho-social health, such as higher than average suicidal thoughts among reindeer herders (Kaiser et al. 2010; Stoor 2016).

The need to "re-image" and reimagine sustainable reindeer husbandry governance

Our analysis demonstrates how the governing systems have been created to address, and continue to address, problem representations and "solutions" defined by actors other than reindeer herders. The governing systems are moreover structured in a manner that restricts the opportunity for herders to take part in meaningful negotiation over needs, goals and visions. As a consequence, they are increasingly seeking alternatives outside the governing systems, while paying the price through, e.g., increasing levels of conflict. Ironically, while the "problems" of reindeer herding have typically been considered internal to the practice itself (e.g., attributed to "irrational herding practices" (Mörkenstam 1999)) "solutions" have often been framed as demanding increased state intervention (Löf 2016). However, the increasing recognition of external drivers and influences means that "solutions" too need to span broader scales and include other actors and governing systems (and thus potentially addressing some of the current fragmentation).

We argue that an underlying reason for this systematic failure is that the governing systems do not address all aspects of reindeer herding. Framing reindeer herding primarily as an industry, coupled with a profit-focused market economy logic that positions actors as interests, appears to be key in driving and maintaining this governing system dynamics. Here, it is important to distinguish between assuring the economy in reindeer herding and assuming profit as the *primary* purpose and defining characteristic (cf. Hinton 2020). Put differently, while herders themselves stress the importance of maintaining good economy in herding practices, there is an important difference in perspective between seeing economy as a tool to sustain a good life, healthy herds and natural pastures (Komu 2020) and regarding economic profit as the main goal for herding practices.

The next logical step is to therefore ask whether it is possible to reimagine and restructure reindeer husbandry governance based on the visions and solutions that currently dominate *or* if the situation demands governing interactions with the potential to *re-image* and *reconsider* both the governed and governing? As noted, the overarching goal for reindeer husbandry governance across Fennoscandia is *to maintain and create conditions for sustainable reindeer husbandry*. The meta-narrative of sustainable development (SD) has become a defining goal across scales. In a nutshell, SD is connected to the ideology of ecological modernization and, essentially, the idea of producing more with less (Arts & Buizer 2009). Overcoming ecological and economic limits is a key objective mediated largely through technological development. The idea of win-win solutions – where nobody is left out – is firmly anchored in the SD discourse (Sarkki et al. 2020), which explains its global application and attraction (Arts & Buizer 2009). However, the complexity and tensions embedded in this concept and discourse are monumental. Just as governance is increasingly recognized as political, so is the idea of SD (Voß & Bornemann 2011). Sustainability is a particularly slippery concept, which not only generates but disguises goal conflicts between its different dimensions. As different actors hold competing definitions of what sustainability entails, the SD discourse can be used to legitimate seemingly unsustainable practices as "*perceptions of sustainability are scale and place specific*" (Nilsson & Larsen, cited in Sörlin 2021, p.327).

Sörlin (2021) concludes that discourses and ideas of sustainability are often shaped by dominant players, with certain presumptions taken for granted, e.g., that extraction or intervention (be it through mining or forestry) is a given. As we have shown, this connects to the idea of adaptation as a solution (see also Löf 2013; 2014). In this case, the narrative of sustainability hides the political aspect of governance and turns it into an administrative and managerial task, e.g., identifying which forests to cut rather than acknowledging limits to co-existence and win-win. There is thus a risk that under the present dominant understanding of what *reindeer herding* is and can be, and what *sustainability* is and can mean, the wrong questions are posed, and the governing systems will probably face more tension and conflict as a result. Voß and Borneman explain that "*the patterns and processes of governance itself have come to be identified as challenges in working toward sustainable development because they define the very capacities by which societies shape and transform themselves*" (2011, p.1). Our analysis of reindeer husbandry governance shows precisely that; as long as herders' views are peripheral in how the governance of reindeer herding is constructed, the conflicts, contestation and loss of legitimacy will continue. *Reimagining* sustainable reindeer herding thus requires us, first, to engage in a process of *re-imaging* the system-to-be-governed, where herders, herding organizations and herding communities must take the lead. Re-imaging requires us to look much deeper than instruments and specific issues. Paraphrasing Johnsen et al. (2015) *do we view the governance task like the state or the herders*? Can it be done differently?

References

Allard, C. (2015). *Renskötselrätt i nordisk belysning*. Göteborg: Makadam Förlag.

Allard, C. & Brännström, M. (2021). Girjas Reindeer Herding Community v. Sweden: Analysing the Merits of the Girjas Case. *Arctic Review*. 12, 56–79.

Anaya, S.J. (2004). *Indigenous peoples in international law*. Oxford: Oxford University Press.

Arts, B. & Buizer, M. (2009). Forests, discourses, institutions: A discursive-institutional analysis of global forest governance. *Forest Policy and Economics*. 11(5–6), 340–347.

Bacchi, C. (2009). *Analysing policy*. French forest: Pearson Higher Education AU.

Bengtsson, B. & Torp, E. (2012). Svensk samerätt: Något om den senaste utvecklingen. In: Øyvind, R. & Henriksen, T. (eds) *Juss i Nord: Hav, fisk og Urfolk*. Tromsö: Gyldendal Akademisk. 136–147.

Benjaminsen, T., Reinert, A.H., Sjaastad, E. & Sara, M.N. (2015). Misreading the Arctic landscape: a political ecology of reindeer, carrying capacities and overstocking in Finnmark, Norway. *Norwegian Journal of Geography*. 69(4), 219–229.

Bevir, M. (2010). Rethinking governmentality: Towards genealogies of governance. *European Journal of Social Theory* 13(4), 423–441.

Brännström, M. (2017). *Skogsbruk och renskötsel på samma mark: En rättsvetenskaplig studie av äganderätten och renskötselrätten*. Skrifter från Juridiska institutionen Umeå universitet. 35:376. PhD. Diss. Umeå University.

Du Plessis, G. (2020). Killing reindeer: A spatial analysis of Nordic states and nomadic forms of life in the arctic. *International Political Sociology*. 14(4), 347–365.

Forbes, B.C., Bölter, M., Müller-Wille, L., Hukkinen, J., Müller, F., Gunslay, N. & Konstantinov, Y. (eds.) (2006). *Reindeer management in Northernmost Europe. Ecological Studies*. 184. Berlin: Springer

Forbes, B.C., Turunen, M.T., Soppela, P., Rasmus, S., Vuojala-Magga, T. & Kitti, H. (2020). Changes in mountain birch forests and reindeer management: comparing different knowledge systems in Sápmi, northern Fennoscandia. *Polar Record*. 55(6), 507–521.

Grönvall, A. & Löf, A. (2020). *Den gränslösa renen, det gränsdragna Sápmi: Om gränsöverskridande renskötsel, statliga regleringar och konsekvenser för Sárevuopmi*. (Skrifter från Várdduo, Centrum för samisk forskning, 26). Umeå: Umeå universitet.

Heikkinen, H.I., Moilanen, O., Nuttall, M. & Sarkki, S. (2011). Managing predators, managing reindeer: contested conceptions of predator policies in Finland's southeast reindeer herding area. *The Polar Record*. 47(3), 218–230.

Hinton, J.B. (2020). Fit for purpose? Clarifying the critical role of profit for sustainability. *Journal of Political Ecology*. 27(1), 236–262.

Holand, Ø., Moen, J., Kumpula, J., Löf, A., Rasmus, S. & Røed, K. (2021). Project ReiGN: Reindeer Husbandry in a Globalizing North–Resilience, Adaptations and Pathways for Actions. In Nord, D. C., (Ed.) *Nordic Perspectives on the Responsible Development of the Arctic: Pathways to Action*. Cham, Springer International Publishing. 227–248.

Hukkinen, J., Heikkinen, H., Raitio, K., & Muller-Wille, L. (2006). Dismantlinag the barriers to entrepreneurship in reindeer management in Finland. *International Journal of Entrepreneurship and Small Business*. 3(6), 705–727.

Johnsen, K.I. (2018). *Conflicting knowledges, competing worldviews: Norwegian governance of Sámi reindeer husbandry in West Finnmark, Norway*. PhD. Diss. Norwegian University of Life Sciences.

Johnsen, K.I., Benjaminsen, T.A. & Eira, I.M.G. (2015). Seeing like the state or like pastoralists? Conflicting narratives on the governance of Sámi reindeer husbandry in Finnmark, Norway. *Norwegian Journal of Geography*. 69(4), 230–241.

Johnsen, K.I., Mathiesen, S.D. & Eira, I.M.G. (2017). Sámi reindeer governance in Norway as competing knowledge systems: A participatory study. *Ecology and Society*. 22(4), 33.
Joks, S., Østmo, L. & Law, J. (2020). Verbing meahcci: Living Sámi lands. *The Sociological Review*. 68(2), 305–321.
Kaiser, N., Sjölander, P., Liljegren, A.E., Jacobsson, L. & Renberg, E.S. (2010). Depression and anxiety in the reindeer-herding Sami population of Sweden. *International Journal of Circumpolar Health*. 69, 383–393.
Keskitalo, A. (2018). Sameretten ved inngangen til 2018. *Kritisk juss*. 44(01), 50–55.
Keskitalo, E.C.H., Horstkotte, T., Kivinen, S., Forbes, B. & Käyhkö, J. (2016). "Generality of mis-fit"? The real-life difficulty of matching scales in an interconnected world. *Ambio*. 45(6), 742–752.
Kivinen, S. (2015). Many a little makes a mickle: Cumulative land cover changes and traditional land use in the Kyrö reindeer herding district, northern Finland. *Applied Geography*. 63, 204–211.
Komu, T. (2020). *Pursuing the good life in the North. Examining the coexistence of reindeer herding, extractive industries and nature-based tourism in northern Fennoscandia*. PhD. Diss. University of Oulu.
Kooiman, J. (2003). *Governing as governance*. London: Sage Publications Ltd.
Kroik, L. & Hellzen, O. (2011). *Helseopplevelse hos reindriftssamene i sør: sørsamers livsmiljø*. (Rapport nr 76). Høgskolen i Nord-Trøndelag.
Kumpula, J., Kurkilahti, M., Helle, T. & Colpaert, A. (2014). Both reindeer management and several other land use factors explain the reduction in ground lichens (Cladonia spp.) in pastures grazed by semi-domesticated reindeer in Finland. *Regional Environmental Change*. 14(2), 541–559.
Kuoljok, K. (2019). Without land we are lost: traditional knowledge, digital technology and power relations. *AlterNative: An International Journal of Indigenous Peoples*. 15(4), 349–358.
Landauer, M. & Komendantova, N. (2018). Participatory environmental governance of infrastructure projects affecting reindeer husbandry in the Arctic. *Journal of Environmental Management*. 223, 385–395.
Landauer, M., Rasmus, S. & Forbes, B.C. (2021). What drives reindeer management in Finland towards social and ecological tipping points? *Regional Environmental Change*. 21(2), 1–16.
Larsen R.K., Raitio K., Stinnerbom M. & Wik Karlson J. (2017). Sami-state collaboration in the governance of cumulative effects assessment: a critical action research approach. *Environmental Impact Assessment Review*. 64, 67–76.
Lawrence, R. (2007). Corporate Social Responsibility, Supply-chains and Saami claims: Tracing the Political in the Finnish Forestry Industry. *Geographical Research*. 45(2), 167–176.
Löf, A. (2013). Examining limits and barriers to climate change adaptation in an Indigenous reindeer herding community. *Climate and Development*. 5(4), 328–339.
Löf, A. (2014). *Challenging Adaptability: Analysing the Governance of Reindeer Husbandry in Sweden*. PhD. Diss. Umeå University.
Löf, A. (2016). Locking in and locking out: a critical analysis of the governance of reindeer husbandry in Sweden. *Critical Policy Studies*. 10(4), 426–447.
Marin, A., Sjaastad, E., Benjaminsen, T.A., Sara, M.N.M. & Borgenvik, E.J.L. (2020). Productivity beyond density: A critique of management models for reindeer pastoralism in Norway. *Pastoralism*. 10, 1–18.
Mathiesen, S.D., Reinert, H. & Turi, E. I. (2009). Adapting to climate change in Sami reindeer herding: the nation-state as problem and solution. In: Adger, W.N., Lorenzoni, I.

& O'Brien, K.L. (eds) *Adapting to Climate Change: Thresholds, Values, Governance*, 417. Cambridge: Cambridge University Press.

Mörkenstam, U. (1999). *Om "Lapparnes privilegier": föreställningar om samiskhet i svensk samepolitik 1883–1997*. PhD. Diss. Stockholm University.

North, D.C. (1990). *Institutions, Institutional Change and Economic Performance*. Cambridge: Cambridge University Press.

Össbo, Å. (2014). *Nya vatten, dunkla speglingar. Industriell kolonialism genom svensk vattenkraftsutbyggnad i renskötselområdet 1910–1968*. PhD. Diss. Umeå University.

Österlin, C. & Raitio, K. (2020). Fragmented Landscapes and Planscapes – The Double Pressure of Increasing Natural Resource Exploitation on Indigenous Sámi Lands in Northern Sweden. *Resources*. 9(9), 104.

Pekkarinen, A.J., Kumpula, J. & Tahvonen, O. (2021). What Drives the Number of Semi-domesticated Reindeer? Pasture Dynamics and Economic Incentives in Fennoscandian Reindeer Husbandry. In: Nord, D.C. (ed.) *Nordic Perspectives on the Responsible Development of the Arctic: Pathways to Action*. Cham: Springer. 249–270.

Persson, S., Harnesk, D. & Islar, M. (2017). "What local people? Examining the Gállok mining conflict and the rights of the Sámi population in terms of justice and power." *Geoforum* 86: 20-29.

Raitio, K. (2012). New Institutional Approach to Collaborative Forest Planning: Methods for Analysis and Lessons for Policy. *Land Use Policy*. 29(2), 309–316.

Raitio, K., & Heikkinen, H. (2003). More initiative, less complaining. The views of administration of development of the institutions that are engaged in reindeer husbandry (in Finnish). Helsinki University of Technology Laboratory of Environment Protection Publication. *Technology, Society, Environment*.

Raitio, K., Allard, C. & Lawrence, R. (2020). Mineral extraction in Swedish Sápmi: The regulatory gap between Sami rights and Sweden's mining permitting practices. *Land Use Policy*. 99, 105001.

Rasmus, S., Kojola, I., Turunen, M., Norberg, H., Kumpula, J. & Ollila, T. (2020). Mission impossible? Pursuing the co-existence of viable predator populations and sustainable reindeer husbandry in Finland. *Journal of Rural Studies*. 80, 135–148.

Risvoll, C. & Hovelsrud, G.K. (2016). Pasture access and adaptive capacity in reindeer herding districts in Nordland, Northern Norway. *The Polar Journal*. 6(1), 87–111.

Risvoll, C. & Kaarhus, R. (2020). Struggling with 'clear zoning'-Dilemmas of carnivore-pastoral coexistence in Nordland, northern Norway. In: Breidlid, A. & Krøvel, R. (eds) *Indigenous knowledges and the sustainable development agenda*. London: Routledge.

Sandström, M. (2020). *Dekoloniseringskonst: Artivism i 2010-talets Sápmi*. PhD. Diss. Umeå University.

Sandström, P., Cory, N., Svensson, J., Hedenås, H., Jougda, L. & Borchert, N. (2016). On the decline of ground lichen forests in the Swedish boreal landscape: Implications for reindeer husbandry and sustainable forest management. *Ambio*. 45(4), 415–429.

Sarkki, S., Heikkinen, H.I. & Löf, A. (2021). Reindeer Herders as Stakeholders or Rights-Holders? Introducing a Social Equity-Based Conceptualization Relevant for Indigenous and Local Communities. In: Nord, D.C. (ed.) *Nordic Perspectives on the Responsible Development of the Arctic: Pathways to Action*. Cham: Springer. 271–292.

Sarkki, S. & Heikkinen, H.I. (2010). Social Movements' Pressure Strategies during Forest Disputes in Finland. *Journal of Natural Resources Policy Research*. 2(3), 281–296.

Sarkki, S., Komu, T., Heikkinen, H.I., García, N.A., Lépy, É. & Herva, V.P. (2016). Applying a Synthetic Approach to the Resilience of Finnish Reindeer Herding as a Changing Livelihood. *Ecology and Society*. 21(4), 14.

Scharpf, F.W. (1999). *Governing in Europe: Effective and Democratic?* Oxford: Oxford University Press.
Sehlin MacNeil, K. (2017). *Extractive Violence on Indigenous Country: Sami and Aboriginal Views on Conflicts and Power Relations with Extractive Industries*. PhD. Diss. Umeå University.
Sjölander-Lindqvist, A., Risvoll, C., Kaarhus, R., Lundberg, A.K. & Sandström, C. (2020). Knowledge claims and struggles in decentralized large carnivore governance: insights from Norway and Sweden. *Frontiers in Ecology and Evolution*. 8, 120.
Sörlin, S. (2021). Is There Such a Thing as 'Best Practice'? Exploring the Extraction/Sustainability Dilemma in the Arctic. In: Nord, D.C. (ed.) *Nordic Perspectives on the Responsible Development of the Arctic: Pathways to Action*. Cham: Springer. 321–348.
Stoor, P. (2016). *Kunskapssammanställning om samers psykosociala ohälsa*. (Rapport från Sámediggi). Gìron/Kiruna.
Torfing, J., Peters, B.G., Pierre, J. & Sørensen, E. (2012). *Interactive governance: advancing the paradigm*. Oxford: Oxford University Press.
Turunen, M., Rasmus, S., Norberg, H., Kumpula, J., Kojola, I. & Ollila, T. (2017). Porot ja pedot–kuinka poronhoidon sopeutuminen petoihin on muuttunut 90 vuodessa? *Suomen riista*. (63), 19–42.
Turunen, M., Rasmus, S., Järvenpää, J. & Kivinen, S. (2020). Relations between forestry and reindeer husbandry in northern Finland–Perspectives of science and practice. *Forest Ecology and Management*. 457, 117677.
Tyler, N.J., Hanssen-Bauer, I., Førland, E.J. & Nellemann, C. (2021). The Shrinking Resource Base of Pastoralism: Saami Reindeer Husbandry in a Climate of Change. *Frontiers in Sustainable Food Systems*. 4, 274.
Ulvevadet, B. (2012), *The governance of Sami reindeer husbandry in Norway: institutional challenges of co-management*. PhD. Diss. University of Tromsø.
Voß, J.P. & Bornemann, B. (2011). The politics of reflexive governance: challenges for designing adaptive management and transition management. *Ecology and Society*. 16(2). [online] URL: www.ecologyandsociety.org/vol16/iss2/art9/
Vuojala-Magga, T. (2012). Adaptation of Sámi reindeer herding: EU regulation and climate change. In: Tennberg, M. (ed.) *Governing the Uncertain: Adaptation and Climate Change in Russia and Finland*. Dordrecht: Springer. 101–122.
Widmark, C. and Sandström, C. (2012). Transaction Costs of Institutional Change in Multiple-Use Commons: The Case of Consultations Between Forestry and Reindeer Husbandry in Northern Sweden. *Journal of Environmental Policy & Planning*. 14(4), 428–449.

APPENDIX

Table 8A.1 Policy literature cited, legislation not included

Finland
Ministry of Justice (2017). Muistio saamelaiskäräjälain 9§:n mukaisesta neuvotteluvelvoiteesta, OM 2/551/2017. https://oikeusministerio.fi/documents/1410853/4735105/SaK%C3%A4L9%C2%A7_Neuvotteluvelvollisuus_Muistio.pdf/1e7c49d2-e911-4cbf-8733-112a00241b36/SaK%C3%A4L9%C2%A7_Neuvotteluvelvollisuus_Muistio.pdf?t=1511959724000
Metsähallitus (2021). Northern and Eastern Finland are special areas for Metsähallitus. www.metsa.fi/en/lands-and-waters/planning-of-state-owned-areas/special-regional-features/ [7.5.2021]

Table 8A.1 Cont.

Finland

Paliskuntain yhdistys (2021). Projects. [Reindeer Herders' Association]. https://paliskunnat.fi/reindeer-herders-association/projects/
Sámediggi (2021). The Sámi in Finland. www.samediggi.fi/sami-info/?lang=en [7.5.2021]
Työ- ja elinkeinoministeriö (TEM) (2019). Finland's updated tourism strategy for 2019–2028. https://tem.fi/en/finland-tourism-strategy
Valtioneuvosto (2019). Porolaidunten käyttösuunnitelmat. Kauden 2020–2030 paliskuntakohtaiset porolaidunsuunnitelmat. https://valtioneuvosto.fi/documents/1410837/12122311/Porolaidunten+k%C3%A4ytt%C3%B6suunnitelmat.pdf/ac53528f-28db-652e-27ab-c86db3ed0023/Porolaidunten+k%C3%A4ytt%C3%B6suunnitelmat.pdf?t=1574674928000

Norway

Ministry of Agriculture and Food (2007). Ot.prp. Nr. 25 (2006–2007), Om lov om reindrift (reindriftsloven).
Ministry of Agriculture and Food (2020). Reindrift www.regjeringen.no/no/tema/mat-fiske-og-landbruk/reindrift/id2339773/ [3.3.2021]
Ministry of Agriculture and Food (2021). Aktuelt fra Landbruks- og matdepartmentet. In National Conference on large carnivores, pastoral animals and society Webinar 25–26 January 2021. www.regjeringen.no/contentassets/90c34e7b253745df80ede83cee9118ee/rovvilt-beitedyr-og-samfunn_-hamar-26.-jan-2021_lmd-4.pdf [3.3.2021]
NOU 2001:35. Forslag til endringer i reindriftsloven.
Prop. 189 S. 2020–2021. Reindriftsavtalen 2021/202.
Riksrevisjonen (2011–2012), Riksrevisjonens undersøkelse av bærekraftig reindrift i Finnmark, Dokument 3:14 (2011–2012), 9.
St.meld. nr. 32 (2016–2017) to the Storting on Reindeer Husbandry policy: Reindeer Husbandry. Old tradition - unique opportunities. www.regjeringen.no/en/aktuelt/reindeer-husbandry.-old-tradition---unique-opportunities/id2548490/

Sweden

Arctic Strategy 2020. Regeringens skrivelse 2020/21:7 Strategi för den arktiska regionen. www.regeringen.se/4a8365/contentassets/000d750cc7d941b98abedf844a07529f/regeringens-skrivelse-2020-21-7.pdf
Committee on Environment and Agriculture (2000). 1999/2000:MJU9. Miljö- och jordbruksutskottets betänkande, Rennäring. https://data.riksdagen.se/fil/480D3C49-B780-417F-97AD-61920E64B89E [4.4.2021]
Ministry of Culture and Democracy (2018). Speech 7 Nov. by Kuhnke, A. www.regeringen.se/tal/2018/02/tal-av-alice-bah-kuhnke-vid-den-officiella-invigningen-av-jubileumsveckan-staare-och-firandet-av-samernas-nationaldag/ [3.5.2021]
Ministry of Culture and Democracy 2015. Samerna – ett folk och urfolk. www.regeringen.se/artiklar/2015/06/samerna---ett-folk-och-urfolk/ [4.4.2021]
Ministry of Enterprise and Innovation 2020. Åtgärder för att stärka rovdjursförvaltningen. www.regeringen.se/pressmeddelanden/2020/11/atgarder-for-att-starka-rovdjursforvaltningen/ [28.4.2021]
Prop. 1992/93:32 Om samerna och samisk kultur m.m.
Prop. 2009/10:80 En reformerad grundlag.

Table 8A.1 Cont.

Sweden
Prop. 2020/21; 64 En konsultationsordning i frågor som rör det samiska folket. Sametinget (2020). Sametingets årsredovisning 2020 [Swedish Sami Parliament's yearly report] www.sametinget.se/156490 [4.4.2021] Sámediggi (2016). Samiskt självbestämmande. [Sámi self-determination] www.sametinget.se/samiskt_sjalvbestammande [20.9.2021] Sámiid Riikkasearvi (2021). Nyhetsbrev februari [Newsletter February] https://www.sapmi.se/sveriges-regering-ser-inte-skogen-for-alla-trad/ SOU 2001:101. En ny rennäringspolitik - öppna samebyar och samverkan med andra markanvändare SvT (Swedish television) (2021). June 23rd www.svt.se/nyheter/inrikes/efter-girjasdom-rennaringslagen-ska-utredas [20.8.2021]

9 Governing maximum reindeer numbers in Fennoscandia

Simo Sarkki, Kathrine Ivsett Johnsen, Annette Löf, Antti-Juhani Pekkarinen, Jouko Kumpula, Sirpa Rasmus, Mia Landauer and Birgitta Åhman

Introduction

Environmental governance is contested and complex, involving divergent and interacting policy, society and science perspectives (Bennett & Satterfield 2018). This chapter explores the governing challenges associated with setting the maximum permitted number of reindeer in Finland, Norway and Sweden. Reindeer herders operate with an estimated herd size (Sara et al. 2016). While herders across Fennoscandia regularly count their reindeer, recording exact numbers can be challenging: some reindeer escape the autumn round-ups, and some are lost during the winter (e.g., due to predation, traffic accidents, starvation and moving to other districts). Many of the calves born in spring are slaughtered in the autumn. Available grazing resources and weather also affect herd sizes, which fluctuate throughout and between years. However, for the past hundred years, state-based government bodies in all three countries have decided on maximum permitted numbers. Structures, processes and mechanisms for planning, implementation and evaluation of herd sizes vary. While reindeer numbers do affect the state of pastures, the governance challenge of defining and imposing a maximum number lies in questions about who should make the decisions, for what purpose and what type of knowledge upon which to ground the decision making. It is thus appropriate to examine whether and how externally defined and imposed maximum numbers may enhance the sustainability of reindeer husbandry and whether the sociocultural dimension is recognized or neglected in the ecological and economic metrics used (Heikkinen et al. 2007).

Reindeer numbers as a sensitive topic

Reindeer numbers and herd structure affect the herders' economy. The interdependence between animals and pastures forms the material basis for reindeer husbandry as a traditional, nature-based livelihood (Tahvonen et al. 2014; Pekkarinen et al. 2015). However, competition with other forms of land use has triggered local and national debates about reindeer numbers, in which different land use interests' claims, accusations, arguments and perspectives compete with those of the herders (Johnsen 2016). Sensitivity to the issues surrounding

DOI: 10.4324/9781003118565-13

reindeer numbers also relates to Indigenous rights of Sámi people to the lands they have inhabited for centuries. In the case of Finland, ethnic Finns also practice reindeer herding and did so long before state intervention on northern lands (Heikkinen 2002).

Fluctuations in reindeer herd sizes have been explained by multi-facetted ecological (e.g., weather and snow conditions), political (e.g., national borders, state-based regulation and other land uses), economic (e.g., demand for reindeer meat) and social (e.g., relationships between herders, and between herders and other land users) aspects (see Pape & Löffler 2012). The customary rights of herders to use the land for their livelihood derive from their historical presence in the areas and are recognized by national laws and regulations, though these are not always implemented (Chapter 8). Moreover, herders hold key expertise and practice-based knowledge about the dynamics between pastures, reindeer and herders (Chapter 7). This knowledge is based on long-term daily experiences of herders who directly witness the socio-environmental changes affecting their livelihood. Herding practices such as herd structure, migration, pasture rotation or use of supplementary feeding have evolved in response to changes in social and environmental conditions and, in turn, have an impact on which pastures can be used, under what circumstances and to what extent (e.g., Kumpula et al. 2011). It can be argued that as both knowledge holders and rights holders, herders' perspectives should be included in decisions about the maximum number of reindeer (see Sarkki et al. 2021; Chapter 8).

From the states' perspectives, setting reindeer numbers depends on the size and ecological condition of pastures. This interdependence often lies at the heart of decision making on maximum numbers. The idea of carrying capacity of winter pastures is critical for setting reindeer numbers in most parts of Fennoscandia, usually measured by the quality and quantity of biomass of terrestrial lichens on winter pastures. Such calculations on pasture–reindeer relations are often used (especially in Norway) to inform decisions on the maximum sustainable number of reindeer. This often relies on ideas informed by the "tragedy of the commons" (Hardin 1968), reflected as a persistent narrative that, without state control, herders will increase their herd sizes to the detriment of all. However, this one-dimensional presentation of reindeer herders' rationale and simplification of the argument is questioned (e.g., Benjaminsen et al. 2015; Johnsen et al. 2015; Marin et al. 2020); issues like supplementary feeding (especially in Finland), herding practices, pasture rotation, cumulative impacts of other land uses and climate change also contribute to the pressure that a particular number of reindeer has on pastures. Consequently, there is a risk that decision makers rely mainly on reductionist views on carrying capacity, while herders' knowledge derives from practice and more complex recognition of landscape processes and interactions (Chapter 7). Herders often point out that their livelihood, culture and rights are compromised by external definitions, regulations and generalized estimates.

Governing maximum number of reindeer in the three countries

In Finland, the Reindeer Husbandry Act states that the Ministry of Agriculture and Forestry (MAF) shall determine the maximum reindeer number for each herding district for each ten-year period, so that "*reindeer grazing does not exceed the sustainable production capacity of winter pastures*" (Reindeer Husbandry Act 848/1990; 21 §). The quantity and quality of pastures and their relationships to reindeer numbers and grazing practices need to be examined and evaluated to inform decisions on sustainable numbers. However, it is important to include the impacts of other forms of land use when assessing the impacts of reindeer grazing on lichens (Sarkki et al. 2013; Kumpula et al. 2014). In particular, commercial forestry affects "*not only the spatial configuration and areas of different pasture patches, but also the grazing pressure at the remaining pasture sites*" (Jaakkola et al. 2013, p. 459). As the impacts of other land uses on reindeer pastures are increasingly taken into consideration, the concept of carrying capacity of winter pastures as a legal instrument to decide maximum number of reindeer is being challenged. For example, Landauer et al. (2021, p. 13) pose further questions regarding "*who has the right to decide what constitutes a desirable system state*".

In Norway, the national Reindeer Husbandry Board (RHB) sets the maximum number of reindeer for each herding district. Regulating numbers, as well as economic incentives for increasing efficiency of meat production, are key tools for ensuring the national policy objective of "a rational, market-oriented industry" (Meld. St. 32 2016–2017, p. 7). There is also an objective to preserve pastures from overgrazing, and therefore the state-regulation of reindeer numbers is informed by estimates of the pastures' "carrying capacity". However, it is a paradox that while herders are forced to reduce their herds to preserve pastures, the authorities allow mineral extraction and wind power development in the same areas (Johnsen 2016; see also Chapter 8). Some scholars and herders also argue that the use of indicators – such as carcass weights – to monitor ecological sustainability excludes the herders' experience-based knowledge about the reindeer and herd dynamics and leads to misreading Arctic landscapes (Benjaminsen et al. 2015; Johnsen et al. 2017; Marin et al. 2020).

In Sweden, the state regulates reindeer numbers at the reindeer herding district level. According to the present Swedish Reindeer Husbandry Act (SFS 1971:437), the County Administrative Boards (CABs) are responsible for determining the highest number of reindeer for each district. Unlike in Finland and Norway, processes for deciding reindeer numbers are rather rare governing interventions and the public discussion on maximum number of reindeer is less intense. The debates on "overgrazing" and "too many reindeer" on sensitive mountain (summer) pastures peaked in the 1990s, resulting, e.g., in changes in the legislation with demands for environmental consideration (Swedish Government Proposition 1995/96:226; Reindeer Husbandry Act §65a). Studies claiming overgrazing in specific areas were also contested and proved to be more of a local problem caused by limiting the natural movement of reindeer by erecting fences (Moen & Danell 2003). This, together with research showing that reindeer grazing had no consistent negative effects on mountain vegetation (Olofsson et al. 2001), put the overgrazing issue on the back burner in Sweden. However, proposals to limit numbers keep resurfacing, while perhaps under more restricted circumstances. Accordingly, maximum numbers are treated more flexibly in Sweden, compared to Finland and Norway, typically serving as a reference rather than an absolute limit (e.g., SOU 2001:101, p.318).

Objectives

Governance approaches pertaining to maximum reindeer numbers differ between the three countries. Here, we highlight different governance challenges and opportunities in the different national contexts: herders' opportunities to participate in decision making on maximum numbers (Finland), clashes between scientific and reindeer herders' knowledge systems (Norway) and flexibility of state-based governance to define and enforce maximum numbers (Sweden).

Background

State policies and regulations for setting a maximum number of reindeer have been influenced by many different objectives. First – especially in Finland and Norway – ecological sustainability has been and is a key objective for regulating herd sizes, operationalized by the concept of carrying capacity. However, framing the criteria for ecologically sustainable herd sizes and use of pastures are a complex and contested process (Chapter 8). Second, since the 1960s and 1970s the objective has been to rationalize reindeer husbandry across Fennoscandia. In Sweden, policies typically favoured owners of large herds and embedded herding in an economic, rather than cultural, paradigm. In Norway, the rationalization policies promoted standardized herd structures and slaughter strategies and optimal reindeer numbers (Paine 1994). In Finland, EU membership since 1995 has directed the nature of reindeer husbandry towards larger herds and more meat production. In Sweden, EU membership has had less impact, and Norway is not an EU member. Third, during the past century across Fennoscandia there has been an objective to control the potential harm caused by reindeer grazing to forestry and agriculture by regulating numbers. However, the situation has moved towards increased recognition of herders' needs and impacts of other land uses on pastures. In Finland, the state forestry enterprise Metsähallitus has improved its practices in this regard. In Sweden, consultations between forestry actors and reindeer herders have a long history but today are mainly guided by market-based governance through Forest Stewardship Council (FSC) certification. However, discrepancies between herders' concerns and other actors' views on reindeer numbers remain. Controlling numbers is still often perceived by state officials as a measure to handle conflicts between reindeer husbandry and other land uses, including forestry, urbanization, mining, predator conservation and renewable energies. Table 9.1 outlines basic information regarding the maximum number of reindeer in the three countries.

Table 9.1 Background for understanding governance of maximum numbers of reindeer in the three countries

Issue	*Finland*	*Norway*	*Sweden*
Who regulates the number of reindeer?	Ministry of Agriculture and Forestry (MAF) sets the numbers for each district and owner in ten-year intervals based on propositions from working groups established by the Ministry.	The reindeer herding districts make proposals and the Reindeer Husbandry Board decides on the maximum reindeer numbers for the districts. If the district complains, the Ministry for Agriculture and Food makes a final decision.	County Administrative Boards (CABs) are responsible for determining the highest number of reindeer for each district.
Rationale for setting maximum reindeer by state-based actors.	"Reindeer grazing should not exceed the sustainable production capacity of winter pastures" (Reindeer Husbandry Act 848/1990; 21 §).	Ecologically sustainable reindeer husbandry, which is understood as a herd size adjusted to available pastures (§60).	Has differed over time. Long-term conservation of grazing resources is the key objective (Reindeer Husbandry Act § 65a), while the impacts of other land uses on pastures are increasingly recognized as important factors to consider (1971:51; SOU 2001:101).
Methods of evaluation/ monitoring.	Inventories every ten years evaluating quantity and quality of ground and arboreal lichen ranges and extent of other land use forms; annual reindeer counts.	Annual reindeer counts and counting inspections; statistics on production.	Annual reindeer counts regulated in the Sámi Parliament ordinances (STFS 2007:2). CABs can inspect the counts.
Actors in setting the max numbers.	(In the latest process) Working group consisting of representatives of MAF, Reindeer Herders' Association, regional authorities, ENGOs, Sámi Parliament, research institute	The herding district boards propose a number; the Reindeer Husbandry Board either approves the number proposed or rejects it and decides an upper reindeer number which it believes to be more sustainable.	Routines differ between CABs in how herders and others are included. Districts are always heard. There are also formalized reindeer husbandry delegations belonging to the northern CABs where general and

(*continued*)

Table 9.1 Cont.

Issue	*Finland*	*Norway*	*Sweden*
	LUKE, state-owned forestry enterprise Metsähallitus and representative of herders.		regional issues concerning herding are discussed but inclusion in decision-making processes varies.
Herders' opportunities to participate.	In the latest process, Reindeer Herders' Association and Sámi Parliament are represented in the working group. A leader of each reindeer herding district called for hearing of the working group.	Herders participate through the reindeer herding district boards.	Herding districts are heard and can appeal decisions. Herders/herding representatives may also be appointed expert investigators in decision processes, but this not formally required and varies from case to case.
Access to state subsidies.	EU subsidies for reindeer owners with 80 or more animals.	Districts with reindeer numbers within the limit and that meet the production criteria to get access to state subsidies.	None related to reindeer numbers.
Sanctions if the maximum number is exceeded.	The regional State Administrative Agency can set a penalty to enforce a reduction in reindeer to the allowed number. Also state subsidies can be cancelled, and sometimes state-forced slaughters may be applied.	Districts with too many reindeer are not granted subsidies. Districts and individual herders can be fined, and the authorities can decide forced slaughter.	While only used occasionally, the CABs can demand reduction of reindeer numbers by issuing injunctions should numbers repeatedly exceed the set limit. They may also impose fines on the districts if demands are not followed.

Three case studies: Examples of different governance practices

Latest process to set maximum permitted number of reindeer in Finland

During the 1980s, reindeer numbers exceeded the maximum allowed in several districts. This was mainly due to favourable winters, but also due to changes in herding practices (calf-slaughter, supplementary feeding and treating parasites).

All these changes led to smaller than typical annual reindeer losses. The deteriorating conditions of the lichen pastures, however, raised more and more concerns, especially when some northern districts simultaneously experienced high reindeer losses due to harsh winters in the late 1980s. In addition, market disturbances and a deteriorating image and demand for reindeer meat due to the 1986 Chernobyl nuclear accident resulted in "too many" reindeer. This led to decreasing numbers of slaughtered reindeer and an increase in herd size. However, despite these complexities, a general conclusion of the reindeer husbandry authorities and researchers at that time was that winter pastures could not sustain the then high number of reindeer. This resulted in a reduction of the maximum number by 25,000 to 203,700 in 2000. In 1990–1991, the set maximum was 228,900, and in the same year the actual number reached 259,611.

In 2018, after keeping the maximum numbers unchanged for 20 years (2000–2020), the MAF started a process to determine the maximum numbers for the years 2020–2030 and set up a working group to prepare a proposal. The 13 members of the reindeer numbers working group (including permanent experts and secretaries) consisted of members of the Reindeer Herders' Association and the Sámi Parliament, and a leader of one Sámi reindeer herding district. Other permanent members were a reindeer researcher from the Natural Resources Institute (LUKE) and members of The Finnish Association for Nature Conservation, Metsähallitus, and The Central Union of Agricultural Producers and Forest Owners (MTK). During the working process, leaders of each reindeer herding district (54 in total) were invited to present their own herding system, suggest a suitable maximum number for their district and give other feedback.

The MAF decided, in accordance with the working group's recommendations, that the maximum number of reindeer in each district basically remains the same as before, but the allowed maximum number of reindeer owned by a single herder be increased to 500. This maximum number for one herder is already in use in the 20 northernmost districts (specific reindeer herding area including Sámi Homeland) but is now applied to all reindeer herding districts. In addition, districts were asked to prepare Pasture Management Plans for the next period, 2022–2030, aimed at reducing grazing pressure on winter pastures.

The final recommendations of the working group were based on scientific and herders' knowledge. However, the justification for the proposal was mainly based on scientific input about winter pastures. The reason for focusing on the state of winter pastures is based on the Reindeer Husbandry Act. However, a synthesis produced by a research group led by LUKE and funded by the MAF shows that it is problematic to regulate maximum reindeer numbers purely on carrying capacity of winter pastures. Both scientists and herders emphasized that the issue is much more complex than simple carrying capacity for at least three reasons: (1) there is no simple shared idea of carrying capacity (sustainable production capacity of winter pastures); (2) following the law would result in absurd situations (especially in southern districts with widespread supplementary feeding); (3) the law does not recognize the effects of other land uses on winter pastures (and thus is in conflict with laws covering reindeer husbandry). Furthermore, herders were not against "empirical, objective and unbiased" scientific information, and appreciated scientific knowledge (even economic-ecological modelling). However, even though herders were heard in the process, we highlight the need to recognize and include herders' knowledge more fully when specifying reindeer numbers. This is because the economic and social factors of reindeer herding, as well as the effects of other land uses on pastures, need to be considered, along with the status of pastures during other seasons (Pekkarinen et al. 2015). In conclusion, the interactions between different land uses, predation, climate change and herding practices (e.g., supplementary feeding) point to complex relationships between reindeer numbers and pasture conditions, bringing into question the use of reductionist logic to determine appropriate numbers.

Norway: Models and indicators for 'rationalizing' reindeer herding

For more than a century, Norwegian government officials have been concerned about "too many reindeer" and "too many herders", especially in West Finnmark, their largest reindeer herding region. In the official view, too many reindeer will lead to overgrazing and degradation of the tundra, and too many herders will jeopardize the economic viability of reindeer husbandry in the north (LMD 2018). This understanding has its background in the rationalization programme introduced in the 1970s with the objective of transforming Sámi reindeer husbandry into a more economically efficient industry (Paine 1994).

To achieve more effective decision making, the Reindeer Husbandry Act of 2007 gives more responsibility to herders in assessing the number of reindeer that could graze the herding districts' pastures and, in cases where the herd sizes are too large, developing reduction plans. In 2008, the Ministry of Agriculture and Food (LMD) established a working group with the mandate to develop criteria and norms for determining sustainable reindeer numbers. The group, which consisted of herders, scientists and government officials, defined indicators that were 'simple, objective and verifiable', such as carcass weights and meat production volumes. In addition, the group described more subjective indicators based on traditional knowledge of the reindeer's well-being, such as the morphology of the antlers and body and the quality of the coat (LMD 2008a). However, in the final *Guidelines for setting ecologically sustainable reindeer numbers*, no traditional knowledge indicators were included (LMD 2008b).

During 2009 and 2010, the herding districts developed proposals for maximum reindeer numbers and reduction plans. The proposals were submitted to the national RHB, which made final decisions on maximum numbers. The first four proposals were approved at the end of 2010, but the LMD overruled the decisions of the RHB and argued that the proposed maximum reindeer numbers would not ensure sustainable herding practices (Johnsen 2016).

The national reindeer husbandry administration developed new guidelines, and the RHB used these to specify new maximum herd sizes. The herding districts were not aware of the new guidelines, but they observed that the new numbers were not consistent with the criteria and norms identified by the working group. As a consequence of the new level, the herding districts in West Finnmark had to reduce their herds by an average of approximately 30%, a reduction twice as large as that originally proposed by the herding districts (Johnsen 2016).

The final decisions on maximum numbers emphasized mathematical models for calculating correlations between reindeer meat production, animal numbers and densities of reindeer. However, these are models based on simplified and generalized versions of reality and do not include the herders' complex, situated and local knowledge of reindeer and pastures, and may even undermine it. The models do not take local differences such as herding practices, geography and disturbances (e.g., predators or human activities) into account.

Sweden: maximum number of reindeer as a flexible governance tool

In Sweden, the process of deciding the highest number of reindeer remains largely unclear. Decisions should be informed by the ecological status of pastures, consideration of 'other interests' and land users, as well as animal welfare (§15, prop. 1971, p. 51; prop. 1992/93:32, SOU 2001, p. 101). Grazing inventories (at district level) alongside previously documented numbers of reindeer (actual presence and historically determined) are recognized as carrying particular weight. However, updated grazing inventories are often lacking and the methods for inventorying are contested. Furthermore, how 'consideration of other interests' (§15) should actually affect decisions about maximum number of reindeer is unclear in several respects: first, what interests this actually refers to (e.g., those not regulated through other legislation such as forestry) and second, to what degree and how different types of consideration are to be weighed against one another (e.g., SOU 2001; Brännström 2017). The consideration criteria, therefore, seem to carry less weight in practice compared to other criteria when determining reindeer numbers. While it is natural to think of limits to reindeer numbers as a restricting governing instrument, it could potentially function as an instrument to *protect* herding districts vis-à-vis encroachment from other actors and interests. For example, the Swedish legislation specifies that the maximum numbers must not be compromised or lowered as a consequence of the actions of competing land users (see, e.g., the §21 in the Forestry Act). That is, felling trees or developing a mine is not considered sufficient reason to reduce numbers. The potential strength as active protection, however, remains untested. The argument has never been used effectively to stop extensive forest felling (SOU 2001, p. 101; Brännström 2017) and other actors, such as mining companies, seem completely unaware of this logic and have openly suggested reducing reindeer numbers as an "adaptation" to proposed mining interventions (Raitio et al. 2020).

The CABs have the opportunity to enforce maximum numbers by issuing injunctions (*föreläggande*) or fines, should the total number of reindeer within a district repeatedly exceed the limit (§15), but this is only rarely used. For example, in Västerbotten county, only one injunction has been issued since their introduction in 1993 (pers. comm. with CAB official). However, there are large differences over time and regions and the use of injunctions may vary between CABs.

As a recent example, in 2017, Västerbotten county determined new maximum numbers for its seven herding districts. The formal maximum numbers had remained constant for over 70 years (since 1945–1946, set by the so-called Lapp administration). Changes were suggested in 1966 after surveying the status of reindeer grazing areas but were not implemented. The discussion resurfaced again in the 1990s after herd sizes increased.

It was not, however, until 2002 when Västerbotten CAB formally decided to revisit the maximum numbers. A so-called reindeer number group was established, consisting of active herders and elders from some of the herding districts concerned. After consulting the herding districts, the group recommended increasing the maximum numbers for all districts except for one, based largely on the herders' own assessment of available pastures and strategies (Idivuoma 2015). However, due to uncertainties regarding administrative borders and overlapping pastures between districts, the proposal was overruled by the administrative court in 2003, demanding that borders were first settled. The CAB, in the meantime, made an interim decision to go with the highest number of suggested reindeer for each district (whether based on the old ordinance from 1945 or the suggestion by the reindeer number group). The borders were finally settled in 2015 after responsibility for determining them had been transferred from the CABs to the Sámi parliament in 2007 and after another round of appeals (one district had, e.g., not been able to participate fully in the process).

In the meantime, a special investigator, a knowledgeable herder from a different region, was appointed by the CAB to provide input for determining new reindeer numbers (Idivuoma 2015). The new investigation generally took the earlier reindeer number group's suggestion at face value and suggested maintaining or increasing maximum numbers, with one exception. The newly settled borders meant that one district lost important winter pastures, which resulted in a recommendation to substantially reduce the maximum numbers there. The CAB's decision, based on the inquiry's recommendations, was again appealed by two of the districts (both arguing that the districts, not the CAB, should have the decision-making power over numbers). The appeals were turned down and the new limits finally came into force in 2017, 15 years after the process was initiated.

This example, despite regional variations, demonstrates some general characteristics. First, the tool in itself is used infrequently and can involve a lengthy bureaucratic process. Second, the process can be undertaken in different ways with varying degrees of participation and recognition of the district's views and practice-based expertise. It *can* be quite inclusive, as this example suggests, but this is not required and, therefore, rather depends on the CAB in question. Third, while the legal basis for determining maximum numbers comes across as hierarchical and controlling, the actual praxis seems to allow more flexibility and recognizes uncertainties in the legal history, resulting in perceived difficulties in determining fixed limits. The authorities thus approach it as a reference value rather than an absolute limit (SOU 2001). If neighbouring districts or other land users complain, intervention may be warranted. Finally, the example demonstrates how reindeer numbers strongly depend on other factors, in this case, borders and access to winter pastures. Together, this explains why many CABs hesitate about the degree to which the governing tool can actually be used and enforced.

Discussion

Though the cases presented above do not allow a robust comparison of specific aspects, they each highlight themes relevant for the governance of reindeer numbers: (i) participation of herders in decision making, (ii) unequal treatment of herders' practice-based knowledge and knowledge relying on optimization and rationalization using simplified scientific models and indicators and (iii) flexibility of governance.

Participation

In Finland, the suggestion was made 30 years ago (MAF 1988) to decentralize governance and give more decision-making power to the herding districts. However, even today, issues of participation are contested. It is likely that this is due to tensions between the fear by some decision makers and scientists of overgrazing and increased recognition that external factors also impact pasture conditions. The Norwegian case illustrates that even when herders were represented in the working group, there was no guarantee of inclusive and transparent decision making. The case from Sweden illustrates that herders can be involved as experts, owners *and* facilitators of knowledge coproduction, which is quite different from Norway's use of mathematical models of carrying capacity. Nevertheless, in all countries, the perception is that reindeer numbers have to be controlled externally.

Knowledge systems

Decision making for rationalizing reindeer husbandry often prioritizes natural science knowledge over reindeer herders' practice- and experience-based knowledge. In Norway's decision making, scientific knowledge is considered objective and superior to the herders' knowledge, which is seen as subjective and biased (Benjaminsen et al. 2015; Johnsen et al. 2017; Marin et al. 2020). In Finland, the legal focus on carrying capacity of winter pastures requires scientific inventories quantifying their state, especially the condition of lichen grounds. The resulting natural science knowledge has a strong role in discussions about reindeer numbers. However, this is problematic as considerable areas with terrestrial or arboreal lichen have been lost, mainly due to activities like forestry, in most districts, and there is extensive supplementary feeding. Furthermore, recently it has become increasingly clear that defining a single constant 'optimal' maximum number of reindeer is very difficult due to ecological, but also economic, sociocultural and other contextual complexities. This realization has focused increased attention on herders' knowledge to inform the discussions. In the Swedish case, the herders' expertise and knowledge system was given a major role in determining new numbers. However, herders still complain that decision-making power lies in the hands of the CABs and not the districts themselves.

Formal recognition by state-based actors of other actors' impacts on grazing conditions varies between countries. In Finland, such recognition is increasing, in Norway the authorities are starting to acknowledge the need to protect reindeer pastures from encroachment, and in Sweden the formal recognition of influence of other land users on pasture condition is strong, yet tools to deal with this in practice are still lacking.

Flexibility

The cases demonstrate both different degrees of rigidity in governing approaches and opportunities for increased flexibility by strengthening aspects of self-governance. Borders, between countries and herding districts, are of central importance in understanding how reindeer numbers are directly and indirectly decided, and enacted. Drawing and closing national borders have been central in the governing of reindeer husbandry since the late 1800s (Lantto 2000). In the Swedish case, while maximum numbers were considered more flexibly than elsewhere, the perceived need and act of defining fixed borders turned out to be the single most important factor affecting the actual 'flexibility' of maximum numbers. In Finland, it was proposed that reindeer densities could be higher in areas with significant meat production, where the number of reindeer owners compared to the number of reindeer is large and where an increase in reindeer numbers does not compromise carrying capacity. Also in Finland, the proposed Pasture Management Plans are a solution to ensure sustainability in a context where the maximum number of reindeer was not changed compared to the previous numbers, over a 10-year time frame. Herders can choose between two out of five options to enhance sustainable use of pastures: (1) reduce the number of reindeer by 7% from the maximum, (2) early slaughter to reduce pressure on winter pastures, (3) further enhancing pasture rotation practices, (4) summer grazing practices to enhance biodiversity values and (5) increasing collaboration between separate herding districts. In Norway, a key issue concerning flexibility is how the indicators and norms used for setting maximum reindeer numbers are the same irrespective of where the district is located in the country and the types of pastures/ecosystems present.

Table 9.2 Key questions on governing the maximum number of reindeer

- How can participation and flexibility be enhanced in decision making concerning reindeer numbers?
- How can herders' experience-based knowledge and Sámi worldviews be included in the governance of reindeer numbers?
- Is there a need to change laws governing maximum numbers of reindeer to increase flexibility to adapt to changing situations and to enhance opportunities for participation of reindeer herders?
- What are the alternatives to the idea of carrying capacity of (winter) pastures upon which to ground decisions about maximum number of reindeer?
- How can scientific knowledge and herders' knowledge be integrated in a mutually constructive way for developing integrated knowledge systems leading to balanced decisions on maximum number of reindeer?
- How do seasonal migration, pasture rotation, use of supplementary feeding and timing of slaughter affect grazing pressures and thereby need to inform decisions on maximum number of reindeer?
- How can impacts of other land users on pasture condition and availability be recognized when defining maximum number of reindeer?

Conclusion

Our analysis and comparison of the cases highlight issues regarding setting maximum reindeer numbers by state-based governance. In particular, further research is needed on how to enhance participation and flexibility, and to find effective ways to integrate herders' knowledge systems into state-based governance. Importantly, more holistic landscape approaches are needed, taking multiple actors, pressures and alternatives into account. Accordingly, we have formulated a number of questions (Table 9.2) that we advise all involved to consider before embarking on new policy processes determining reindeer numbers.

References

Benjaminsen, T. A., Reinert, H., Sjaastad, E. & Sara, M. N. (2015). Misreading the Arctic landscape: A political ecology of reindeer, carrying capacities, and overstocking in Finnmark, Norway. *Norsk Geografisk Tidsskrift*. 69, 219–229.

Bennett, N. J., & Satterfield, T. (2018). Environmental governance: A practical framework to guide design, evaluation and analysis. *Conservation Letters*. 11, e12600.

Brännström, M. (2017). Skogsbruk och renskötsel på samma mark – En rättsvetenskaplig studie av äganderätten och renskötselrätten. Ph.D. diss., Umeå University.

Hardin, G. (1968). The Tragedy of the Commons. *Science*. 162(3859), 1243–1248.

Heikkinen, H. I., Lakomäki, S. & Baldridge J. (2007). The dimensions of sustainability and the neoentrepreneurial adaptation strategies in reindeer herding in Finland. *The Journal of Ecological Anthropology*. 11(1), 25–42.

Heikkinen, H. (2002). *Sopeutumisen mallit – Poronhoidon adaptaatio jälkiteolliseen toimintaympäristöön Suomen läntisellä poronhoitoalueella 1980–2000*. Ph. D. diss. University of Oulu.

Idivuoma, P.G. (2015). Utredningsförslag högsta renantal samebyarna i Västerbottens län.

Jaakkola, L.M., Heiskanen, M.M., Lensu, A.M. & Kuitunen, M.T. (2013). Consequences of forest landscape changes for the availability of winter pastures to reindeer (Rangifer tarandus tarandus) from 1953 to 2003 in Kuusamo, northeast Finland. *Boreal Environment Research*. 18(6), 459–472.

Johnsen, K.I. (2016). Medbestemmelse, makt og mistillit i reindriftsforvaltningen. In: T. A. Benjaminsen, T. A., Eira I. M. G. & Sara, M. N. (eds.) *Samisk reindrift, norske myter.* Bergen: Fagbokforlaget. 195–220.

Johnsen, K.I., Benjaminsen, T.A. & Eira, I.M.G. (2015). Seeing like the state or like pastoralists? Conflicting narratives on the governance of Sámi reindeer husbandry in Finnmark, Norway. *Norsk Geografisk Tidsskrift – Norwegian Journal of Geography*. 69(4), 230–241.

Johnsen, K.I., Mathiesen, S.D. & Eira, I.M.G. (2017). Sámi reindeer governance in Norway as competing knowledge systems: a participatory study. *Ecology and Society*. 22(4), 33.

Kumpula, J., Kurkilahti, M., Helle, T. & Colpaert, A. (2014). Both reindeer management and several other land use factors explain the reduction in ground lichens (Cladonia spp.) in pastures grazed by semi-domesticated reindeer in Finland. *Regional Environmental Change*. 14(2), 541–559.

Kumpula, J., Stark, S., Holand, Ø. (2011). Seasonal grazing effects by semi-domesticated reindeer on subarctic mountain birch forests. *Polar Biology*. 34(3), 441–453.

Landauer, M., Rasmus, S. & Forbes, B.C. (2021). What drives reindeer management in Finland towards social and ecological tipping points? *Regional Environmental Change*. 21(2), 1–16.

Lantto, P. (2000). *Tiden börjar på nytt. En analys av samernas etnopolitiska mobilisering i Sverige 1900–1950.* Umeå: Kulturgräns norr.

LMD (2008a). *Kriterier/indikatorer på økologisk bærekraftig reintall: Rapport fra arbeidsgruppe opprettet av LMD.* Landbruks- og matdepartementet (Ministry of Agriculture and Food) Retrieved from www.regjeringen.no/globalassets/upload/lmd/vedlegg/brosjyrer_veiledere_rapporter/rapport_okologisk_reintallskriterier.pdf

LMD (2008b). *Veileder for fastsetting av økologisk bærekraftig reintall.* Landbruks- og matdepartementet Retrieved from www.regjeringen.no/globalassets/upload/lmd/vedlegg/brosjyrer_veiledere_rapporter/veileder_fastsetting_okologisk_baerekraftig_reintall_des_2008.pdf

LMD (2018). Økologisk, kulturell og økonomisk bærekraft. Retrieved from www.regjeringen.no/no/tema/mat-fiske-og-landbruk/reindrift/okologisk-kulturell-og-okonomisk-barekraft/id2339776/

Marin, A., Sjaastad, E., Benjaminsen, T.A., Sara, M.N.M. & Borgenvik, E.J.L. (2020). Productivity beyond density: A critique of management models for reindeer pastoralism in Norway. *Pastoralism*. 10, 1–18.

Meld. St. 32 (2016–2017). Reindrift: Lang tradisjon – unike muligheter. vol 5 April 2017. Landbruks- og matdepartementet, Oslo.

Moen, J. & Danell, Ö. (2003). Reindeer in the Swedish mountains: An assessment of grazing impacts. *Ambio*. 32(6), 397–402.

Olofsson, J., Kitti, H., Rautainen, P., Stark, S. & Oksanen, L. (2001). Effects of summer grazing by reindeer on composition of vegetation, productivity and nitrogen cycling. *Ecography*. 24(1), 13–24.

Paine, R. (1994). *Herds of the tundra: A portrait of Saami reindeer pastoralism.* Washingtion, DC: Smithsonian Institution Press.

Pape, R. & Löffler, J. (2012). Climate change, land use conflicts, predation and ecological degradation as challenges for reindeer husbandry in northern Europe: what do we really know after half a century of research? *Ambio.* 41(5), 421–434.

Pekkarinen, A.-J., Kumpula, J. & Tahvonen, O. (2015). Reindeer management and winter pastures in the presence of supplementary feeding and government subsidies. *Ecological Modelling,* 312, 256–271.

Raitio, K., Allard, C., Lawrence, R. (2020). Mineral extraction in Swedish Sápmi: The regulatory gap between Sami rights and Sweden's mining permitting practices. *Land Use Policy,* 99, 105001.

Reindeer Husbandry Act 848/1990; 21 §. www.finlex.fi/fi/laki/kaannokset/1990/en19900848.pdf

Sara, M.N., Eira, I.M.G., Bjørklund, I. & Oskal, A. (2016). "Hvordan skal vi forstå reintall?" In: Benjaminsen, T.A., Eira, I.M.G. & Sara, M.N. (eds.) *Samisk reindrift, norske myter.* Bergen: Fagbokforlaget. 51–68.

Sarkki, S., Heikkinen, H.I. & Karjalainen, T. P. (2013). Sensitivity in transdisciplinary projects: Case of reindeer management in northern Finland. *Land Use Policy.* 34, 183–192.

Sarkki, S., Heikkinen, H.I. & Löf, A. (2021). Reindeer herders as stakeholders or right-holders? Introducing a social equity-based conceptualization relevant for indigenous and local communities. In: Nord D. (ed) *Nordic Perspectives on the Responsible Development of the Arctic: Pathways to Action.* Springer Polar Sciences. Cambridge: Springer. 271–292.

SFS 1971:437. Rennäringslagen [Swedish Reindeer Husbandry Act].

SOU (2001). *Förhållandet mellan samebyarnas och andras markanvändning.* Stockholm.

Swedish Government Proposition 1995/96:226. Hållbar utveckling i landets fjällområden [Sustainable development in the mountain areas].

Tahvonen, O., Kumpula, J. & Pekkarinen, A.-J. (2014). Optimal harvesting of an age structured, two-sex herbivore-plant system. *Ecological Modelling.* 272, 348–361.

Part IV

Challenges for productivity, health and adaptation of reindeer

10 The productive herd

Past, present and perspectives

Øystein Holand, Asko Mäki-Tanila, Thomas Kvalnes, Kirsi Muuttoranta, Amélie Paoli, Jaakko Pietarinen, Robert B. Weladji and Birgitta Åhman

Introduction

The essence of reindeer pastoralism is the conversion of natural resources into animal products to satisfy human needs. However, the ecological niche of reindeer pastoralists is hard to manipulate. To meet production goals in the short and long term and secure the animals' survival, adjusting herd size and composition to ecological conditions become essential. These goals are dynamic and must be placed in a social-ecological context (e.g., Paine 1994; Næss 2020). Obviously, the transition from traditional pastoralism to present-day market-oriented meat production has created tensions between and within herders' communities. The authorities' push for rationalization by use of economic incentives and regulations created frustration among the herders (e.g., Hausner et al. 2011; Riseth et al. 2019) and is still ongoing (Chapter 9). It is therefore imperative to investigate how herders traditionally managed and selected animals to meet their production goals and to explore the transition to the current market-oriented meat production and its implication for herd size and composition, as well as slaughter regimes and selection.

Herd dynamics and production depend on an array of abiotic and biotic factors: quality and seasonal balance of pastures, climatic conditions, anthropogenic disturbances and predator pressure, as well as management practices. Animal density and herd composition are influenced by slaughter and selection practices (Danell & Petersson 1994). Even the appropriate measures to describe a herd's (meat) productivity can be disputed (Marin et al. 2020). Indeed, ecological and social factors affecting production are often overlooked. Most of the production-oriented research has focused on variation of productivity within a country (Lundqvist 2007; Tveraa et al. 2007; Riseth 2009; Næss 2020; Marin et al. 2020, but see Helle & Kojola 2008). Obviously, an analysis of differences in meat productivity both between and within countries is due.

A warmer and wetter climate in Fennoscandia may have detrimental effects on reindeer well-being. Increased rain-on-snow events, which make the ground vegetation layer inaccessible (e.g. Tyler et al. 2021), may drain the animals' body reserves. Further, a mismatch between the peak of resource demands by

DOI: 10.4324/9781003118565-15

reproducing animals and the peak of forage availability has been suggested in *Rangifer* (Post & Forchhammer 2008). The relative role of phenotypic plasticity and microevolutionary response to a warmer and wetter environment with frequent fluctuations remains largely unquantified in semi-domesticated reindeer (but see Holand et al. 2020). We, therefore, end this chapter by discussing challenges and strategies for maintaining viable and productive herds in a changing north.

Past

In the early phase of reindeer pastoralism, reindeer supplied an array of products from both live and slaughtered animals. However, in the 1800s, these small multipurpose herds evolved in many regions into large herds with seasonal migrations, with a focus on meat production. This became the main form of pastoralism among Sámi herders in Fennoscandia (Chapter 1). Transportation of household equipment between seasonal dwelling sites became essential. Before entering the transportation squad, selected males had to be trained and finally castrated at 3–4 years old. The traditional Sámi strategy was to keep reproductive females in good condition and predominantly slaughter adult males, including castrates. The proportion of males in the herd could reach 50% (Skuncke 1964).

The Sámi herders' rationality

The herd represents the owners' and their household's investment for their future. The herders' rationale was to withstand unpredictable events by keeping large and robust herds (Ingold 1980). This was achieved by having a herd that contained a high proportion of the age classes with low mortality. The drive for a larger herd often led to overaccumulation across consecutive good years. This was followed by crashes in reindeer numbers as the pastures were not able to permanently bear high stocking rates. Such crashes were caused by extreme winter conditions resulting in starvation often amplified by disease outbreaks (Päiviö 2006). The fluctuations in herd size were also modified by social interactions between herders, as movements to mitigate poor range conditions relied on reciprocal access to pastures and mutual agreements among neighbouring herding groups (Chapter 7).

Viable and productive herds depend on sufficient pasture resources, escape habitats and functional corridors during all seasons. Likewise, a diverse and functional herd adapted to a fluctuating environment is a prerequisite for using the natural resources optimally and thus confers resilience to the herd, inherent in the "beautiful herd" (*čappá eallu*) concept (Magga 2006).

An optimal herd composition was, and still is, the foremost goal in herd management (Oskal 1999). The owners' success depends on their ability to assess the performance of individual reindeer as well as the performance of their herd and the herds of other members of their herding group. In addition to favouring specific phenotypic traits such as large-bodied animals, selecting for specific behaviours, such as maternal care or following behaviour contributed to the domestication process.

New ideas enter the scene

As early as the late 1930s, Soviet researchers (e.g., Dobrotvorsky 1938) started to advocate for changes in herd composition to increase meat production. They suggested restructuring herds towards a higher proportion of reproductive females and introduced the slaughtering of calves. These ideas entered Fennoscandia in the late 1950s. Varo (1964; 1972) analysed phenotypic and genetic variation in important production traits and promoted calf slaughter to utilize the high growth potential of the juvenile segment of the population. This new practice was also advocated by Norwegian and Swedish researchers (e.g., Skuncke 1964; Skjenneberg & Slagsvold 1968). However, without the transportation revolution brought about by the availability of snowmobiles in the 1960s, the new herd composition and slaughter regime would have been hard to implement.

Herd investigation and transformation

In Norway, the transformation towards new herd composition aimed at maximizing production started among the Sámi in the South. In the 1960s, the herders in several *sijtes* agreed on internal resource distribution, including a maximum number of animals per household (Fjellheim 1999). This was possible given the small number of herders and enabled them to join forces to withstand external pressures (Næss 2020). The aim was to develop a herd structure and slaughter strategy that combined tradition-based practices with modern production theory.

Reproduction rates, as well as the timing of conception, were investigated and related to female slaughter weight and male age structure during the rut. Female body mass was identified as a quantitative and objective proxy for reproductive performance (Figure 10.1). Lenvik and Fjellheim (1988) documented a high correlation for autumn body mass between calves and yearlings. Therefore, phenotypic selection based on female calves' autumn body mass would improve their body mass at 1.5 years of age. The female's pre-rut body mass at 1.5 years should be 60 kg to give birth to viable calves in the following spring (Lenvik 1988, Figure 10.1). Lenvik argued that a male:female ratio of 1:10 and the use of 1.5 years old males with a body mass larger than 60 kg during rut would secure full conception. Realizing the animals' growth potential required an appropriate pasture balance (Lenvik 1988). Where such conditions were met, prime age females reached a spring body mass of around 70 kg and reproduced each year with minimal calf mortality (Figure 10.1). Females that were too large were regarded as disadvantageous as they required more energy, and hence more winter forage without any production gain.

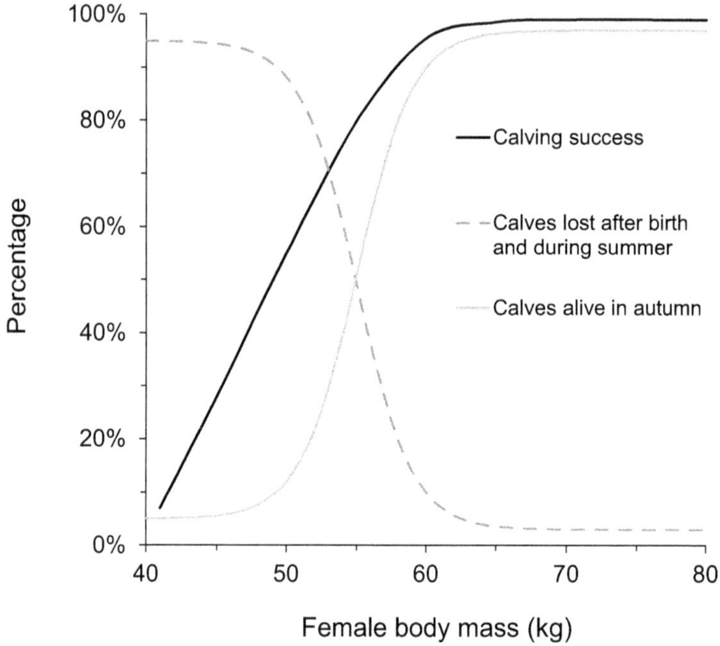

Figure 10.1 Female calving success and mortality of their calves in relation to the females' own spring body mass (redrawn after Lenvik (1990)).

Changes in the herd's sex ratio and age structure took place by trial and error during the 1970s and early 1980s in many Sámi districts in southern Norway (Figure 10.2). Selection improved animal condition and meant that most 1.5-year-old females were able to reproduce regularly for about 10 years. The proportion of reproducing females increased in the winter herd, accompanied by higher culling (phenotypic selection) of female calves. Among the male calves, the selection was even stronger. Around 80% of the male calves were slaughtered in autumn and early winter. The transformation was fuelled by a calf slaughter bonus since 1977, agreed annually by the government and the Norwegian herder's association (NRL). As a result, the strategy started to spread to northern Norway.

Longitudinal data generate new knowledge

Ruvhten Sijte (Jämtland, Sweden) started an individual recording programme in the early 1980s, which is still running. Petersson and Danell (1993) demonstrated that the body mass of calves in autumn had a substantial genetic component and therefore phenotypic selection would yield an appreciable response. However, they underlined that the long-term advantages from such

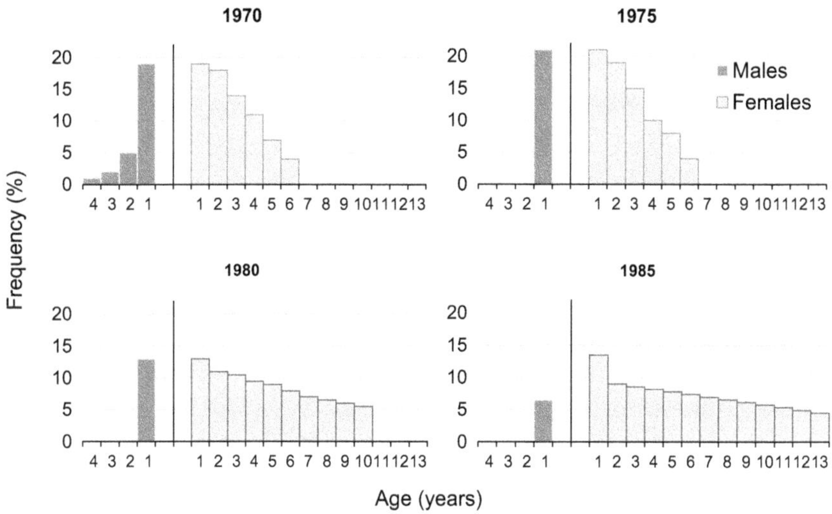

Figure 10.2 The gradual change in winter herd sex and age composition during the 1970s and early 1980s in Gåebrien sijte (redrawn after Lenvik (1990)).

a selection regime should be further dissected in relation to longevity, lifetime reproductive success and maternal traits. Rönnegård and Danell (2003) used an expansion of the same data set and found an increase of 0.35 kg in body mass of calves in autumns after 11 years of selection. By comparing the rate of inbreeding, including the maternal effect in their analyses, Rönnegård et al. (2003) concluded that there is no risk of losing genetic variation if the herd size exceeds 2,000 animals.

Based on detailed life history data collected since 1969 from the Finnish Reindeer Herders' Association experimental reindeer herd at Kutuharju (northern Finland), Eloranta and Nieminen (1986) confirmed the effects of maternal age and female body mass on calf birth weight and survival. Weladji et al. (2008) explored the reproduction costs of females in the Kutuharju herd. Comparing age-specific survival and reproduction between four reproductive states, they found no cost of gestation and lactation in terms of future reproduction and survival. Females that successfully raised their first calves early remained successful throughout their life, suggesting the existence of substantial variation in female maternal quality (Weladji et al. 2006). However, supplementary feeding during winter, first implemented in 1986, might conceal reproductive costs.

In the same herd, Muuttoranta et al. (2013; 2014) investigated the genetic variation in traits related to calf growth. Genetic variation was substantial in the traits related to birth weight and even birth date (see also Holand et al. 2020). They also found that the female's body mass is inversely related to her maternal care. However, there is too little data to investigate the variation in traits prone to non-genetic variation and for understanding the consequences on fitness traits of artificial selection for meat production.

Herders' selection practice

Muuttoranta and Mäki-Tanila (2011) surveyed the attitudes of the Finnish and Sámi leaders of the herding districts in Finland towards the use of selection to improve production. The leaders considered selection and optimization of herd composition to be the most important management operations.

Calf slaughter is a common strategy among the herders to modify the herds' age distribution and sex ratio for higher productivity and profitability. The leaders interviewed were aware of female age affecting calf body mass and survival. During autumn slaughter, the most vigorous calves are left alive to increase the vigour of the herd and future generations. The main selection criteria in Finland are calf health, vigour, body size and muscularity. The calf's temperament had only a small effect on selection. The leaders indicated an inverse relationship between an individual's own and the mother's effects on calf growth, as reported by Muuttoranta et al. (2014).

About half of the districts marked all their breeding animals individually. One-third had an individual bookkeeping system. Indeed, it is essential that the pedigree information can be related to phenotype measurement, such as calves' autumn body mass. Muuttoranta and Mäki-Tanila (2011) reported that calves are never weighed at the autumn round-up, and selection is based on visual assessments. This would obviously compromise quantification of the selection efficiency although it is most practical in terms of time requirements and infrastructure.

Production

Herd productivity depends on factors such as quality and seasonal balance of pastures, climatic conditions, anthropogenic disturbances and predator pressure, as well as management practices; animal density and herd composition are influenced by slaughter and selection practices (Figure 10.3). The net profit of production is a function of production costs and revenues (Chapter 11). While costs are not directly related to herd size, revenues depend on meat production volume, sale of by-products (e.g., skins and antlers) and market conditions. Indeed, product prices vary seasonally, between age classes and with quality.

Herd dynamics and meat production measures

Primary production of forage resources suitable for reindeer grazing depends on biotic and abiotic conditions, with high spatial and temporal variations. Furthermore, stochastic weather and climate events (in particular snow conditions) have a strong impact on the population dynamics of reindeer (Helle & Kojola 2008; Bårdsen et al. 2017). This implies that a dynamic stocking rate,

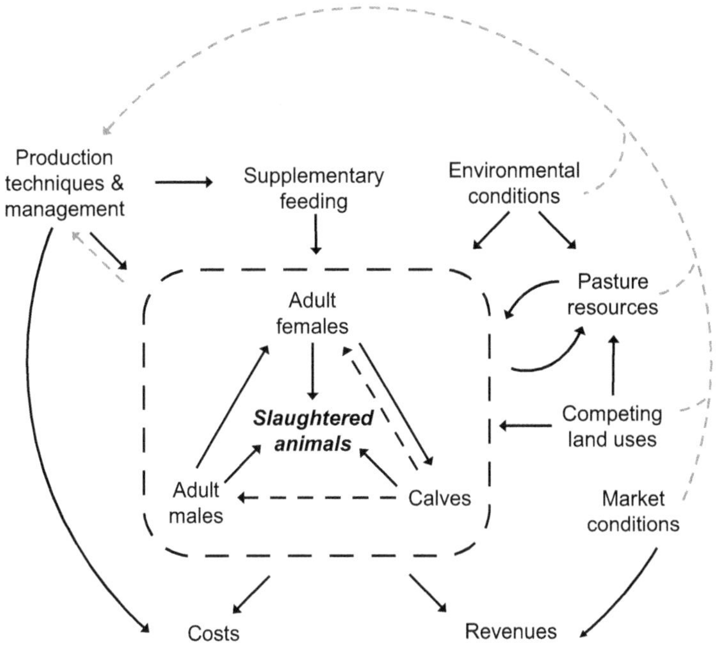

Figure 10.3 Herd dynamics and production depend on herd composition and slaughter strategy, indicated by links between animal classes within the dashed box; these are influenced by an array of external factors and natural conditions. Outside the box, solid arrows represent direct effects whereas dashed lines indicate possible influences (redrawn after Danell and Petersson (1994)).

herd composition and harvesting approach would be appropriate to respond to density-dependent and independent factors that can influence population dynamics (Bårdsen et al. 2017). Given a highly female skewed winter herd, the stochastic simulation predicts that maximum annual meat production is reached when around 70% of the calves are slaughtered (Bårdsen et al. 2017). However, such herds have high growth potential and will run into food shortages if not harvested appropriately. An optimal harvest also depends on its timing, to utilize the high summer growth rate and to minimize mortality during winter. If the availability of winter grazing resources is limited, the slaughter should take place before the herd enters the winter pasture.

The females' risk-averse life history strategy (Bårdsen et al. 2008) is essential to understand the herd dynamics, which are primarily driven by female body mass and condition. The changes in a female's condition are reflected in the performance of her offspring (Rönnegård et al. 2003). Indeed, the autumn calf carcass weights and the herd's autumn female:calf ratio are two measurable proxies for females' condition. Further, these two parameters indicate the herd's welfare status and ability to withstand extreme winter events.

The growing integration of reindeer pastoralism into the market economy in the latter half of the twentieth century rationalized production and meant that most of it passed through approved slaughterhouses. In Norway, Sweden and Finland, the recording of slaughter data for carcass weight, sex, age class and carcass quality was established. Thus, reliable and systematic data have been available since the late twentieth century in these countries. The summed statistics do not fully reflect the total production, leaving out some privately sold meat and subsistence use. However, the proportion moved through official channels is high, estimated to average around 90%, because many of the subsidies are connected to official slaughter records. The slaughterhouse data, combined with information on herd size and size of pasture area, allow several aspects of the production to be quantified and analysed at different spatial and temporal scales: total production, production per animal in the winter herd after slaughter or per pasture area.

Herd adjustment

Given sufficient pasture resources, almost all adult females give birth to a calf every year. However, climatic variation has a strong influence on the females' productivity and the growth of their calves. Production per animal in the winter herd, therefore, reflects the winter herd composition and slaughter strategy. Herd production per pasture area can be used to adjust herd size to available grazing resources. Both measures are proxies and must be adapted locally. Adapting herd size to available grazing resources requires careful monitoring of the state of pastures. In particular, terrestrial lichens, the principal winter grazing resources, are very sensitive to rapid increases in herd size and are thus a bottleneck for sustainable herd size. In parts of Finnmark, a strong inverse relationship between lichen biomass and reindeer herd size has been shown (Tømmervik et al. 2014). In several herding districts in Finland where there is no pronounced separation between summer and winter grazing areas, trampling of the unprotected lichen cover during summer has strongly decreased its availability as a winter grazing resource (Kumpula et al. 2014). Navigating the balance between-herd size and forage availability, therefore, requires decisions regarding herding and slaughter strategy, as well as herd composition at the level of herding units. This is particularly the case as strategies regarding herd composition and slaughter by individual reindeer owners may vary, thus influencing the inter-owner productivity (Weladji et al. 2002).

Between-herd comparison

Comparisons can be made at different spatial scales, such as herding district, region and country level. Variation in productivity may be due to pasture quality and size of the seasonal pastures, herd size, climate and herd management, including herding, driving and round-up practices. Comparisons, therefore, need to be based on the production per animal, or pasture area. Potential effects of encroachment on pastures and disturbance by other forms of land use must also be considered (Chapter 4). This may reduce the usable pasture area, and if the herd is not reduced proportionally, the stocking rate will increase. Disturbance may also influence forage acquisition negatively and hence growth. We acknowledge that overgrazing and overstocking are debated concepts (Mysterud 2006) and are often misinterpreted and taken out of context. Indeed, these concepts and the regulation enforcement practised have to be put into a political ecology context as discussed in Chapter 9.

Production in the three countries – similarities and differences

In the following, we present statistics from the official national records: the Finnish Reindeer Herders' Association reports (Paliskuntain yhdistys) 2010–2020, the Norwegian Government reports (Landbruksdirektoratet) 2010–2020 and the Swedish Sámi Parliament database.

The average annual total production in the last decade has been highest in Finland, followed by Norway and then Sweden (Table 10.1). Finland stands out even more when measured as production per km^2 and the production per animal in the winter herd. The high productivity in Finland cannot be explained by range quality, slaughter practice or winter herd composition, which is rather similar at the country level (Table 10.1). We suggest it is primarily related to the extensive use of supplementary feeding during winter in Finland, which secures high female fertility and calf survival. Nieminen (2010) estimated that, in Finland, ca. 100 kg of pellets were fed per reindeer each winter. In addition, he estimated that the amount of roughage fed was similar in energy content but considerably higher in weight per animal. In Sweden and Norway, supplementary feeding is not currently a regular practice to the same degree as in Finland. However, the use of this practice has increased in both countries in the last decade (Chapter 12).

Table 10.1 Yearly average (2010–2020) production statistics for Finland, Norway and Sweden Coefficient of variation (%) in brackets. Herd composition during winter and slaughter composition are divided into male/female/calf (M/F/Calf)

Country	*Av. tot. prod. (tons)*	*Av. prod (kg) per animal in winter herd*	*Av. prod. (kg)/km^2 available pasture*	*Winter herd composition (%) M/F/Calf*	*Slaughter composition (%) M/F/Calf*
Finland	1957 (10%)	10.2 (11%)	17.4 (10%)	7/78/15	8/16/76
Norway	1526 (10%)	6.8 (12%)	10.5 (10%)	6/78/16	11/10/79
Sweden	1366 (9%)	5.5 (9%)	6.0 (9%)	8/68/24	10/20/70

Sources: Landbruksdirektoratet (2021); Paliskuntain yhdistys (2010–2020); Swedish Sámi Parliament (2021).

Higher productivity in Norway compared to Sweden (per animal in the winter herd) may partly be explained by the higher proportion of females in the winter herd and the higher percentage of slaughtered calves (Table 10.1). Further, the lower production per area in Sweden may be attributed to a larger proportion of the potential pasture area not being available. Indeed, estimates of the Swedish total pasture area vary. We have used Sandström's (2015) estimate of 226,000 km^2.

Considerably larger populations of predators, compared to the other two countries (Chapter 6), are another potential explanation for low productivity in Sweden, but see Bårdsen et al. (2017). According to Hobbs et al. (2014), every reproduction of lynx or wolverine may reduce the annual harvest by, on average, around 100 reindeer. With the present Swedish reindeer populations, this corresponds to 20,000–25,000 reindeer and will obviously impact the population dynamics and slaughter off-take. The impact due to bears is unknown, but predation of neonatal calves may be high when bears are present in the calving area (Sivertsen 2017).

In all three countries, the total production of the winter population has varied considerably over the last hundred years (Chapter 1). Since 2012, the winter populations of the three countries have been rather stable, as has the total production (Figure 10.4). This may indicate an appropriate stocking rate in all three countries combined with no extreme winter events in the last decade. However, winter 2020 was difficult in many parts of northern Fennoscandia and resulted in extensive emergency feeding. Nevertheless, losses to starvation

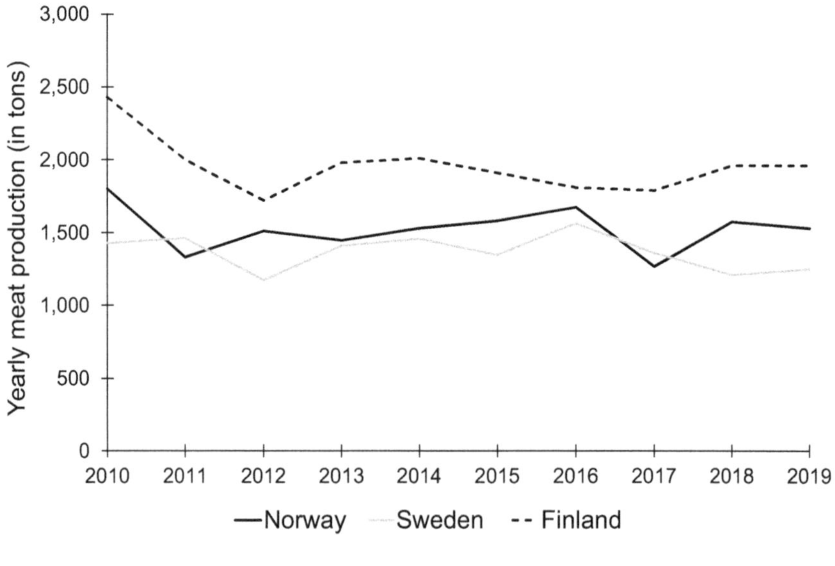

Figure 10.4 Yearly total meat production (in tons) in Finland, Norway and Sweden passing through official slaughterhouses in the last decade.

were high and preliminary data from Norway suggest a pronounced decrease in the number of animals slaughtered in autumn 2020.

Variation within the countries

In all three countries, the herding regions mostly follow administrative borders, meaning that the inter-regional comparison, based on official statistics, should be interpreted with caution, because they may hide larger within-region variation in natural conditions as well as management practices.

In Finland, the reindeer herding area is divided into three main regions (Chapter 1). The average production per animal in the winter herd during the last decade has been rather similar across these regions, probably reflecting the almost identical winter herd composition and slaughter strategy (Table 10.2) as well as feeding practices. The lower productivity per pasture area in the southern part seems partly related to intensive forestry practices and disturbance. In Finland, the southern districts have considerably less variation in meat production between years, most likely due to higher supplementary feeding.

In Norway, the differences in productivity between regions are pronounced (Table 10.3). Herd composition and slaughter strategy are similar, probably due to high calf slaughter subsidies, and cannot explain the variation. The proportion of slaughtered animals for home consumption and local sale and thus outside the official channels seems to be highest in the north and may explain some of the variations. The production per animal in the winter herd is highest in the southern regions, where there is also lower between-year variation (Table 10.3).

Table 10.2 Yearly average (2010–2020) production statistics for different reindeer herding regions in Finland. Coefficient of variation (%) in brackets. Average winter herd composition divided into percentage of male/female/calf (M/F/Calf) and average slaughter composition (M/F/Calf) for the same period are also given.

Main regions	*Av. tot. prod. (tons)*	*Av. prod. (kg) per winter head*	*Av. prod. (kg)/ km² available pasture*	*Winter herd comp. (%) M/F/ Calf*	*Slaughter comp. (%) M/F/Calf*
Sámi homeland area	845 (18%)	10.8 (17%)	17.8 (18)	7/80/13	10/17/73
Special reindeer herding area excluding the Sámi homeland	356 (14%)	8.7 (20%)	18.9 (14%)	8/76/16	7/17/76
Southern part of the reindeer herding area	755 (8%)	10.0 (9%)	13.3 (8%)	7/78/15	7/16/77

Source: Paliskuntain yhdistys (2010–2020).

Table 10.3 Yearly average (2010–2020) production statistics for different reindeer herding regions in Norway. Coefficient of variation (%) in brackets. Average winter herd composition divided into percentage of male/female/calf (M/F/C) and average slaughter composition for the same period are also given.

Regions	*Av. tot. prod. (tons)*	*Av. prod. (kg) per winter head*	*Av. prod. (kg)/ km² available pasture*	*Winter herd comp. (%) M/F/ Calf*	*Slaughter comp. (%) M/F/ Calf*
Øst-Finnmark	500 (22%)	6.9 (23%)	16.3 (22%)	6/78/16	9/10/81
Vest-Finnmark	456 (24%)	5.1 (23%)	17.2 (24%)	6/78/16	8/11/81
Troms	37 (24%)	3.1 (24%)	2.0 (24%)	10/71/19	17/12/71
Nordland	58 (10%)	4.1 (12%)	1.8 (10%)	10/70/20	19/15/66
Nord-Trøndelag	107 (15%)	10.8 (14%)	4.8 (15%)	6/76/18	16/15/69
Sør-Trøndelag/ Hedmark	166 (12%)	13.1 (11%)	19.3 (12%)	4/77/19	14/12/74
Non-Sámi	213 (7%)	16.9 (6%)	34.7 (7%)	5/73/22	18/14/68

Source: Landbruksdirektoratet (2021).

In the south, the herds have been kept rather stable based on internal regulation and high production output since the late 1970s. A paradox is that here the disturbances and fragmentations are considerable. For production per area, the non-Sámi region stands out, producing almost 35 kg per km^2. The Sámi southern (Sør-Trøndelag/Hedmark) and northern (Øst- and Vest-Finnmark) regions produce 16–19 kg per km^2. The considerable differences between South and North in production per animal in the winter herd suggest that the production potential in Finnmark is high. Indeed, some districts in Øst-Finnmark match the per winter head productivity of the non-Sámi area. Næss (2020) argued that internal competition and lack of trust, within and between districts, have contributed to what he calls an "Assurance game" resulting in herd accumulation and size fluctuation in parts of Finnmark. The new regulations implemented in the early 2010s seem to have stabilized herd size but have induced mistrust between herders and the authorities and increased internal conflicts (Chapter 9). In Troms and Nordland, the productivity is low (Table 10.3). Tveraa et al. (2007) argued that the shortage of winter pastures in these two regions limits production, even though the summer pastures are excellent. The high predator pressure may also contribute to the low output.

The variation in productivity between the reindeer herding regions in Sweden is pronounced (Table 10.4). The low productivity in the mountain reindeer herding districts (RHDs) in northern Norrbotten can be partly explained by a large number of herders having few reindeer and a large personal outtake, not accounted for in the official statistics. The same is true for the concession RHDs. The highest productivity measured per winter head and per km^2 pasture area is in the mountain RHDs in southern Jämtland (Table 10.4) and may be partly explained by the high female percentage in

Table 10.4 Yearly average (2010–2019) production statistics for different reindeer herding regions (RHDs) in Sweden. Coefficient of variation (%) in brackets. Average winter herd composition divided into percentage of male/female/calf (M/F/Calf) and average slaughter composition (M/F/Calf) for the same period are also given.

Regions	*Av. tot. prod. (tons)*	*Av. prod. (kg) per winter head*	*Av. prod. (kg)/ km² pasture*	*Winter herd comp. (%) M/F/ Calf*	*Slaughter comp. (%) M/F/ Calf*
Mountain RHDs					
Norrbotten county North	166 (22%)	2.2 (21%)	4.3 (22%)	9/63/28	11/14/75
Norrbotten county South	264 (16%)	5.9 (16%)	6.8 (16%)	9/68/23	14/25/61
Västerbotten county	302 (14%)	6.6 (17%)	6.3 (14%)	9/68/23	13/22/65
Jämtland county North	106 (27%)	5.3 (26%)	2.6 (27%)	7/69/24	12/18/70
Jämtland county South	313 (17%)	12.5 (16%)	16.4 (17%)	6/77/17	8/16/76
Forest RHDs	174 (15%)	5.8 (14%)	6.4 (15%)	8/70/22	5/21/74
Concession RHDs	42 (19%)	3.7 (16%)	3.1 (19%)	8/69/23	11/17/72

Source: Swedish Sámi Parliament (2021).

the winter herd. Lundqvist (2007) argued that different animal densities and the length of the growing season contribute to the variation. Supplementary feeding seems not to be the main explanation for differences between regions, as feeding is least common in the southernmost region. As mentioned previously, predator pressure has been shown to affect slaughter volume (Hobbs et al. 2012) and may therefore be an important explanation for differences in productivity between regions.

Perspectives

Building on the knowledge and practice from the past and present relating to herd dynamics and productivity, we discuss how climate change and globalization may affect the viability of reindeer pastoralism from a production point of view.

Climate change

Rapid climate change may threaten the stability and functioning of Arctic ecosystems. As the Arctic is warming, it has been widely observed that shrubs expand their distribution, abundance and size in the hitherto treeless areas, contributing to regional warming due to increased absorption of solar radiation and other ecosystem effects (Verma et al. 2020). High summer grazing pressure may slow down "shrubification" (Verma et al. 2020) and hence prevent a decreased albedo. However, a too high stocking rate may reduce the production output per winter head. This may increase the enteric methane emission per kg meat produced and hence the herds' CO_2 footprint.

There is a lack of knowledge about the ecological and evolutionary adaptations of large northern herbivores to a changing and fluctuating environment. Climate change is anticipated to increase winter precipitation in the form of snow, at least in the mid-term, but also lead to increased risk of rain-on-snow events which render the ground vegetation layer inaccessible (e.g., Tyler et al. 2021). This will reinforce reliance on supplementary winter feed to prevent starvation (Horstkotte et al. 2020). However, extensive winter feeding may change the animals' natural foraging behaviour and weaken their forage acquisition skills. More frequent occurrence of ice crusts may also have implications for herd size and age and sex composition as strong animals in good condition (including males and castrates) are better able to break ice layers. The effects of fragmentation of ranges and habitat loss, expressed through disturbance of their foraging and movement pattern and habitat use (Chapter 4), amplify these negative consequences. Positive effects can also be anticipated. Climate warming and increased precipitation will prolong the growing season, reducing the length of the winter. This may improve the animals' body condition (Weladji & Holand 2006; Tveraa et al. 2013) and hence their ability to cope with harsh winter conditions. However, during extremely warm events reindeer become "heat trapped", jeopardizing their heat balance. They must allocate time and energy to thermoregulate and hence are not able to fully realize their growth potential.

Parturition and mating behaviour of reindeer have evolved along with their migratory behaviour, feeding specialization and social structure of large and mobile herds. Ultimately, the mating synchrony reflects their need to time births according to the onset of vegetation green-up during spring for optimal survival of young. Due to climate change, there is increasing concern regarding the potential for a mismatch between the peak of resource demands by reproducing animals and the peak of forage availability that individuals rely on to ensure the survival of their young. However, testing the so-called Match/Mismatch Hypothesis has yielded contradictory results in *Rangifer* (Post & Forchhammer 2008; Tveraa et al. 2013; Paoli et al. 2019).

Using the Kutuharju life history dataset, Paoli et al. (2018) found that climate change may affect reindeer reproductive phenology. The calving season advanced by approximately seven days in the period from 1970 to 2015 (see also Holand et al. 2020). Advanced birth dates were correlated with lower precipitation, mainly in the form of snow and a reduced snow cover in April and warmer temperatures in April–May. No increasing mismatch between parturition time and the earlier emergence of spring was found, suggesting the reindeer to be a highly plastic species (Paoli et al. 2019). However, simultaneously the calving synchrony in this herd has weakened, indicating that the climatic trend also affects the variation in females' calving date.

The relative role of plasticity and microevolutionary change remains largely unquantified in reindeer. However, Holand et al. (2020) estimated breeding values and showed that earlier parturition in the Kutuharju herd has a substantial micro-evolution component. They also found directional and stabilizing selection towards a combination of earlier birth date and heavier birth mass, with an intermediate optimum, and that these traits have a negative genetic correlation. This indicates that in reindeer there is an optimal trade-off between birth weight and birth date as the environment changes. A stochastic environment induces a selection pressure for highly plastic phenotypes, but as environmental changes progress plasticity may not be sufficient or too costly and there will be selection for genetic changes. This emphasizes the importance of maintaining genetic and phenotypic variation and has implications for selection of animals for production.

Advances in selection schemes

To achieve and maintain an optimal herd composition requires systematic slaughter, which automatically leads to changes in the herds' genetic composition. The herders implement balancing selection based on calf growth and reindeer mother's capacity to care for the calf. The coordinated development and utilization of genetic variation are, to a large extent, overlooked in domestic reindeer management. In particular, little is known about the consequences of selection. Animal breeders have developed models to answer such questions (Willham 1963). We have developed a method to find out what kind and how much data is needed to analyse the outcome and consequences of selection in reindeer herds (Pietarinen et al. 2018; Pietarinen & Mäki-Tanila 2020). The method allows the comparison of cases based on genetic parameters, information sources and selection intensity.

Let us assume that, for a selected trait in reindeer, there is a genetic component in the phenotypic variation for both direct and maternal effects, and the effects are inversely related.

We will base the analysis on the information available for a reindeer calf during the autumn slaughter, when the slaughter decision is made. If the selection is based only on the calf's weight, the change in maternal effect would be negative. We need additional data on the females' performance to improve the responses of the selection strategy. Because reindeer are uniparous and, over their whole life, produce few progenies, we need information on the relatives and, in particular, on the sire's relatives (Figure 10.5). Even if there is an inverse relationship between direct and maternal effects, with suitable information about relatives, it is possible to improve both effects simultaneously.

As reindeer production occurs in a harsh and fluctuating environment, ongoing natural selection needs to be accounted for in the herders' long-term selection for optimized production. Extensive data bodies on traits like survival and reproduction are needed to assess the relationship between production and fitness traits. Natural selection among male reindeer can be very intensive, strongly affecting the outcome of artificial selection by the herders. To maintain the genetic variation within the herd, the minimum number of breeding males has to be 50–100. With 10% selection among male candidates,

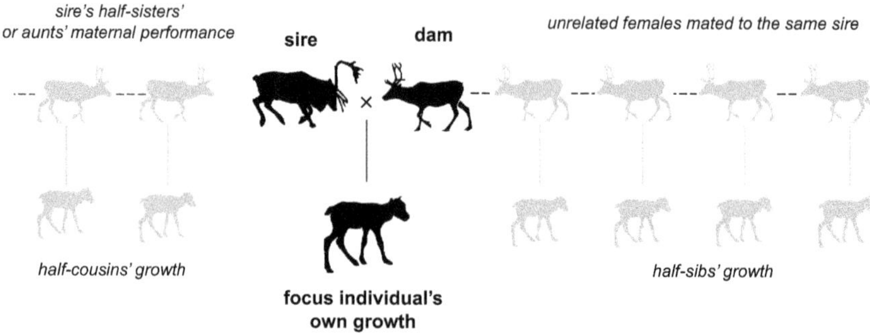

Figure 10.5 The possible core type of collateral relatives to be used in evaluating reindeer calves for maternally affected traits.

the optimal number of selection candidates is 1,000–2,000 with the other half being females (Rönnegård et al. 2003).

New technology – implications

The snowmobile revolution in the 1960s changed the mode of production. Lately, GPS collars and drones have been introduced to ease daily herding and surveillance. Virtual fencing technology has the potential to improve herding of free-ranging animals (Campbell et al. 2021). Such remote real-time control of the herd may mitigate conflicts with other land users and reduce predator losses but may decouple the close association between herders and their animals. On the other hand, supplementary feeding during winter will strengthen that association. However, extensive feeding will have consequences for breeding, as survival traits adapted to extreme environmental conditions will be deemphasized. Animal breeding now relies more and more on genomic information. Genomic selection would require a cheap and dense panel of DNA markers and an extensive number of genotyped and phenotyped individuals which could be read by ID chips. Building the marker panels is feasible as the reindeer genome has been sequenced (Li et al. 2017).

New technologies have associated costs and will have management consequences and cascade into larger herds and reduce the labour force demand. This works in tandem with the authorities' goals to professionalize and rationalize production and will result in less diversified pastoralism. However, new technology and innovation may generate new job opportunities, e.g., microprocessing of meat and new niche products.

Conclusions

Herd size and composition, and slaughter and breeding strategies are interrelated and reflect the herders' production goals. These goals will change over time, influenced by natural, social and economic conditions and the technology available. The governments have employed regulations and subsidies to rationalize the sector by stimulating meat production and market integration. This has influenced herd composition and slaughter strategy, as well as herd size.

Given limited winter pasture resources, a herd composition dominated by productive females and not exceeding the size that can be supported by available winter resources will contribute to reduced winter herd mortality, a high proportion of calves in the summer herd and a high percentage of calves to be slaughtered in autumn. Many herders follow this adaptive management strategy and consider the relevant traits that contribute to high productivity. However, herd size, as well as the herd's sex and age composition, may have an intrinsic value for the herders which is not compatible with the optimization of meat production. This is manifested in the large variation in productivity between herding districts and regions.

Integrating resilience to environmental variability into reindeer breeding programmes may improve the capacity of the production system to tackle anticipated detrimental extreme climatic events. A successful reindeer breeding scheme, including genomic selection, relies on quantifying, utilizing and securing genetic variation by the build-up of large phenotypic and genomic data sets, not only for economically important production traits but also for less heritable traits which contribute to the animals' fitness.

To understand the adaptation of reindeer to climate change, either extensive site-specific longitudinal data or the exposure of genetically related animals to different environmental conditions is necessary. Both types of data would give answers to pertinent questions on how the reindeer population may cope with a warming climate.

Losses of land, especially winter pastures, exacerbated by high predation and human disturbance, reduce production output and increase herding costs. This may result in reliance on supplementary winter feeding. By securing access to forage during winter through supplementary feeding, females may allocate less resources to survival and more to reproduction. This may result in two management regimes: one based on intensive feeding and one based on extensive year-round ranching.

References

Bårdsen, B.-J., Fauchald, P., Tveraa, T., Langeland, K., Yoccoz, N. & Ims, R. A. (2008). Experimental evidence of a risk-sensitive reproductive allocation in a long-lived mammal. *Ecology*. 89(3), 829–837.

Bårdsen, B.-J., Bergland, H., Stien, A. & Tveraa, T. (2017). *Effekten av høsting påproduksjon og lønnsomhet i reindriften* (in Norwegian, English summary). NINA Rapport 999, Tromsø.

Campbell, D. L. M., Marini, D., Lea, J. M., Keshavarzi, H., Dyall, T. R. & Lee, C. (2021). The application of virtual fencing technology effectively herds cattle and sheep. *Animal Production Science*. 61(13), 1393–1402.

Danell, Ö. & Petersson, C. J. (1994). A comprehensive transition matrix model for projecting production and resource consumption in reindeer herds. *Rangifer*. 14(3), 99–112.

Dobrotvorsky, I. M. (1938). Growth and development of reindeer calves in the conditions of the Malozemelsk tundra (In Russian, English summary). *Trans. Inst. Polar. Agric. Anim. Husb., Fish. Hunt. Ind. Ser. Reindeer Ind.* 3, 93–98.

Eloranta, E. & Nieminen, M. (1986). Calving of the experimental reindeer herd in Kaamanen 1970–85. *Rangifer*. 6(2), 115–121.

Fjellheim, S. (1999). *Samer i Rørostraktene* (in Norwegian). Snåsa: Fjellheim.

Hausner, V. H., Fauchald, P., Tveraa, T., Pedersen, E., Jernsletten, J-L., Ulvevadet, B., Ims, R. A., Yoccoz, N. G. & Bråthen K. A. (2011). The ghost of development past: the impact of economic security policies on Saami pastoral ecosystems. *Ecology and Society*. 16(3), e26268920.

Helle, T. & Kojola, I. (2008). Demographics in an alpine reindeer herd: effects of density and winter weather. *Ecography*. 31, 221–230.

Hobbs, N. T., Andrén, H., Persson, J., Aronsson, M. & Chapron, G. (2012). Native predators reduce harvest of reindeer by Sámi pastoralists. *Ecological Applications*. 22(5), 1640–1654.

Holand, H., Kvalnes, T., Røed, K. H., Holand, Ø. Sæther, B-E. & Kumpula, J. (2020). Stabilizing selection and adaptive evolution in a combination of two traits in an arctic ungulate. *Evolution*. 74(1), 103–115.

Horstkotte, T., Lépy, É. & Risvoll, C. (eds). (2020). *Supplementary feeding in reindeer husbandry: Results from a workshop with reindeer herders and researchers from Norway, Sweden and Finland*. Report, Umeå University.

Ingold, T. (1980). *Hunters, Pastoralists and Ranchers: Reindeer Economics and Their Transformations*. New York: Cambridge University Press.

Kumpula, J., Kurkilahti, M., Helle, T. & Colpaert, A. (2014). Both reindeer management and several other land use factors explain the reduction in ground lichens (*Cladonia* spp.) in pastures grazed by semi-domesticated reindeer in Finland. *Regional Environmental Change*. 14(2), 541–559.

Landbruksdirektoratet (2021). Ressursregnskapet for reindriften 2010–2020. www.landbruksdirektoratet.no/nb/nyhetsrom/aktuelle-tema/ressursregnskapet-for-reindriftsnaeringen [20.08.2021]

Lenvik, D. (1988). *Utvalgsstrategi i reinflokken* (in Norwegian). PhD thesis, Norges Landbrukshøgsskole, Ås.

Lenvik, D. (1990). Flokkstrukturering – tiltak for lønnsom og ressurstilpasset reindrift. (in Norwegian). *Rangifer*. Special Issue 4, 21–35.

Lenvik, D. & Fjellheim, A. (1988). Utvalgsstrategi i reinflokken 2. Ungsimlenes vekt ved 18 måneder relatert til vekten ved 2 og 6 måneder (in Norwegian, English summary). *Norsk landbruksforskning*. 1, 263–274.

Li, Z., Lin, Z., Ba, H., Chen, L., Yang, Y., Wang, K., Qiu, O., Wang, W. & Li, G. (2017). Draft genome of the reindeer (*Rangifer tarandus*). *GigaScience*. 6(12), egix102.

Lundqvist, H. (2007). Ecological cost-benefit modelling of herbivore habitat quality degradation due to range fragmentation. *Transactions in GIS*. 11(5), 745–763.

Magga, O. H. (2006). Diversity in Saami terminology for reindeer, snow, and ice. *International Social Science Journal*. 58(187), 25–34.

Marin, A., Sjaastad, E., Benjaminsen, T., Sara, M. N. & Langfeldt Borgenvik, E. J. (2020). Productivity beyond density: A critique of management models for reindeer pastoralism in Norway. *Pastoralism*. 10(9), e001643.

Muuttoranta, K. & Mäki-Tanila, A. (2011). Selection decisions among reindeer herders in Finland. *Rangifer.* 31(1), 129–138.

Muuttoranta, K., Tapio, M., Holand, Ø., Røed, K. H., Nieminen, M. & Mäki-Tanila, A. (2013). Genetic and environmental factors affecting the birth date and birth weight of reindeer calves. *Rangifer.* 33(1), 25–35.

Muuttoranta, K., Tapio, M., Holand, Ø., Røed, K. H., Nieminen, M. & Mäki-Tanila, A. (2014). Genetic variation in meat production related traits in reindeer (*Rangifer t. tarandus*). *Rangifer.* 34(1), 21–36.

Mysterud, A. (2006). The concept of overgrazing and its role in management of large herbivores. *Wildlife Biology.* 12(2), 129–141.

Næss, M. W. (2020). Cultural group selection and the evolution of reindeer herding in Norway. *Human Ecology.* 48, 279–291.

Nieminen, M. (2010). Why supplementary feeding of reindeer in Finland? *Rangifer Report.* 14, 41.

Oskal, A. I. (1999). Tradisjonelle vurderinger av livdyr (in Norwegian). *Rangifer Report.* 3, 121–124.Paine, R. (1994). *Herds of the tundra: a portrait of Saami reindeer pastoralism.* London: Smithsonian Institution Press.

Päiviö, N. J. (2006). Sirkas sameby – om konsekvenser av beitekatastrofer (in Norwegian). *Ottar.* 1, 10–17.

Paliskuntain yhdistys (2010–2020). Tilastoja poronhoitovuodelta (in Finnish). Extracted from Poromies.

Paoli, A, Weladji, R. B., Holand, Ø. & Kumpula, J. (2018). Winter and spring climatic conditions influence timing and synchrony of calving in reindeer. *Plos One.* 13(4), e0195603.

Paoli, A., Weladji, R. B., Holand, Ø. & Kumpula, J. (2019) Early-life conditions determine the between-individual heterogeneity in plasticity of calving date in reindeer. *Journal of Animal Ecology.* 89(2), 370–383.

Petersson, C. J. & Danell, B. (1993). Causes of variation in growth rate of reindeer calves. *Rangifer.* 13(2), 105–116.

Pietarinen, J. & Mäki-Tanila, A. (2020). The efficiency of multi-generation selection on maternal traits, with implications for reindeer. *Agricultural and Food Science.* 29(5), 395–404.

Pietarinen, J., Muuttoranta, K. & Mäki-Tanila, A. (2018). Pseudo-BLUP methodology for maternally affected traits. *Proceedings of the World Congress on Genetics Applied to Livestock Production, Auckland, New Zealand.* e11.869.

Post, E. & Forchammer, M. C. (2008). Climate change reduces reproductive success of an Arctic herbivore through trophic mismatch. *Philosophical Transactions of the Royal Society B.* 363(1501), 2367–2373.

Riseth, J. Å., Tømmervik, H. & Forbes, B.C. (2019) Sustainable and resilient reindeer herding. In: Tryland, M. & Kutz, S. J. (eds). *Reindeer and Caribou: Health and Disease.* 23–43. New York: CRC Press.

Rönnegård, L. & Danell, Ö. (2003). Genetic response to selection on reindeer calf weights. *Rangifer. 23*(1), 13–20.

Rönnegård, L., Woolliams, J. A. & Danell, Ö. (2003). Breeding schemes in reindeer husbandry. *Rangifer.* 23(2), 45–55.

Sandström, P. (2015). *A Toolbox for Co-production of Knowledge and Improved Land Use Dialogues.* PhD. diss. Swedish University of Agricultural Sciences. Umeå.

Sivertsen, T. R. (2017). *Risk of Brown Bear Predation on Semi-Domesticated Reindeer Calves. Predation Patterns, Brown Bear – Reindeer Interactions and Landscape Heterogeneity.* PhD. Diss. Uppsala: Swedish University of Agricultural Sciences.

Skjenneberg, S. & Slagsvold, L. (1968). *Reindriften og dens naturgrunnlag* (in Norwegian). Oslo: Universitetsforlaget.

Skuncke, F. (1964). *Rennäringens ekonomi, skötsel, avkastning och markvärden* (in Swedish). Lappväsendet, Renforskningen, Meddelande 9, Uppsala.

Swedish Sámi Parliament (2021). Statistik rennnäringen. www.sametinget.se/renstatistik%20%5b [20.08.2021]

Tveraa, T., Fauchald, P., Yoccoz, N. G., Anker Ims, R., Aanes, R. & Høgda, K. A. (2007). What regulate and limit reindeer populations in Norway? *Oikos.* 116(4), 706–715.

Tveraa, T., Stien, A., Bårdsen, B.-J. & Fauchald, P. (2013). Population densities, vegetation green-up, and plant productivity: impacts on reproductive success and juvenile body mass in reindeer. *PLoS One.* 8(2), e56450.

Tyler, N. J. C., Hanssen-Bauer I., Førland, E. & Nellemann, C. (2021). The shrinking resource base of pastoralism: Saami reindeer husbandry in a climate of change. *Frontiers in Sustainable Food Systems*. 4, e585685.

Tømmervik, H., Bjerke, J. W., Laustsen, K., Johansen, B. E. & Karlsen, S. R. (2014). *Overvåking av vinterbeiter i Indre Finnmark 2013 – resultater fra feltrutene* (in Norwegian, English summary). NINA Rapport 1066.

Varo, R. M. (1964). Tutkimuksia poron jalostusmahdollisuuksista. (in Finnish, English Summary). *Annales Agriculturae Fenniae.* 3, 296–310.

Varo, R. M. (1972). Investigations on the possibilities of reindeer breeding II. *Journal of the Scientific Agricultural Society of Finland.* 44(4), 234–248.

Verma, M., Schulte to Bühne, H., Lopes, M., Ehrich, D., Sokovnina S, Hofhuis, S.P. & Pettorelli, N. (2020). Can reindeer husbandry management slow down the shrubification of the Arctic? *Journal of Environmental Management.* 267, e110636.

Weladji, R.B., & Holand, Ø. (2006). Influences of large-scale climatic variability on reindeer population dynamics: implications for reindeer husbandry in Norway. *Climate Research.* 32(2), 119–127.

Weladji, R. B., Holand, Ø., Steinheim, G. & Hansen, H. (2002). Effect of "owners" selection strategies on autumn weight in reindeer (*Rangifer t. tarandus*) calves. *Rangifer.* 22(2), 107–113.

Weladji, R.B., Gaillard, J-M., Yoccoz, N. G., Holand, Ø., Mysterud, A., Loison, A., Nieminen, M. & Stenseth, N.C. (2006). Good reindeer mothers live longer and become better raising offspring. *Proceedings of the Royal Society of London, Series B.* 273(1591), 1239–1244.

Weladji, R. B., Loison, A., Gaillard, J-M., Holand, Ø., Mysterud, A., Yoccoz, N. G., Nieminen, M. & Stenseth, N. C. (2008). Heterogeneity in individual quality overrides costs of reproduction in female reindeer. *Oecologia.* 156, 237–247.

Willham, R. L. (1963). The covariance between relatives for characters composed of components contributed by related individuals. *Biometrics.* 19(1), 18–27.

11 Bioeconomics of reindeer husbandry in Fennoscandia

Antti-Juhani Pekkarinen, Jouko Kumpula, Øystein Holand, Birgitta Åhman and Olli Tahvonen

Introduction

Most studies in reindeer husbandry research have concentrated on the biology or ecology of reindeer (Pape & Löffler 2012). Understanding of the economics of reindeer husbandry and economically optimal pasture use has been lacking. Interdisciplinary analysis of reindeer husbandry requires a clear understanding of reindeer ecology, pasture use and its economics. Indeed, reindeer husbandry systems should be studied as a whole (Pape & Löffler 2012). In these systems, ecology and economics are in constant dynamic interaction. An appropriate way to analyse the dynamics of these interactions is the use of mathematical system models and bioeconomic analysis (Schmolke et al. 2010; Pekkarinen 2018).

Bioeconomics is the study of economically optimal utilization (including other values besides monetary income) of biological resources. Bioeconomic models solved by dynamic optimization are at the centre of bioeconomic research (Clark 1976). Interdisciplinary bioeconomic models include a description of the ecology and economics of the studied system. The level of detail required from the model depends on the system being studied as well as on the questions asked. Simple models are easier to analyse and are, therefore, valuable for educational purposes and for analysing the basic driving forces of system dynamics. Among other things, they have been used for analysing hypothesized "tragedy of the commons" situations in Fennoscandian reindeer husbandry and thus showing how unmanaged use of the common pasture resources could affect the reindeer husbandry system (Johannesen & Skonhoft 2009; Skonhoft et al. 2017).

However, in this chapter our focus is on studying the optimal utilization of reindeer populations and their pastures in Fennoscandia. To achieve this, we concentrate on models that aim to describe the main properties of the real reindeer husbandry systems in detail. Thus, to study the slaughtering and feeding decisions made by herders, we need models that can describe the age and sex structure of the population, diet choice and the use of natural food resources and supplementary feeding.

One of the key aspects determining the productivity of a reindeer husbandry system is how reindeer herds utilize their pastures (Pape & Löffler 2012).

DOI: 10.4324/9781003118565-16

Winter lichen pastures are considered to be the limiting factor for the growth rate and productivity of most reindeer populations (Kumpula 2001b). Thus, to analyse sustainable lichen biomass levels with bioeconomic model of reindeer husbandry, lichen dynamics must be included. Including lichen dynamics also makes it possible to estimate whether or not lichen pastures are currently overgrazed, i.e., is there enough lichen on pastures to fulfil the nutritional needs of reindeer during winter. In addition, the recovery from overgrazed pastures can be studied using dynamic reindeer–lichen models.

The first bioeconomic model for the Scandinavian reindeer–lichen system was a model with two state variables, produced by Virtala (1992). Moxnes et al. (2001) adopted a similar approach in their model and included a detailed description of energy intake from various natural energy resources. They included summer pastures and lichen wastage but no description of population structure. Skonhoft et al. (2017) and Johannessen et al. (2019) specified a stage-structured reindeer population model to study the effects of predation. However, their model does not include pasture resource dynamics or sufficient description of the age structure of the reindeer population which would enable the analysis of optimal slaughter strategies.

Reindeer as well as their pastures can be viewed as biological resources affecting economic profitability. Thus, both should be included in any detailed bioeconomic analysis of Fennoscandian reindeer husbandry system. In addition, an age- and sex-structured modelling framework provides insights into optimal herd structure and slaughter strategy, which cannot be fully studied with biomass models or with simplified stage structure. None of the models mentioned above includes all these features. Thus, in this chapter we utilize an age- and sex-structured reindeer-pasture model created by Tahvonen et al. (2014) and Pekkarinen et al. (2015), to analyse sustainable herding practices and pasture use under various economic and ecological conditions.

Reindeer herding practices (e.g., slaughter strategy, use of pastures, supplementary feeding) vary between and within Fennoscandian countries. These differences in herding practices are often adaptations to local conditions. Economic–ecological analysis sheds light on the reasons behind different management decisions under varying conditions. In this chapter, we analyse how variations in economic and ecological conditions affect economically sustainable reindeer husbandry. We consider economically sustainable adaptations and herding practices as well as economically optimal solutions under different conditions. We generate economically optimal model solutions to analyse optimal reindeer numbers, lichen biomass, feeding strategies, structure of the reindeer population, slaughter strategy and the effects of different subsidy systems.

We begin by defining three hypothetical reindeer herding districts that represent herding conditions from mountainous areas with migratory pasture rotation systems to forested areas with stationary herding systems. These three hypothetical herding districts represent the typical variation in conditions between and within Fennoscandian countries. We then generate economically optimal model solutions using the parameter values for each of the three hypothetical districts and demonstrate how costs, prices, interest rate (the marginal rate of return from alternative investments, e.g., other natural resources or stock markets) and government subsidies affect economically sustainable reindeer husbandry. We study which slaughter strategies, lichen biomass levels, feeding strategies and reindeer population sizes give the highest net revenues over the long term under varying economic and ecological conditions. We also ask how different subsidy systems used in Fennoscandian reindeer husbandry direct economically sustainable reindeer husbandry. Finally, we discuss and compare economic incentives, winter pasture conditions and impacts of government subsidies in Nordic countries in the light of our model analysis.

Bioeconomic reindeer husbandry model

In this chapter, we utilize a bioeconomic reindeer husbandry model presented in Tahvonen et al. (2014) and Pekkarinen et al. (2015). The model includes four sub-models: population, energy intake, lichen and economic. General descriptions of each are presented in the following sections, but for complete mathematical descriptions and optimization codes, see the original publications.

Population sub-model

The reindeer population sub-model includes 17 female and 13 male age classes and a description of the population dynamics. The number of reindeer in age class s, in sex class i, in a year t is denoted by $x^i_{s,t}$, $s = 0,1,\ldots,n_i, i = f,m,t = -1,0,1,\ldots$, where f and m denote males and females, respectively. The model year starts immediately after the autumn slaughter, at the beginning of the winter period. The population structure evolves according to:

$$x^i_{1,t+1} = \left(1 - m^i_0\right)u_i x_{0,t} - h^i_{0,t}, i = f,m,t = 0,1,\ldots$$

$$x^i_{s+1,t+1} = \left[1 - m^i_s\left(wd_t\right)\right]x^i_{s,t} - h^i_{s,t}, s = 1,2,\ldots,n_i, i = f,m,t = 0,1,\ldots$$

where m^i_0 is the summer mortality of calves and $h^i_{s,t}$, $s = 0,1,\ldots,$ $i = f,m,t = 0,1,\ldots$ denotes the number of reindeer harvested from age and sex classes at the end of the period. The share of calves (age class 0) belonging to sex class i is denoted by u_i, $i = f, m$.

Reproduction is specified by a modified harmonic mean mating system (Bessa-Gomes et al. 2010) which accounts for the polygynous features of reindeer reproduction. The number of calves born during spring in year t is given as:

$$x_0 = \sum_{s=1}^{n_f} \beta_{t-1} f_s\left(wd_t\right)\left[1 - m^f_s\left(wd_t\right)\right]x^f_{s,t}, \quad t = 0,1,\ldots,$$

where β_{t-1} gives the fraction of females mated at the end of period t-1 and $f_s\left(wd_t\right)$ is the average number of calves per female in age class s. Winter mortalities are denoted by $m^i_s\left(wd_t\right)$. Winter food availability and the associated energy intake in relation to energy need during winter define an individual's weight change during winter (wd_t) and its effects on mortality and reproduction (Tahvonen et al. 2014). Thus, low energy intake decreases spring weight, which in turn reduces the number of calves born and calf birth weight. In addition, significant weight loss during winter increases mortality. The weight change during winter is a function of average daily energy intake during winter, which is calculated by the energy intake sub-model.

Energy intake sub-model

The energy intake sub-model defines the daily energy intake and diet choice during winter (Pekkarinen et al. 2015). The diet choice between arboreal lichens, ground lichens and other resources excavated from beneath the snow (dwarf shrubs, mosses and graminoids) and supplementary feed follows the principles of the optimal foraging theory (e.g., Stephens & Krebs 1986). Thus, reindeer are assumed to choose the combination of energy resources that gives the highest energy intake relative to the time taken for foraging. In addition, reindeer living on natural pastures are assumed to have a preference for natural food resources over supplementary feed (Danell et al. 1994). The amount of supplementary feed given (kg/ha) is decided by the herders and is thus a control (optimized) variable. Arboreal lichen availability and consumption are affected by the availability of old forests and their arboreal lichen biomass.

Lichen sub-model

The lichen sub-model describes the growth and consumption of ground lichens. Lichen biomass (kg/ha) in year t (at the beginning of winter period) is denoted by z_t and lichen growth during summer by $G\left(z_t - l_t^{wi} - l_t^{sp}\right)$. The development of lichen biomass is given as:

$$z_{t+1} = z_t - l_t^{wi} - l_t^{sp} - l_t^{su} + G\left(z_t - l_t^{wi} - l_t^{sp}\right) - l_t^{au}, \quad t = 0,1,\ldots,$$

where $l_t^e, t = 0,1,\ldots, e = wi, sp, su, au$ denote the consumption of lichen (kg/ha) during season e and wi, sp, su and au denote winter, spring, summer and autumn seasons. The total lichen consumption during the different seasons depends on the age- and sex class-specific energy requirements and daily energy intake from lichen. Daily energy intake from lichen is specified in the energy intake sub-model and depends on the relative availabilities of all energy resources and on the size and structure of the reindeer population. To account for the total reduction in lichen caused by grazing reindeer, the model also includes the wastage of lichen by reindeer, in addition to what is ingested and converted to energy. This wastage is mainly the result of trampling and dropping of lichen by reindeer. Pekkarinen et al. (2017) estimated two wastage functions (constant and linear) to describe the situation in northernmost Finland. In this study, we use the constant wastage function as it is simpler and reduces computing time compared to the linear wastage function.

Growth of ground lichens $\left(G\left(z_t - l_t^{wi} - l_t^{sp}\right)\right)$ depends on the lichen biomass after consumption during winter and spring. In addition, lichen pasture type affects lichen growth. Following the formulation presented in Pekkarinen et al. (2015), the annual lichen growth in mountain heaths is 40% of that in old or mature pine forest. Lichen production in young pine forests, logging areas and mountain birch forests is assumed to be 60% of that in old or mature pine forest. The growth function for mature and old pine forests is based on a long-term monitoring study (see more details in Tahvonen et al. 2014). Carrying capacity (undisturbed maximum biomass) of lichen is 6,400 kg/ha. Lichen biomass of 2,300 kg/ha produces the maximum annual lichen growth, which is 142 kg/ha/year in old or mature pine forest.

Economic sub-model

The economic sub-model includes prices, costs and descriptions of subsidy systems analysed. In addition, it describes the objective function and optimization method. In this study, we use Knitro optimization software (version 12.2) and the AMPL programming language (Byrd et al. 2006) for all calculations and optimizations. For economic optimization, we assume that the reindeer herding district maximizes the present value of net revenues, given by:

$$J = \sum_{t=0}^{\infty} (R_t - C_t)^{\alpha} \left(\frac{1}{1+r}\right)^t,$$

where R_t is the annual revenues from slaughtering and C_t is the total annual costs. Total costs include the constant and variable management cost, slaughter costs and feeding costs. The decision variables are the number of animals chosen for slaughter from the age and sex classes and the quantity of supplementary food given.

The assumption of maximizing net present value for herding districts is a simplification of the complex social, cultural and economic objectives that herders experience in reality. However, these other objectives are often difficult to quantify. In addition, this assumption allows us to study a clearly defined question about how to manage a reindeer herding system in order to obtain the highest possible monetary value over an infinite time horizon. Monetary costs of changes in the herding environment and alternative management actions can then be calculated using this same approach.

Tahvonen et al. (2014) showed that their model solutions converge into an economically optimal steady state or cycle around that steady state depending on the linearity of the objective. The difference in the present values of net revenues between the solutions calculated using a linear ($\alpha = 1$) and non-linear ($0 < \alpha < 1$) objective function is minor. We use the non-linear objective in this study because high fluctuations in annual revenues would be problematic in actual reindeer herding livelihoods. Using a non-linear objective (in this study $\alpha = 0.8$) also means that we assume that herders prefer a steady income flow.

The objective function is maximized subject to the model presented in detail in Pekkarinen et al. (2015) and in the model extensions presented in Pekkarinen et al. (2017). The initial state of the system is given. The optimization codes are available as supplementary material in the original publications (Tahvonen et al. 2014; Pekkarinen et al. 2015), on the website of the Economic-Ecological Optimization Group (www2.helsinki.fi/en/researchgroups/economic-ecological-optimization-group/codes), and upon request.

Economic and ecological conditions within and between Fennoscandian countries

Most of the features and parameters in the model are based on Finnish data from the northernmost forest-dominated herding districts. However, to study how variation between and within Fennoscandian countries affects sustainable reindeer husbandry, we define three hypothetical herding districts (mountainous, mixed and forest herding districts). These districts represent the typical variation in conditions that we are interested in.

Defining mountainous, mixed and forest herding districts

Figure 11.1 illustrates how pastures and movement of reindeer differ between Fennoscandian countries. On average, winter pastures in Norway are more commonly found in open mountainous areas, while in Finland most winter lichen pastures are in forested areas. In most parts of Norway, reindeer migrate between winter and summer pastures. In Finland, it is more common to have a stationary system where reindeer have access to the same pastures throughout

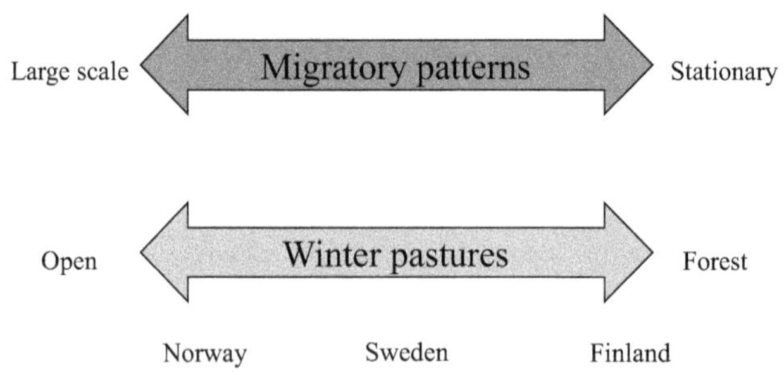

Figure 11.1 Illustration of the differences between reindeer herding in Norway, Sweden and Finland. The figure shows typical features in the countries, but most can also be found within each country.

the year. However, even without pasture rotation, reindeer typically select different pastures during summer and winter. Sweden is located between Norway and Finland both geographically and in terms of typical pasture types and pasture use. Most herding districts in Sweden have large-scale migratory patterns, similar to Norway, and winter pastures in the forest, like in Finland.

In this study, we demonstrate how these different migratory systems and pasture conditions affect economically sustainable reindeer husbandry. We define and parameterize three hypothetical districts to represent typical conditions in mountainous, mixed and forest-dominated districts. We define Mountainous districts, as districts where winter pastures are in open mountainous areas and reindeer migrate between winter and summer pastures. These features are typical in Norway. Forest districts represent districts without pasture rotation and with pastures in forested areas. This is common in Finland but also in some parts of Sweden. Mixed districts are districts where pasture rotation is used, and winter pastures are located in both forested and mountainous areas. Mixed districts include features common in Sweden but also in some areas of Finland.

Although the parameterization of the three districts follows the gradients presented in Figure 11.1, these districts do not directly describe any specific herding district or country. Most of the features of these herding districts can be found in all countries even though they are more common in others. For example, mountainous winter pasture areas are typical in northernmost Norway, but in some of the Finnish districts the majority of winter pastures are also in mountainous areas. Mountainous areas are also typical Sweden, but they are used as summer pastures, and winter pastures are located in forests. In Norway and Sweden, migratory pasture rotation systems are common, but some Finnish districts also have a seasonal pasture rotation system, controlled by means of fences.

Table 11.1 shows the parameter values describing the pasture conditions in these three hypothetical herding districts. The total land area of each district is set to be 3,000 km^2. In the Forest district, the area of lichen pastures available for reindeer during winter is 1,000 km^2. In herding districts with mountainous and mixed lichen pastures, a seasonal pasture rotation is used and the area of lichen pastures available for winter grazing is assumed to be 400 km^2. Winter lichen pastures in mountainous herding districts are in mountain heaths and in mountain birch forests (including other similar vegetation types). Lichen pastures in forest districts are assumed to be in forests at various stages of succession (old, mature, young, logging area). Arboreal lichen pastures are only in old or mature forests.

Lichen pasture type affects lichen growth in the model used in this chapter. The maximum annual lichen growth in old or mature pine forest is 142 kg/ha/year and lower in mountainous pastures and in younger forests. Thus, the maximum annual lichen growth for the three hypothetical herding districts is 71, 92 and 114 kg/ha/year for mountainous, mixed and forest districts, respectively. For optimization, the minimum lichen biomass is restricted to 200 kg/ha. This ensures that optimal solutions lie within the use range of the model. In addition,

Table 11.1 Parameter values describing the pasture conditions in the three hypothetical herding districts

Pasture rotation system	*Mountainous*	*Mixed*	*Forest*
	Seasonal migration	*Seasonal migration*	*No pasture rotation*
Total land area of herding district, km^2	3,000	3,000	3,000
Area of winter lichen pastures, km^2	400	400	1000
• in mountain heaths, %	50	25	
• in mountain birch forests, %	50	25	
• in young forests, %		25	50
• in old forests, %		25	50
Area of arboreal lichen pastures, km^2[a]	0	200	1000

Note:
a Includes only those old/mature coniferous forests where the availability of arboreal lichens is considered to be sufficient (6 kg/ha or more on average).

with extremely low lichen biomasses, the associated reindeer density is very high. At very high population densities, other density-dependent factors besides winter food limitation would begin restricting population growth. However, these effects are not included in the bioeconomic model used.

Costs and prices

Pekkarinen et al. (2020a) calculated the unit costs and producer meat prices for the 20 northernmost herding districts in Finland for the years 2015–2016, based on data from the Reindeer Herders' Association. They found that the average annual variable management costs were approximately €40 (per reindeer in the winter population) and the slaughter costs were €22 (per slaughtered reindeer). The fixed management costs were €1.6 (per ha of the total land area used by the reindeer herding cooperative) and the estimated producer meat price was €10 (per kg of meat). In this study, we use these same costs, although costs and prices actually vary between the countries. Keeping the costs and prices constant, we can analyse how different pasture conditions in Fennoscandia affect economically optimal model solutions.

The different pasture conditions are represented by the three hypothetical districts. To study how these pasture conditions alone affect economically optimal solutions, we keep economic parameters the same between the hypothetical districts. However, because costs and prices vary between the years and areas, we also derive the solutions with different costs, meat prices and subsidy systems. We study how changing management costs, slaughter costs, feeding costs and meat price affect model solutions.

Indeed, costs and prices vary between Fennoscandian countries. For example, the producer meat price is lower in Sweden (Sametinget 2020) than in Finland. In contrast, slaughter costs for reindeer herders are small in Sweden since these are mainly covered by the slaughter company. In addition, herding costs vary within and between the countries depending on the characteristic of herding districts. We do not change the level of fixed management cost, as it does not affect economically optimal herding strategies, even though it changes the absolute level of annual net revenues. In this model, fixed costs depend only on the size of herding district and thus remain at a fixed level no matter how the reindeer population is managed.

The costs of supplementary feeding depend on the price of supplementary feed and on the costs of delivering the feed to winter pasture areas. In this study, we use €0.5 per kg as an estimate for the costs of supplementary feeding. The price of commercial supplementary feed accounts for about half of the costs and the other half is for transporting and distributing feed to winter pasture areas. We vary the costs of supplementary feeding to study how lower costs would change slaughter strategies, herding strategies and optimal pasture use.

Subsidy systems in Finland, Sweden and Norway

Government subsidies aim to support local livelihoods, while regulation is used to reduce the possible harmful effects of these livelihoods. The use of natural resources is often strongly regulated and subsidized. In addition to the direct effects, subsidies and regulation also affect the economically optimal ways to manage these natural resources. In this study, we describe different subsidy systems used in Fennoscandian reindeer husbandry and study how they can affect economically rational reindeer management.

All Fennoscandian countries have subsidy systems for reindeer husbandry. The Finnish government subsidizes reindeer herders according to the size of their reindeer herds during winter. Thus, reindeer owners with large enough herds are subsidized by €28.5 per reindeer. In Sweden, a subsidy is paid for meat production. Reindeer herders are paid €1.45 per kilo carcass weight for calves and €0.9 per kilo for reindeer over one year of age. Payment is made for reindeer slaughtered at approved slaughterhouses. The slaughterhouse sends a list of slaughtered reindeer to the Sámi Parliament, which pays the subsidy to the owner of the slaughtered reindeer. The subsidy is paid to all reindeer owners irrespective of the number of live or slaughtered animals.

Norway has combined several subsidy systems with the intention of developing reindeer herding in directions considered favourable in different situations. The system is complex and more than ten different subsidies are paid according to different requirements. They can be divided into three main categories: (1) operating subsidies, mainly covering districts' common administrative costs as well as costs for welfare and social security, (2) production incentives for high productivity and calf slaughter and (3) innovation and infrastructure support. In addition, some subsidies are allocated for compensation and preventive measures. Although, the Norwegian system is more complex than the ones used in Sweden or Finland, it includes similar elements. There are similarities especially with the Swedish system, as many of the subsidies increase with increasing meat production and slaughter rate.

In this study, we analyse the incentives associated with different subsidy systems. We focus on the two main systems used in Scandinavia: meat production subsidy and reindeer subsidy. We define the former as a subsidy paid to herders per kilo of meat produced. This subsidy system is the main one used in Sweden and many of the Norwegian subsidies have similar features. We define reindeer subsidy as a subsidy paid per reindeer in the winter population. A reindeer subsidy system is used in Finland, and some of the Norwegian subsidies have similar elements as they increase with increasing management costs (the logic holds when management costs increase with increasing herd size).

For bioeconomic model calculations, the effects of reindeer subsidy are equal to the lower management costs (€ per reindeer). To study the consequences of this reindeer subsidy system, we decrease the variable management costs in the model by €28.5 per reindeer (the sum paid in Finland). Similarly, the effects of meat production subsidy are equivalent to the meat price being higher (€ per kg of meat produced). To study the effects of meat production subsidy we increase the meat price by €1.6 per kg. This is somewhat higher than the current subsidy level in Sweden. However, with €1.6 per kg, the total sum of subsidies paid in our optimal model solutions is equal to the total subsidies paid if the reindeer subsidy is €28.5 per animal. This way we can compare the incentives created by these systems while keeping the total costs to the government and the total sum of subsidies paid to the herders the same between the subsidy systems.

Results and discussion

Dynamic and steady-state solutions

Dynamic models, like the model used in this chapter, include time as a variable. Such models can be used for studying how systems develop over time, but also what type (if any) of long-term steady states the system can reach. To fully understand a reindeer herding system using bioeconomic analysis, we need to study both steady states and dynamic transition solutions.

Steady-state analysis describes the long-term stability and balance between reindeer numbers and pastures. According to previous model solutions (Tahvonen et al. 2014; Pekkarinen et al. 2015) and empirical observations on isolated islands (Klein 1968), natural stable steady states are typically not found in reindeer–lichen systems without harvesting by humans, predation or significant alternative energy resources. In uncontrolled situations, reindeer numbers tend to increase to a very high level, consuming their lichen resources. Because the low growth rate of lichen cannot compensate for the increased consumption, reindeer populations may crash and possibly even face local extinction (Tahvonen et al. 2014; Pekkarinen et al. 2015). However, human influence and sustainable management may lead to more stable situations. Thus, analysis of economically sustainable long-term steady states considerably increases our understanding of these systems.

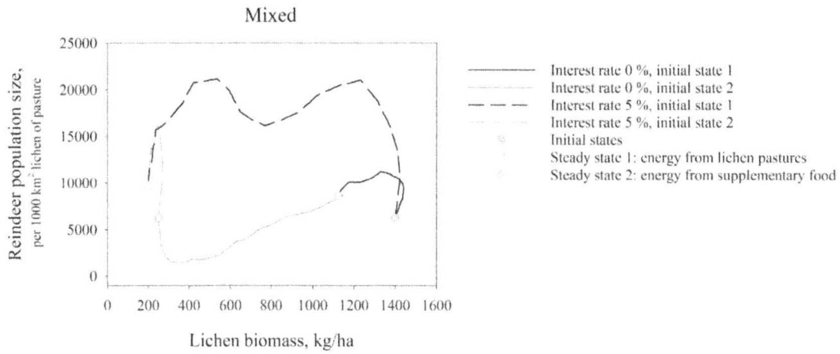

Figure 11.2 Examples of economically optimal dynamic solutions and steady states in different situations. Four dynamic solutions from two initial states that lead to two steady states are shown. Solid lines represent solutions leading to steady state 1, where reindeer herding is based on natural pastures. In these solutions the interest rate is 0%. Dashed lines represent solutions leading to steady state 2, where reindeer herding is based on intensive supplementary feeding. In these solutions the interest rate is 5%.

In addition to steady-state analysis, dynamic solutions are needed for solving transitions from various initial states to these steady states. This is especially important in reindeer herding systems where the transition to a steady state may take a long time because of the slow recovery of lichen pastures and fairly long lifespan of reindeer. In addition, dynamic solutions are necessary for achieving optimal steady-state solutions with a positive interest rate.

Figure 11.2 shows an example of four dynamic transition solutions from two initial states leading to two optimal steady states. These solutions are derived for a mixed herding district (see Table 11.1) using the reindeer–lichen model presented in this chapter. The initial state on the left-hand side represents a situation with low initial lichen biomass. In contrast, the initial state on the right-hand side has a higher lichen biomass. The solutions show economically optimal transitions from these initial states to the two steady states. In steady-state 1, economically optimal management of the reindeer population is based on natural pastures. In this example, it is economically optimal to direct the system towards this state when the interest rate is low (0%). With a high interest rate (5%), the development towards steady-state 2 gives a higher present value of net revenues. In steady-state 2, supplementary feeding is the main energy resource for reindeer and the lichen biomass level is very low.

Optimal slaughter strategies and population structures

Tahvonen et al. (2014) found that in the Finnish reindeer husbandry system, it is economically optimal to rely on intensive calf slaughter and on the minimum effective proportion of adult males. This same applies to the solutions in Figure 11.2, which are calculated using the costs and prices in Finnish reindeer husbandry. Thus, in these solutions, the population structure and slaughter strategy (relative to population size) remain similar, although the population size of reindeer, lichen biomass and the main energy resource of reindeer differ greatly. Figure 11.3 (Slaughter strategy I) shows this steady-state population structure and slaughter strategy. More than 60% of female calves and more than 95% of male calves are slaughtered during their first autumn. Adult females are kept alive until the age of 9.5 years and adult males until the age of 5.5 years. The number of adult males is kept as low as possible without significantly reducing the fertilization rate of females and the reproduction rate of the population.

With current prices and costs, slaughter strategy I (Figure 11.3) becomes optimal. However, lower management costs, lower meat price or high reindeer

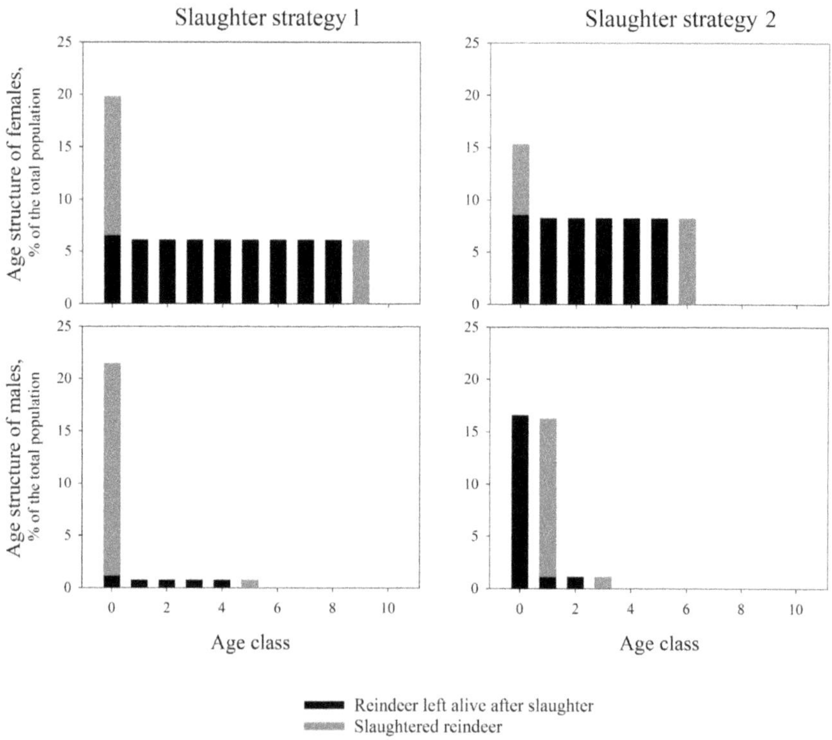

Figure 11.3 Optimal steady-state age and sex structures and slaughter strategies without any other mortality factors (predators, deceases, traffic) besides winter food limitation. Calf slaughter dominates in Slaughter strategy I, whereas in strategy II most of the reindeer slaughtered are adults (1.5 years or older). Strategy I is optimal with current costs and prices in Finland, but Strategy II may be come optimal with lower meat prices, variable management costs or higher reindeer subsidies.

subsidy can change the optimal slaughter strategy towards adult slaughter. The example of an alternative optimal slaughter strategy is presented in Figure 11.3 (slaughter strategy II). In this example, management costs and meat price are low and thus it becomes optimal to use this alternative approach. With this slaughter strategy, more than 80% of the slaughtered reindeer are adults (at least 1.5 years old) and less than 20% of the calves are slaughtered during their first autumn. Thus, the proportion of adults in the population and adults slaughtered are much higher in strategy II than in strategy I. In addition, the total number of reindeer is higher and thus lichen biomass is lower in strategy II. Lower lichen biomass implies reduced calf production and calf weights. Thus, slaughter strategy II is not based on maximizing calf production, calf weight or calf slaughter. Instead, it is based on higher reindeer numbers and higher proportional weight increase from calf to adult than in strategy I. When meat production is mainly based on adult slaughter (e.g., in slaughter strategy II), it is optimal to slaughter adults earlier (from younger age classes) compared to situations in which the adult population is mainly used for reproduction (slaughter strategy I). In slaughter strategy II, females are slaughtered at the age of 6.5, because after that their weight no longer increases.

In the solutions presented in this chapter, winter food limitation is the only mortality factor for adults (2% of calves are assumed to die during summer). As it is not economically rational to let reindeer starve, natural mortality in optimal steady states is close to zero. Including other mortality factors (predators, diseases and traffic) may change optimal herd structure and the slaughter strategy. Indeed, Pekkarinen et al. (2020a) showed that high predation pressure reduces the relative importance of calf slaughter. A high density of grey wolves changes the optimal slaughter strategy of adults towards younger age classes, but high brown bear density does not have the same effect. The difference is caused by the differences in age class-specific predation mortalities. Grey wolves, and also lynx and wolverine, predate all age classes more equally, whereas brown bears mostly target calves or young reindeer during summer before autumn slaughter (see Chapter 6). Incentives associated with different predator compensation systems may alter these solutions.

The effects of costs, meat price and subsidies on optimal slaughter strategies

Solutions presented in Figure 11.2 are derived using prices and costs estimated for northernmost Finland in the years 2015–2016. However, costs and prices differ within and between Fennoscandian countries. The choice between the two types of slaughter strategy presented in Figure 11.3 depends on the level of variable management costs and meat price. Government subsidies can affect these as reindeer subsidy reduces costs per reindeer and production subsidies increase the revenues gained per kilogram of meat. Figure 11.4 shows various combinations of meat prices (including meat production subsidies) and management costs (including the effects of reindeer subsidies) and the corresponding optimal slaughter strategies. It shows that in districts where winter pastures are

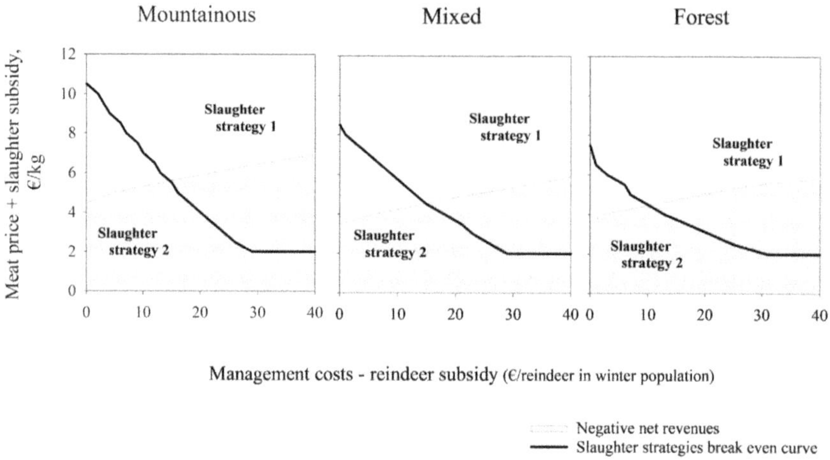

Figure 11.4 Effects of meat price (including meat production subsidy) and management costs (minus reindeer subsidy) on the choice of slaughter strategy in optimal steady states with 0% interest rate. Reindeer subsidy favours Slaughter strategy II, as it reduces costs per reindeer. Meat production subsidy increases meat price and thus favours strategy I. Calf slaughter dominates strategy I, whereas in strategy II most of the reindeer slaughtered are adults (1.5 years or older).

less productive (mountainous districts), it is more often beneficial to postpone slaughter until most reindeer are adults (slaughter strategy II). In those cases, calf production and calf weights are lower. This may have been the situation in some mountainous districts in Norway in the past, where calf weights were low due to poor pasture conditions and supplementary feeding was not used. However, nowadays incentives have been implemented in Norway to reduce pressure on the winter pastures, which has resulted in a higher proportion of calf slaughter. In contrast, in the Forest district, intensive calf slaughter is optimal even with a lower meat price. This has been typical in Finnish and Swedish districts with productive ground lichen and arboreal lichen pastures in old and mature forests. In southern districts in Finland where pastures are less productive due to forestry, supplementary feeding has been used to ensure higher calf weight and calf production.

Figure 11.4 also shows that, in order to achieve positive net revenues, a slightly higher meat price is required in Mountainous districts than in Forest districts. However, the costs may also vary between the districts depending on pasture and winter conditions. Overall, the differences in required meat price are relatively small. These solutions suggest that to gain positive net revenues, meat price (+ meat production subsidy) must be at least €4–€7 per kg. However, at that level the revenues from meat production only just cover the costs of reindeer husbandry.

Optimal model solutions under different pasture conditions and interest rates

Pasture conditions and migratory patterns vary within and between Fennoscandian countries (Figure 11.1). In addition, the interest rate available to herders may differ between areas, individuals and time. Table 11.2 shows economically optimal model solutions for the three hypothetical herding districts which represent different pasture conditions in Fennoscandia. With lower interest rates, the lichen biomass is clearly lower in the Forest district compared to mountainous and mixed districts. This is mainly due to a lack of pasture rotation in the Forest district. Without pasture rotation, lichen is not protected from grazing and trampling during snow-free periods. In addition, high availability of arboreal lichens in the Forest district helps reindeer to survive and reproduce even with lower ground lichen availability. Because of these two factors, it is not beneficial to invest in higher lichen biomass.

With higher interest rates, it becomes optimal to use intensive supplementary feeding, as demonstrated in Figure 11.2. In that case, lichen biomass falls to a very low level and reindeer rely on a mixed diet during winter, gaining energy from resources excavated from beneath the snow, supplementary feed and also arboreal lichens if available.

Table 11.2 and Figure 11.4 also show that, according to our model solutions, mountainous districts are less productive than mixed or forest districts. This is due to more favourable conditions for ground lichens in old and mature pine forests (Kumpula et al. 2014) and because of high arboreal lichen availability in

Table 11.2 Optimal steady state lichen biomass, number of reindeer and annual net revenues in different types of herding districts with interest rates from 0 to 5%.

	Interest rate	*Mountainous*	*Mixed*	*Forest*
Lichen biomass, kg/ha	0	1180	1144	479
	1	862	859	403
	3	210*	620	214*
	5	201*	202*	210*
Number of reindeer	0	6352	8628	8275
	1	6580	8860	8506
	3	7924	9944	9362
	5	8496	11132	9411
Annual net revenues, €	0	356240	653600	567900
	1	351120	652000	566400
	3	101000	570440	524200
	5	65120	293880	524000

Note:

* The main energy resource for reindeer in these solutions is supplementary feed. Thus, lichen biomass falls to the lowest possible level. To ensure that the optimization solutions lie within the use range of the model, the minimum lichen biomass in model solutions is set to 200 kg/ha.

old and mature forests (Esseen et al. 1996). Pasture conditions are most favourable in the forest district, but seasonal pasture rotation increases the productivity in mixed and mountainous districts. Thus, under the model assumptions, the least productive systems would be mountainous systems without pasture rotation. However, this solution is highly dependent on the daily digging area (reindeer excavate resources from beneath the snow). In our model daily digging area is assumed to be on average 30 m^2 (Kumpula 2001a). However, our preliminary results suggest that if the average digging area in mountainous pastures is larger, e.g. due to more favourable snow conditions, it may increase the productivity of reindeer husbandry (Pekkarinen et al. 2020b) compared to forest pastures with a lower average digging area. However, more research on average digging areas and availability of food resources beneath the snow in various conditions is needed to validate this result.

Pasture degradation and the associated high reindeer numbers are one of the main concerns in northernmost Scandinavia and a pressing topic in reindeer husbandry research (Pape & Löffler 2012). According to our model solutions, a higher interest rate and favourable pasture conditions are associated with a higher number of reindeer (Table 11.2), when reindeer herding districts maximize their long-term net revenues. In addition, various other possible reasons exist for high reindeer population densities, which can lead to increased grazing pressure and possibly to overgrazing. For example, Johannesen and Skonhoft (2011) found that herders may keep large herds to gain higher social status within the community. In addition, "tragedy of the commons" scenarios may result in high reindeer densities (Johannesen & Skonhoft 2009), when pasture use is not limited by the herding district or government. According to Næss and Bårdsen (2010), in a randomly variable environment, large herds may also be used as a risk-reduction strategy. However, Pekkarinen et al. (unpublished) found that poor pasture conditions caused by high reindeer density may expose reindeer husbandry to greater negative effects of randomly variable winter conditions. In addition, reduced pasture area or quality, e.g., due to forestry, may result in increased grazing pressure even if reindeer numbers remain the same (Pekkarinen et al. 2021). Thus, it is not only the number of reindeer that determines the sustainability of reindeer husbandry but the balance between the grazing resources and reindeer density. Consumer-resource models, like the one used in this study, are an appropriate method for analysing this relationship and thus including a dynamic description of the grazing resource is crucial in studying reindeer husbandry systems.

Optimal use of supplementary feed

In addition to optimal slaughter strategy and optimal lichen biomass levels, detailed bioeconomic models can be used to study whether it is optimal to rely on intensive supplementary feeding. Pekkarinen et al. (2015) found that, assuming average winter conditions, optimal steady-state solutions are typically based either on the use of natural pastures or on intensive supplementary feeding. The choice between these two strategies depends on economic and ecological factors. In addition, it is economically sensible to use supplementary feeding during a transition phase when restoring heavily grazed lichen pastures. In randomly variable winter conditions, supplementary feeding also becomes economically optimal during those winters when weather and snow conditions significantly restrict the use of natural pastures (Pekkarinen et al. unpublished).

Figure 11.5 gives an example of optimal solutions in two different situations. Both solutions start from the same initial state, but due to the different prices

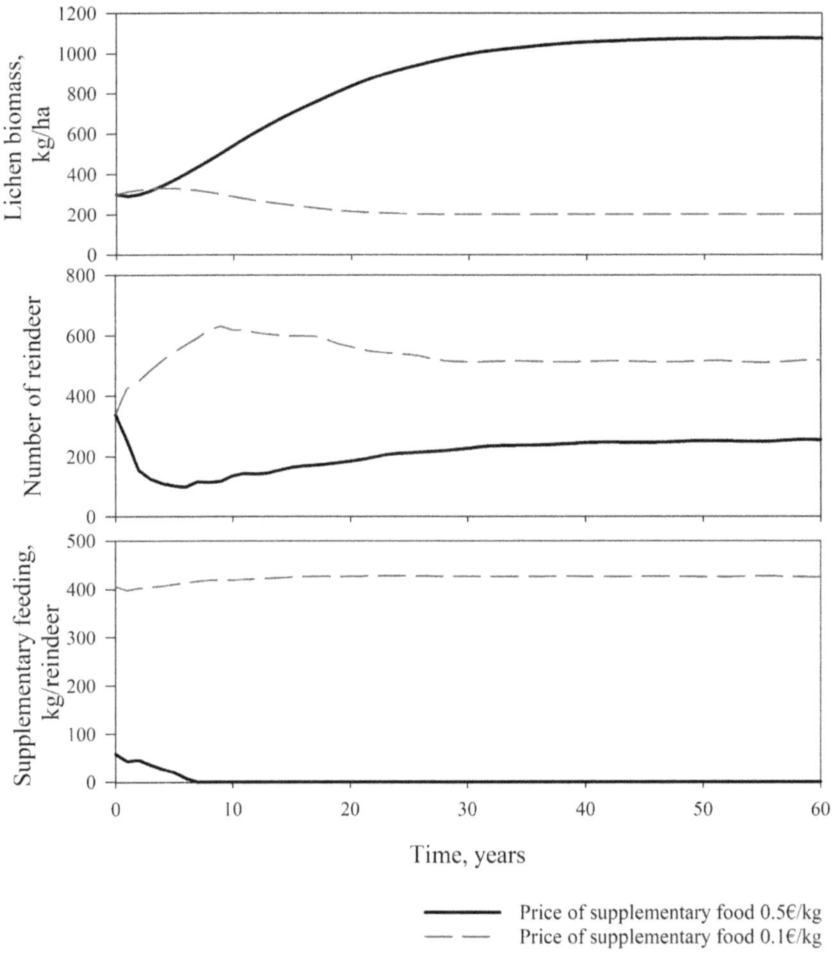

Figure 11.5 Dynamic optimal model solutions starting from relatively low lichen biomass. Solutions represented by solid lines are based on the estimated current costs of supplementary feeding (€0.5 per kg). Solutions represented by dashed lines are based on very low feeding costs (€0.1 per kg).

of supplementary feed, optimal solutions lead to two different steady states. In the solution based on €0.5 per kg feeding costs, supplementary feeding is used during the first years of the transition towards steady state, but not once the steady state is reached. However, if feeding is very inexpensive (€0.1 per kg) or the interest rate is high (see Figure 11.2) it becomes economically rational to base reindeer herding on intensive feeding.

It is also noteworthy that with higher feeding costs, the initial situation can be regarded as representing overgrazing as the pastures do not support economically sustainable production and supplementary feeding is not profitable. However, in a situation with low feeding costs, it might not be reasonable to consider the initial situation as representing overgrazing, at least from the perspective of reindeer husbandry, because the economically viable lichen biomass is lower than the lichen biomass in the initial state. This clearly demonstrates that the questions of overgrazing and sustainable levels of lichen pastures are not purely ecological concepts but are also affected by economics and by management objectives (see Mysterud 2006 for discussion on overgrazing in general).

The choice of feeding strategy under different interest rates, prices and costs

Figures 11.2 and 11.5 show that whether or not it is economically rational to use intensive supplementary feeding depends on interest rates and on the price of supplementary feed. In addition, the availability of different winter pastures and their condition affect whether supplementary feeding is appropriate. Figure 11.6 shows the effects of interest rate and feeding costs on the use of supplementary feeding for the optimal steady states in the three different

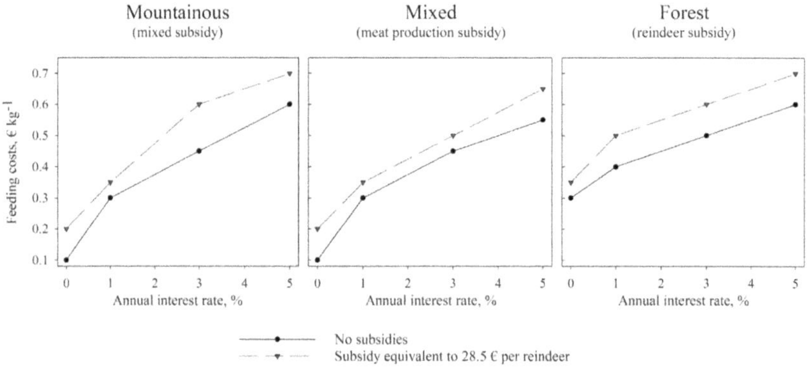

Figure 11.6 Effect of interest rate and feeding costs on the use of supplementary feeding in the optimal steady states. Curves represent the feeding costs for which it is optimal to offer supplementary feed as a main winter energy resource for reindeer. Feeding is not used in situations located above the curves, and reindeer management relies on natural pastures. Below the curves, intensive supplementary feeding is used, resulting in very low lichen densities.

types of herding districts. Compared to mountainous or mixed districts, feeding in the forest district becomes optimal with higher feeding costs, especially when the interest rate is low. However, the effect of high interest rates seems to outweigh the effect of pasture conditions and feeding becomes optimal with similar feeding costs in all of the three herding districts.

Figure 11.6 also shows the effect of different subsidy schemes on the optimality of supplementary feeding. The level of the subsidy is calibrated so that the direct costs of subsidies for the governments are equal in all subsidy schemes. Thus, for the forest district we use €28.5 per reindeer (reindeer subsidy), for the mixed district we use €1.6 per kg of meat produced (meat production subsidy) and for the mountainous district we use €14.25 per reindeer and €0.8 per kg of meat produced (mixed subsidy). This shows that the reindeer-based subsidy gives a slightly higher incentive for the use of supplementary feeding than the meat production subsidy. However, differences appear to be small and all three subsidy schemes favour the use of intensive supplementary feeding.

Conclusions

In this chapter, we have demonstrated how bioeconomic analysis, economic–ecological system models and economic optimization can be used as efficient tools to study the dynamics of complex reindeer husbandry systems. We have shown that economically optimal model solutions depend on various economic and ecological factors. As an example, economically optimal lichen biomass can vary significantly depending on interest rate, costs, prices, lichen pasture productivity, availability of other natural energy resources and government subsidies.

Pekkarinen et al.'s (2015) model solutions show that, in an undisturbed herding environment, reindeer husbandry relying on natural pastures is, in most cases, more profitable than reindeer husbandry based on intensive supplementary feeding. However, according to their recent study (Pekkarinen et al. 2021), most of the current changes, pressures and economic incentives affecting Fennoscandian reindeer husbandry seem to favour lower lichen biomass and the use of supplementary feeding. In the southern part of the Finnish reindeer husbandry area, in particular, intensive supplementary feeding and low lichen biomass are common. Our solutions suggest that this is economically rational as intensive forestry has reduced the area and productivity of winter pastures. In addition, government subsidies seem to favour larger herds and supplementary feeding, especially in Finland.

In Norway and Sweden, reindeer herding districts use seasonal migratory pasture rotation systems. Previous research has shown that pasture rotation protects valuable winter lichen pastures from excessive consumption during snow-free periods (Kumpula et al. 2014). However, pasture rotation is more difficult to arrange in smaller and fragmented herding districts, which are common in southern areas of Finland. According to our results, when a pasture rotation system cannot be used, it may become economically sensible to let lichen biomass fall to a lower level and rely more on other natural food resources or, in some cases, on supplementary feeding.

Most herding districts in Fennoscandian countries rely on intensive calf slaughter and on a minimum effective proportion of adult males in the winter population. According to our solutions, this slaughter strategy is the most economically productive in most cases. In addition, government subsidies in Sweden and Norway promote calf slaughter. However, the economically optimal slaughter strategy depends on economic and ecological factors, which have changed in the past and will change in the future. Pekkarinen et al. (2020a) showed that, in some cases, high predation pressure may change the optimal slaughter strategy and reduce the importance of calf slaughter. Similarly, in this chapter we demonstrated that lower meat price and management costs or higher reindeer subsidy may shift the optimal slaughter strategy from calves to adult reindeer.

Reindeer herding practices and herding conditions vary within and between Fennoscandia countries. Similarly, bioeconomic analysis of reindeer husbandry systems presented in this chapter shows that economically optimal solutions depend on various ecological and economic factors. Thus, different situations in reindeer husbandry require different herd structures, slaughtering strategies, reindeer densities, feeding strategies and pasture use. Our analysis suggests that many of the differences seen in practical reindeer husbandry may be economically rational adaptations to local conditions.

References

Bessa-Comes, C., Legendre, S. & Clobert, J. (2010). Discrete two-sex models of population dynamics: on modelling the mating function. *Oecologica*. 36, 439–445.

Byrd, R.H., Nocedal, J. & Waltz, R.A. (2006). KNITRO: An integrated package for nonlinear optimization. In: di Pillo, G. & Roma, M. (eds) *Large-Scale Nonlinear Optimization*. New York: Springer. 35–59.

Clark, C. (1976). *Mathematical bioeconomics: The optimal management of renewable resources*. New York: Wiley.

Danell, K., Utsi, P.M., Palo, R.T. & Eriksson, O. (1994). Food plant selection by reindeer during winter in relation to plant quality. *Ecography*. 17(2), 153–158.

Esseen, P.A., Renhorn, K.E. & Pettersson, R.B. (1996). Epiphytic lichen biomass in managed and old-growth boreal forests: effect of branch quality. *Ecological Applications*. 6(1), 228–238.

Johannesen, A.B. & Skonhoft, A. (2009). Local common property exploitation with rewards. *Land Economics*. 85(4), 637–654.

Johannesen, A.B. & Skonhoft, A. (2011). Livestock as insurance and social status: Evidence from reindeer herding in Norway. *Environmental and Resource Economics*. 48(4), 679–694.

Johannesen, A.B., Olaussen, J.O. & Skonhoft, A. (2019). Livestock and carnivores: Economic and ecological interactions. *Environmental and Resource Economics*. 74(1), 295–317.

Klein, D.R. (1968). The introduction, increase, and crash of reindeer on St. Matthew Island. *The Journal of Wildlife Management*. 32, 350–367.

Kumpula, J. (2001a). Winter grazing of reindeer in woodland lichen pasture: Effect of lichen availability on the condition of reindeer. *Small Ruminant Research*. 39(2), 121–130.

Kumpula, J. (2001b). Productivity of the semi-domesticated reindeer (Rangifer t. tarandus L.) stock and carrying capacity of pastures in Finland during 1960–90s. PhD. diss. University of Oulu.

Kumpula, J., Kurkilahti, M., Helle, T. & Colpaert, A. (2014). Both reindeer management and several other land use factors explain the reduction in ground lichens (Cladonia spp.) in pastures grazed by semi-domesticated reindeer in Finland. *Regional Environmental Change*. 14(2), 541–559.

Moxnes, E., Danell, Ö., Gaare, E. & Kumpula, J. (2001). Optimal strategies for the use of reindeer rangelands. *Ecological Modelling*. 145(2), 225–241.

Mysterud, A. (2006). The concept of overgrazing and its role in management of large herbivores. *Wildlife Biology*. 12, 129–141.

Næss, M.W. & Bårdsen, B.J. (2010). Environmental stochasticity and long-term livestock viability – herd-accumulation as a risk reducing strategy. *Human Ecology*. 38(1), 3–17.

Pape, R. & Löffler, J. (2012). Climate change, land use conflicts, predation and ecological degradation as challenges for reindeer husbandry in northern Europe: What do we really know after half a century of research? *Ambio*. 41(5), 421–434.

Pekkarinen, A.-J. (2018). Ecology and economics of reindeer herding systems. PhD. diss. University of Helsinki.

Pekkarinen, A.-J., Kumpula, J. & Tahvonen, O. (2015). Reindeer management and winter pastures in the presence of supplementary feeding and government subsidies. *Ecological Modelling*. 312, 256–271.

Pekkarinen A.-J., Kumpula J. & Tahvonen O. (2017). Parameterization and validation of an ungulate-pasture model. *Ecology and Evolution*. 7(20), 8282–8302.

Pekkarinen, A-J., Kumpula, J. & Tahvonen, O. (2020a). Predation costs and compensations in reindeer husbandry. *Wildlife Biology*. 2020(3), 1–14.

Pekkarinen, A-J., Kumpula, J. & Tahvonen, O. (2020b). Bioekonominen analyysi poronhoidon ja talvilaidunten tilan taloudellisesta kestävyydestä Suomessa. In: Kumpula, J. & Siitari, S. (eds) *Kestävä biotalous porolaitumilla -hankkeen osaraportit, johtopäätökset ja toimenpide-ehdotukset*. Luonnonvara- ja biotalouden tutkimus 29/2020. Helsinki: Luonnonvarakeskus. 6–27.

Pekkarinen, A-J., Kumpula, J. & Tahvonen, O. (2021). What Drives the Number of Semi-domesticated Reindeer? Pasture Dynamics and Economic Incentives in Fennoscandian Reindeer Husbandry. In: Nord, D.C. (eds) *Nordic Perspectives on the Responsible Development of the Arctic*. Springer Polar Sciences. Cham: Springer. 249–270.

Sametinget (2020). *Rennäringens tillstånd 2019*. Kiruna: Sametinget.

Schmolke, A., Thorbek, P., DeAngelis, D.L. & Grimm, V. (2010). Ecological models supporting environmental decision making: A strategy for the future. *Trends in Ecology and Evolution*. 25(8), 479–486.

Skonhoft, A., Johannesen, A.B. & Olaussen, J.O. (2017). On the tragedy of the commons: When predation and livestock loss may improve the economic lot of herders. *Ambio*. 46(6), 644–654.

Stephens, D. & Krebs, J. (1986). *Foraging Theory*. Princeton, NJ.: Princeton University Press.

Tahvonen, O., Kumpula, J. & Pekkarinen, A.J. (2014). Optimal harvesting of an age-structured, two-sex herbivore–plant system. *Ecological Modelling*. 272, 348–361.

Virtala, M. (1992). Optimal harvesting of a plant-herbivore system: Lichen and reindeer in northern Finland. *Ecological Modelling*. 60(3–4), 233–255.

12 Role of supplementary feeding in reindeer husbandry

Birgitta Åhman, Minna Turunen, Jouko Kumpula, Camilla Risvoll, Tim Horstkotte, Élise Lépy and Svein Morten Eilertsen

Introduction

Providing reindeer with supplementary feed (hereafter referred to as "feeding") in winter is a strategy practised by reindeer herders when natural grazing resources are unavailable or the grazing conditions are challenging for various reasons. Causes and extent of this strategy have changed over time and differ across the reindeer herding area (Åhman et al. 2018). Winter feeding is practised across the whole of Fennoscandia and has increased during recent decades. A main reason is the combination of the impacts of climate change on the accessibility of natural grazing resources and the loss of land to other forms of land use (Chapters 4 and 5). Climate change has resulted in more frequent winters with deep and hard snow, and thaw-refreezing events creating ice crusts in the snow or ice layers on the ground. Such events can prevent reindeer from gaining access to forage resources on the ground, mainly terrestrial lichens. Loss of valuable land for grazing due to industrial land use (i.e., forestry, wind power parks or mining), and infrastructure for recreation and tourism, results in higher grazing pressure on the remaining pastures. Increased presence of predators (Chapter 6) also makes it difficult to use certain areas for reindeer grazing. These factors together increase the risk of there being situations when there is not enough available natural forage, and feeding is then the only solution to provide sufficient food for the reindeer. Feeding is also undertaken in other situations, not directly linked to lack of natural forage but, e.g., to facilitate handling or protect reindeer from predators.

Nowadays, it is less problematic than previously to feed large herds of reindeer due to the availability of factory-made feeds specifically formulated for reindeer, together with motorization, increased infrastructure and growing knowledge and practical experience among herders. However, feeding is both costly and laborious and may involve health risks for the animals (Chapter 13).

In this chapter, we describe how feeding of reindeer is practised in Finland, Sweden and Norway, and what the overall consequences are for reindeer husbandry. We summarize scientific results regarding positive and negative effects of feeding, review how feeding, as a recurring practice in reindeer husbandry, is perceived by herders, what concerns they raise in relation to feeding, discuss

DOI: 10.4324/9781003118565-17

benefits and drawbacks of feeding and how it may help or hinder reindeer husbandry in the future.

Feeding of reindeer in Fennoscandia

Reindeer have probably been provided with supplementary feed by their owners since early domestication, in particular the animals that were kept for transportation and milking. Isotope studies of archaeological bone samples of reindeer dated to the 13th century suggest that feeding may have already been a part of the reindeer herders' practices at that time (Salmi et al. 2020). There is also documentary evidence of winter feeding practices by Forest Sámi in Finland, dating back to the period of intensive herding before the 18th century, when domestic reindeer were fed, e.g., bread, lichens and hay (Itkonen 1948; Kortesalmi 2007). Widespread practices during difficult winters were to provide emergency feed for reindeer by pulling arboreal lichens off trees, cutting down lichen-rich trees and breaking hard snow cover with shovels to make digging easier for the reindeer (Itkonen 1948; Helle and Jaakkola 2008; Berg 2011). Historical remains (tree stumps) from cutting trees rich in lichens, dating back to the early 1800s, can still be found within the Swedish reindeer herding area (Berg 2011). The majority of the remains date from the period 1844 to 1880, while the latest were from the 1930s. In Finland, lichen tree cutting was most common from the late 19th century until the Second World War (Itkonen 1948; Kortesalmi 2007).

There were large losses of reindeer due to mass starvation in the 1960s and 1970s, during a series of winters with difficult snow conditions affecting all three Fennoscandian countries (Helle & Säntti 1982; Vuojala-Magga et al. 2011; Berg et al. 2011; Riseth et al. 2016). This encouraged increased use of hay for winter feeding across large parts of the Finnish reindeer herding area (Helle & Saastamoinen 1979; Helle & Säntti 1982; Helle & Jaakkola 2008) and promoted the development of grain-based feeds formulated for reindeer in all three countries (numerous feeding experiments reviewed by Staaland & Sletten 1991). Composition and nutrient content of feeds for reindeer became an important topic for research (e.g., Skjenneberg & Slagsvold 1968). Loss of grazing areas due to modern forestry from the 1950s onwards (Finland and Sweden), together with other industrial activities and infrastructure development (Helle & Jaakkola 2008; Uboni et al. 2020; Riseth et al. 2016), gradually increased the need to feed reindeer in winter. Shrinking winter pastures increased reindeer densities in some locations and intensified grazing pressure on the remaining land, causing additional negative long-term effects on lichen pastures (Jaakkola et al. 2013; Kumpula et al. 2014).

The present use of feeding differs between the Fennoscandian countries, and also between regions. Temporary feeding of reindeer, targeted at strategic events such as gathering and migration, has been a rather common practice in all countries for several decades (Staaland & Sletten 1991). Feeding in order to prevent acute starvation ("emergency feeding") is also practised in all countries.

In particular, in the southern parts of the Finnish reindeer herding area, feeding during several winter months has become a regular part of the management system (Turunen & Vuojala-Magga 2014), while this has been more limited in Northern Finland and still rather sporadic in Sweden and Norway.

Finland

In Finland, regular winter feeding, either on pasture (Figure 12.1) or in enclosures (Figure 12.2), became a part of the herding system in the southern and central reindeer herding districts (RHDs) in the late 1980s and mid-1990s, mainly as a result of the detrimental impact of forestry on ground and arboreal lichen resources (Turunen et al. 2020). Since then the practice has also expanded towards the northern part of the Finnish reindeer herding area, although to a limited extent and mainly as a means to keep the herd in control or provide extra energy and nutrients for reindeer on winter pasture, especially when there are difficult snow conditions. In addition, seasonal migration and pasture rotation of reindeer generally occur over smaller areas in Finland compared to Sweden and Norway. In some RHDs this has led to increased grazing and trampling on lichen ranges during the snow-free season (Kumpula et al. 2011,

Figure 12.1 Free-ranging reindeer fed hay in the field.
Photo: Jouko Kumpula.

Figure 12.2 Reindeer eating from a feeding crib.
Photo: Jouko Kumpula.

2014), which might have contributed to the need to provide supplementary feed to reindeer.

Winter feeding of reindeer in Finland developed in combination with small-scale agriculture in the southern parts of the reindeer management area (Helle & Jaakkola 2008). The fact that herders in this area have been able to grow their own hay and had experience of feeding other animals has facilitated the development of feeding practices for reindeer (Turunen & Vuojala-Magga 2014). The gradual extension of winter feeding from south to north in Finland was influenced by increasing encroachment of forestry, and also by the Field Reservation Scheme, enacted in 1969. This scheme was established to cope with and reduce overproduction in the dairy sector. Thus, the state paid farmers to leave fields uncultivated. However, they were still allowed to use hay grown on these fields for feeding reindeer. Production of hay was further stimulated by subsidies when Finland joined the EU in 1995. As a result, most of the meadows and hay fields within the Finnish reindeer herding area are presently used almost entirely for production of hay for reindeer (Helle & Jaakkola 2008).

Reindeer in most parts of the Finnish reindeer herding area are thus regularly fed hay, grass silage, pellets or a combination of these during winter. The purpose is to keep reindeer in enough good condition and prevent starvation, as well as reduce loss to predators by keeping reindeer under control on pastures or gathered around feeding stations, and help to achieve or maintain pregnant females at an adequate nutritional status over winter. Reindeer are fed both in enclosures, where all or most forage is provided by the herder, and on pasture, where they can get part of their forage from natural vegetation. The duration of feeding depends on the annual weather and snow conditions. In favourable winters, with thin snow and no ground icing, reindeer can be fed for two or three months, whereas in years with a deep snow cover, ground icing or both, the feeding period may last for four or five months (Turunen & Vuojala-Magga 2014). Providing supplementary feed on natural pasture (Figure 12.1), as it is generally practised in the northern part of the Finnish reindeer herding area, usually lasts for one to three months in late winter, but lasted up to five months in the exceptionally harsh winter of 2019/2020 (Kumpula et al. 2020). Feeding of pregnant females during calving, combined with marking of the newborn calves before the animals are released onto natural pasture for the summer, is common in some of the central and northern RHDs (Turunen & Vuojala-Magga 2014).

Sweden and Norway

The situation regarding reindeer feeding in Sweden and Norway differs from Finland, although there is a large variation depending on region, RHD and individual herder (Landbruksdirektoratet 2017; Persson 2018). Although the reasons for feeding are similar to those cited for Finland, feeding is not as common and few herders regularly feed their reindeer during a large part of the winter. In Sweden, feeding during migration and gathering has been commonly practised for decades in many RHDs, while this practice seems to be less common in Norway. In both countries the need for so-called emergency feeding, to prevent starvation, has however increased during recent years, due to there being more winters with unfavourable weather and loss of land for grazing (Åhman et al. 2018; Horstkotte et al. 2020).

In cases when pastures are locked due to unfavourable snow conditions, reindeer herders usually try to move the animals to areas with better grazing conditions. Alternatively, they allow the reindeer to spread over larger areas in small herds, to search for suitable grazing conditions on their own. However, the presence of predators makes herders reluctant to use this strategy, and during periods with locked pastures they may, therefore, prefer to keep the reindeer in enclosures or in tight herds and provide supplementary feed.

Judging from the sale of commercial feeds for reindeer in Sweden (Uboni et al. 2020), the extent of feeding grew rapidly from the mid-1980s onwards. According to a survey of reindeer herders in Sweden, about half of the herders occasionally fed their reindeer during the late 1990s (Statistics Sweden 1999). Since then, the sale of feed for reindeer has increased, although it varies considerably between years.

The use of commercial reindeer feeds has previously been rather limited in Norway. Until recent years, a considerable part of factory-made reindeer feed was imported from Finland or Sweden. The increased need for feeding has, however, promoted domestic production in Norway. So far, these feeds are mainly used in the northern parts of the Norwegian reindeer herding area (Nordland, Troms and Finnmark).

In certain areas in both Sweden and Norway, radioactive fallout from the Chernobyl nuclear power plant accident in 1986 resulted in contaminated pastures and high levels of radioactive caesium in reindeer. This forced herders to feed reindeer in order to reduce contamination of the meat before slaughter (Åhman 1999, Skuterud et al. 2016). The cost of this type of feeding is refunded by the respective state and is still happening in a few RHDs in both countries (Wiklund et al. 2018).

Unlike in Finland, it is unusual in both Sweden and Norway for reindeer herders to own agricultural land and produce their own forage. Thus, herders mostly have to buy all the feed that they use. High costs prevent most herders from feeding on a large scale, if not absolutely necessary in order to save reindeer from starvation. Herders also find that it is difficult to obtain the desirable quality of hay or silage (Persson 2018). Lack of knowledge about dietary requirements of reindeer among both herders and the farmers who produce forage is regarded as a problem, especially as reindeer have different requirements than, e.g., cattle and sheep (Åhman et al. 2018). Some herders are therefore reluctant to use hay or silage at all, and rely solely on factory-made grain-based pellets, even though this kind of feed is less like the natural food eaten by reindeer, and therefore generally associated with more digestive problems (see below and Chapter 13).

Feeds and feeding practices

Commercial grain-based feeds for reindeer (hereafter referred to as "concentrates", although they can be used as the only feed for reindeer) are commonly used in all three countries and offered by a number of feed producers. All concentrates are nowadays manufactured from milled ingredients in the form of pellets. The ingredients are similar to those used in concentrates for other ruminants, although the relative proportion of ingredients differs. Concentrates for reindeer thus contain various types of grain (including by-products), by-products from the sugar industry (beet pulp, molasses), some sources of extra fat (e.g., rapeseed) and protein (e.g., distillers draff) as well as added minerals and vitamins. The composition is based on the numerous feeding experiments that were undertaken mainly during the 1960s and 1970s, and practical experiences of reindeer feeding since then. The nutritional quality of all concentrates for reindeer is rather similar, though some feeds are adapted for feeding reindeer prior to slaughter and contain, e.g., more protein (up to 14%), compared to basic feeds that are formulated primarily to prevent starvation and keep the reindeer in adequate condition over winter (usually containing 10–12% protein). Some feeds contain additives that are aimed at preventing ruminal acidosis, which is a common problem in reindeer when they have to switch from natural pasture to a grain-based diet (Åhman et al. 2018; Chapter 13).

According to Saarni and Nieminen (2011), the annual use of reindeer feed (including both hay/silage and pellets) in Finland is in the order of about 100–120 kg dry feed per live reindeer in the winter herd. However, the use of feed varies considerably between areas and years. In Sweden, two main producers of grain-based feeds for reindeer have been active on the market since the 1980s. The annual sales for these companies fluctuate between years, corresponding to 20–60 kg per reindeer in the winter herd (Uboni et al. 2020). This is combined with unknown amounts of silage and hay. In Norway, less than 1,000 tons of reindeer feed used to be produced annually (corresponding to an average of 4 kg per reindeer in the winter herd). There were, however, increasing imports from Sweden and Finland from the year 2000 onwards, which stimulated national production of concentrates for reindeer. In 2019, the production was about 10 kg per reindeer, and in the extreme winter of 2019/2020, with exceptionally deep snow, production reached 25 kg per reindeer (Landbruksdirektoratet 2020).

Grass silage and hay are commonly used to feed reindeer. Grass pellets are also sometimes used. Baled silage came into use during the 1980s and replaced much of the use of dry hay. Silage often contains less fibre (cellulose) than hay due to the harvesting and conservation process. This is important for reindeer as mixed feeders, since their digestive system cannot handle as much fibre as that of typical grazers such as cattle and sheep (Åhman et al. 2018).

Reindeer lichens, an important part of the natural winter diet of reindeer, are commonly used to complement concentrates and hay or silage, although in limited amounts. Reindeer greatly prefer lichens, and they are therefore used in order to accustom the reindeer to supplementary feeds and to being fed. Lichens can also be used in case of digestive problems linked to lack of adaptation to commercial feeds. Lichens are usually bought, or collected by the herders themselves, from areas outside the reindeer herding area. With a large herd or planned feeding for a longer period, lichens can only be a minor part of the feed given to the reindeer. In northern Norway, there is currently a shortage of lichens, due to the recent outbreak of Chronic Wasting Disease (CWD), that prevents herders from collecting (or buying) lichen from southern Norway, which was previously the tradition (see Norwegian Food Authority Regulation).

Feeding practices differ between herders (Figures 12.3 and 12.4). Some employ work-demanding systems, where the feed is spread manually on the snow for free-ranging reindeer, using no equipment other than a snowmobile, sled and shovel or pitchfork. Others have advanced systems, where reindeer are kept in enclosures and concentrates are provided in cribs, using various kinds of modern equipment and machinery which are common in agriculture (like silos for concentrates and bale cutters for silage). Sometimes the reindeer are kept in large enclosures with access to some natural pastures as well, while in other situations the animals have to rely entirely on the feed provided by the herder. Herders who feed free-ranging reindeer often need to transport concentrates (in cribs or sleighs behind their snowmobile) or bales of silage or hay over long

Figure 12.3 Distributing feed for reindeer.
Photo: Minna Turunen.

Figure 12.4 A reindeer eating grain-based feed (pellets) spread out on the snowmobile track.
Photo: Minna Turunen.

distances (several kilometres). Access to water is essential when reindeer are fed dry feed. Water can be provided in troughs (heated so that it does not freeze), or via a natural stream or river running through the feeding area. The availability of clean snow is often sufficient for reindeer fed on pasture.

Recommendations regarding feeding practices for reindeer are available in all three countries (e.g., Maijala et al. 2013; Rockström & Åhman 2017; Eilertsen & Winje 2017). Feeding of reindeer, and how to avoid or reduce the risks involved, is also a recurring topic at seminars and courses for reindeer herders. Nevertheless, there are many knowledge gaps, and herders are demanding more coproduction of knowledge and exchange of practical know-how between herders with various levels of experience (Horstkotte et al. 2020, Landbruksdirektoratet 2020).

Effects on reproduction and herd productivity

Body condition, i.e., body reserves of fat and protein, is a key factor for survival and reproduction in reindeer, as in other animals (Åhman & White 2018). When natural forage resources are limited, feeding is therefore generally expected to have positive effects on survival, reproduction and population growth, and thus on the productivity of the reindeer herd.

It is well documented that female body mass (BM) in autumn is positively correlated to the chance of the female getting pregnant and calving the following spring (e.g., Cameron et al. 1993; Rönnegård et al. 2003). Poor winter grazing conditions may nevertheless have negative effects on calving success (Vuojala-Magga et al. 2011), which can be counteracted by feeding (Ballesteros et al. 2013). Rognmo et al. (1983) demonstrated positive effects of enhanced energy and protein intake on foetal growth rate and early survival, and Eloranta & Nieminen (1986) showed that female BM prior to calving correlated with calf birth weight and early survival. Female BM in spring is also positively correlated to milk production (Jacobsen et al. 1981), which in turn strongly correlates to the early growth rate of the calf (White & Luick 1984). Consequently, the calves of reindeer that have been offered supplementary feed are often heavier than those of reindeer that have had to rely on natural pastures only (Säkkinen et al. 1999; Maijala & Nieminen 2001), which may positively affect their future performance.

According to earlier studies in Finland, one main benefit of introducing regular winter feeding has been increased and stabilized meat production (Kojola & Helle 1991, Helle & Kojola 1993, 1994; Kumpula et al. 1998). However, expanded and intensified feeding is currently a major financial burden for herders. The comparatively regular and extensive feeding practices in Finland seem nonetheless necessary to keep up the present productivity level in the face of the declining and unpredictable environment for winter grazing (Kumpula et al. 2002; Pekkarinen et al. 2015).

In Sweden, the use of feeding in the reindeer management system partly explains why population size has remained relatively stable, despite a considerable loss or reduced quality of winter pastures, and increasing occurrences of difficult winter weather and snow conditions (Uboni et al. 2020). The productivity (meat production relative to herd size) is nevertheless low compared to Finland, and probably less affected by feeding (Chapter 10).

Negative consequences and risks associated with feeding

Feeding is obviously associated with high financial costs, not only for the feed itself but also for facilities, equipment, machinery and fuel. It is also associated with a change of workload (other, and often heavier, work). Expenses for feeding may be one of the major costs in some RHDs (Rantamäki-Lahtinen 2008). However, feeding sometimes is the only profitable option for herders. The economic gain from feeding varies considerably depending on, e.g., pasture quality, meat prices and the subsidy system (see Chapter 11). It is thus difficult to assess the potential final economic gain of planned feeding and often hard for a herder to decide whether to feed or not based solely on economic grounds.

There are several health risks for reindeer associated with feeding (Åhman et al. 2018; Tryland et al. 2019). Many of the health problems are linked to poor adaptation of the digestive system to a novel diet. Others are related to keeping animals within restricted areas, particularly when they are fed in enclosures, and the associated increased risk of spreading infectious diseases. Stress in relation to handling also increases the risk of impaired health. Health problems related to feeding are further discussed in Chapter 13.

Feeding on pasture may result in increased grazing and trampling pressure on vegetation and soil due to high animal densities around feeding stations. Furthermore, left-over silage or hay has the potential to affect the natural vegetation. A two-year experiment in Finland (Turunen et al. 2013) showed that frequent feeding may cause changes in the soil and in vegetation composition. In the longer run, this can lead to a gradual shift from a nutrient-poor forest (sub-xeric heath forest type) towards a more nutrient-rich type of forest (mesic heath forest type). No invasive plant species were, however, introduced by spreading grass silage and hay, but the cover and height of the naturally occurring wavy hair grass (*Deschampsia flexuosa*) increased, while the cover of some mosses, lichens and shrubs declined.

In addition to the direct effects of feeding, there are potential long-term effects on reindeer grazing behaviour and on the reindeer management system as a whole, which are further discussed below in relation to herders' perspectives on feeding.

Economic support

As mentioned previously, reindeer herders in Finland can get EU subsidies for using fields to produce hay for reindeer, in accordance with the size of their fields. The Finnish "Act on compensation for damage caused to reindeer husbandry" (987/2011 and 655/2016) has not been used earlier to cover extra costs associated with feeding due to exceptionally difficult weather. Now, this is happening in relation to the exceptionally hard winter of 2019/2020 (Kumpula et al. 2020).

Reindeer herders in Sweden can apply for government support for feeding through the Sámi Parliament (so-called katastrofskadeskydd/disaster relief), when snow conditions make it impossible for the reindeer to dig for forage on the ground (STFS 2019:1). Compensation can then be paid to the RHD for up to 50% of the verified costs for buying feed for the reindeer. It has also become increasingly common that compensation is paid for encroachments on grazing areas caused by other forms of land use, like mines or wind parks, to cover feeding costs (Lawrence & Larsen 2017). Payment for feeding as a means to protect reindeer from predators has hitherto been uncommon. However, during the winter of 2019/2020, several RHDs were paid for feeding reindeer when they had to be fenced in order to protect them from wolves (https://sverigesradio.se/artikel/7383770).

The Norwegian government provides financial support for feeding under certain circumstances (Landbruksdirektoratet 2017). One is to cover feeding expenses for RHDs that are unable to utilize parts of their traditional winter pastures in Sweden due to the termination in 2005 of the former Reindeer Grazing Convention between Norway and Sweden. Herders can also apply for financial support to prevent losses to predators through the funding scheme "Prevention and conflict reducing measures" (Forebyggende og konfliktreduserende tiltak – FKT) administered by the regional governments. The money can be used for feeding in order to aid the tending of the herd. In Norway, RHDs are building "disaster funds" ("katastrofefond") which can be used for buying feed when winter pastures are unavailable. Up to 2 NOK per reindeer per day can be used for this purpose (Landbruksdirektoratet 2020). During the catastrophic winter of 2019/2020, all money in the funds was used up early in the winter, but after negotiations with the government, an extra 30 million NOK was added to the funds.

Herders' perceptions of feeding

Discussions between reindeer herders from Finland, Sweden and Norway at a workshop in 2018 (Horstkotte et al. 2020) revealed several short- and long-term concerns related to the feeding of reindeer. The health and welfare of the animals is obviously a worry for the herders, and more facts on this topic were requested. Many herders pointed out the value of exchanging knowledge and experiences with other herders in order to learn how to avoid health problems related to feeding. In line with this, reindeer herders in Finland interviewed by Turunen and Vuojala-Magga (2014) emphasized the importance of knowledge and experience, and the value of learning through the transfer of knowledge between herders to achieve good results when feeding reindeer.

High costs and increased workload, and the difficulty of making trade-offs between costs and benefits of feeding, were also highlighted by the herders at the workshop (Horstkotte et al. 2020). However, herders focused in particular on the long-term effects that the growing need to feed the reindeer could have on the reindeer management system as a whole. Herders place a strong emphasis on the fact that reindeer herding needs to be based on the use of natural pastures in order to be ecologically, economically and culturally sustainable (Chapter 5).

Herders see a potential risk of losing access to pastureland if government bodies, developers or society at large, without knowledge about reindeer husbandry, regard feeding in fenced areas as a solution and a substitute for natural pastures. However, in reality, supplementary feeding is only treating symptoms of absent or inaccessible forage resources (Horstkotte et al. 2020). Herders are also concerned about the risk of losing traditional experience-based knowledge about reindeer's use of winter pastures. Knowledge about reindeer, nature and landscape is learned through experience and participation in herding activities. Consequently, if herders need to spend more time at the feeding troughs than in the landscape with their herd, there is a risk that such knowledge is not transferred to future generations (a concern also reported by Risvoll & Hovelsrud 2016).

Another risk mentioned by herders (Horstkotte et al. 2020) is that meat and other products from reindeer will no longer be regarded as purely "nature-based" and "traditional". In spite of the fact that reindeer are generally slaughtered in autumn, after several months solely on natural pasture, and that the majority of the slaughtered reindeer are calves, this might have negative effects on the market.

Changes in the reindeer's behaviour and tameness were other concerns emphasized (Horstkotte et al. 2020). These issues were also raised in earlier surveys and interviews with reindeer herders (e.g., Helle & Jaakkola 2008; Vuojala-Magga et al. 2011; Turunen & Vuojala-Magga 2014; Persson 2018). Increased interactions between humans and reindeer during feeding make reindeer more tame and thus often easier to handle. Reindeer that are used to feeding can be attracted by forage and get other reindeer to follow, which helps to get the herd into round-up fences. On the other hand, reindeer that become too habituated to humans may lose their fear of vehicles and people, and therefore become more difficult to gather and move. Herders are also concerned about reduced vigilance making reindeer more susceptible to traffic accidents and predators (Turunen & Vuojala-Magga 2014).

Moreover, herders see a risk of effects on the reindeer's normal grazing and migration behaviour. For example, there is a concern that reindeer that get used to feeding will lose their willingness or ability to search for forage on their own (including the skill of digging through snow). If so, this will compromise their capability to live and survive on natural pastures (Turunen et al. 2016; Persson 2018; Horstkotte et al. 2020). Another concern is that reindeer feeding may make reindeer unnecessarily fat during a season when they are normally rather lean. This is especially important in pregnant females and can lead to problems at calving.

Future role of feeding

Feeding has evidently become an integral part of the reindeer management system in many areas of Fennoscandia. Feeding for short periods (a few days), and in specific situations, seldom creates any major problems and seems to be generally accepted by herders. The use of supplementary feed in acute situations when there is the risk of starvation and during winters with long periods of extreme weather conditions are strategies that have helped to stabilize reindeer numbers and maintain productivity (Helle & Kojola 2006; Uboni et al. 2020). The financial costs, however, are high, and there are obvious health risks associated with the change of diet, especially if the reindeer are already weakened. Access to land and flexibility to move to better areas are key factors for reducing the need for feeding. However, if encroachments on land and altered grazing conditions due to climate change continue to increase, reindeer herding will most certainly become even more sensitive to unfavourable weather.

The system with several months of feeding every winter is a management practice that is much disputed by herders but regarded as the only solution for survival by many herders in the regions where it is commonly applied. There are evidently many economic arguments against a reindeer herding system that depends strongly on feeding, but bio-economic model analyses show that it may sometimes be the only economically profitable solution for the herder (Pekkarinen et al. 2015). However, key arguments against feeding are that the right to grazing resources for reindeer may be questioned as a result, that knowledge and skills related to the utilization of land and how to herd reindeer will be lost, and that the reindeer will lose their ability to survive on natural forage in winter. If these things happen, it would be hard to return to a herding system based on the use of natural pastures all year round.

It has been repeatedly pointed out that the increased use of feeding is not a choice preferred by herders but rather forced on them by external factors related to encroachments by others on the land (including predator policy) and climate change (e.g., Risvoll et al 2016; Turunen et al. 2016; Horstkotte et al. 2020). There is a risk that frequent use of feeding will create an undesirable transition in the reindeer management system that will be difficult to escape from (Landauer et al. 2021; Chapter 14).

The future role of feeding is very much dependent on state policy related to the right to land and the relationship between reindeer husbandry and the surrounding society. Particularly in Sweden and Finland, restoration of degraded pastures through alternative forms of forest management, compared to the current practices, could improve the long-term availability of natural forage (Chapter 4). In Finland, a reorganization of seasonal grazing patterns, to protect lichen pastures from trampling and grazing during the snow-free season, could also promote forage availability.

Escaping or preventing the trap of systemic feeding is a complex task that requires action beyond the agency of reindeer herders themselves. Instead, the wider impact of different actors within the reindeer herding area, including state legislation, needs to be considered to avoid pushing reindeer herders in the direction of feeding. However, climate challenges will persist, and feeding will therefore continue to be a necessary response as crisis relief.

References

Åhman, B. (1999). Transfer of radiocaesium via reindeer meat to man – effects of countermeasures applied in Sweden following the Chernobyl accident. *Journal of Environmental Radioactivity*. 46, 113–120.

Åhman, B., Finstad, G.L. & Josefsen, T.D. (2018). Feeding and associated health problems. In: Tryland, M. & Kutz, S.J. (eds.) *Reindeer and caribou: Health and disease*. Boca Raton: Taylor & Francis. 135–156.

Åhman, B. & White, R.G. (2018). Rangifer diet and nutritional needs. In: Tryland, M. & Kutz, S.J. (eds) *Reindeer and caribou: Health and disease*. Boca Raton: Taylor & Francis. 107–134.

Ballesteros, M., Bårdsen B.J., Fauchald, P., Langeland, K., Stien, A. & Tveraa, T. (2013) Combined effects of long-term feeding, population density and vegetation green-up on reindeer demography. *Ecosphere*. 4, 45.

Berg, A., Gunnarsson, B. & Östlund, L. (2011). "At this point, the lichens in the trees are their only means of survival": A history of tree cutting for winter reindeer fodder by Sami people in Northern Sweden. *Environment and History*. 17, 265–289.

Cameron, R.D., Smith, W.T., Fancy, S.G., Gerhart, K.L. & White, R.G. (1993). Calving success of female caribou in relation to body weight. *Canadian Journal of Zoology*. 71, 480–486.

Eilertsen, S.M. & Winje E. (2017). Tilleggsfôring av reinsdyr. *NIBIO POP*. 3(43).

Eloranta, E. & Nieminen, M. (1986). Calving of the experimental reindeer herd in Kaamanen during 1970–85. *Rangifer Special Issue*. 1, 115–121.

Helle, T.P. & Jaakkola, L.M. (2008). Transitions in herd management of semi-domesticated reindeer in northern Finland. *Annales Zoologici Fennici*. 45, 81–101.

Helle, T. & Kojola, I. (1993). Reproduction and mortality of Finnish semi-domesticated reindeer in relation to density and management strategies. *Arctic*. 46, 72–77.

Helle, T. & Kojola, I. (1994). Body mass variation in semidomesticated reindeer. *Canadian Journal of Zoology*. 72, 681–688.

Helle, T. & Saastamoinen, O. (1979). The winter use of food resources of semi-domesticated reindeer in northern Finland. *Communicationes Instituti Forestalis Fenniae*. 95(6), 1–27.

Helle, T. & Säntti, V. (1982) Vinterkatastrofer inom renskötseln i Finland: Förluster och deras förebyggnade (Winter-catastrophes in the reindeer husbandry of Finland: Losses and their prevention). *Rangifer*. 2(1), 2–8.

Horstkotte, T., Lépy, É., Risvoll, C., Eilertsen, S.M., Heikkinen, H.I., Hovelsrud, G.K., Landauer, M., Löf, A., Omazic, A., Paulsen, M., Pekkarinen, A.-J., Sarkki, S., Tryland, M. & Åhman, B. (2020). *Supplementary feeding in reindeer husbandry – Results from a work-shop with reindeer herders and researchers from Norway, Sweden and Finland*. Umeå: Umeå University.

Itkonen, T.I. (1948). *Suomen Lappalaiset. Osa 1. Suomen Lappalaiset vuoteen 1945. Osa 2. Poronhoito*. Porvoo: WSOY.

Jaakkola, L.M., Heiskanen, M.M., Lensu, A.M. & Kuitunen, M.T. (2013). Consequences of forest landscape changes for the availability of winter pastures to reindeer (*Rangifer tarandus tarandus*) from 1953 to 2003 in Kuusamo, northeast Finland. *Boreal Environment Research*. 18, 459–472.

Jacobsen, E., Hove, K., Bjarghov, R.S. & Skjenneberg, S. (1981). Supplementary feeding of female reindeer on a lichen diet during the last part of pregnancy. Effects on plasma composition, milk production and calf growth. *Acta Agriculturæ Scandinavica*. 31, 81–86.

Kojola, I. & Helle, T. (1991). Productivity of semi-domesticated reindeer in Finland. *Rangifer*. 11, 53–63.

Kojola, I. & Helle, T. (1993). Calf harvest and reproductive rate of reindeer in Finland. *Journal of Wildlife Management*. 57, 451–453.

Kojola, I., Helle, T., Niskanen, M. & Aikio, P. (1995). Effects of lichen biomass on winter diet, body mass and reproduction of semi-domesticated reindeer *Rangifer t. tarandus* in Finland. *Wildlife Biology*. 1, 33–38.

Kortesalmi, J.J. (2007). *Poronhoidon synty ja kehitys Suomessa*. Suomalaisen Kirjallisuuden Seuran Toimituksia 1149. Tampere: Tammer-Paino Oy.

Kumpula, J., Colpaert, A. & Nieminen, M. (1998). Reproduction and productivity of semidomesticated reindeer in northern Finland. *Canadian Journal of Zoology*. 76, 269–277.

Kumpula, J., Colpaert, A. & Nieminen, M. (2002): Productivity factors of the Finnish semi-domesticated reindeer (*Rangifer t. tarandus*) stock during 1990s. *Rangifer*. 22, 3–12.

Kumpula, J., Stark, S. & Holand, Ø. 2011. Seasonal grazing effects by semi-domesticated reindeer on subarctic mountain birch forests. *Polar Biology*. 34(3), 441–453.

Kumpula, J., Jokinen, M., Siitari, J. & Siitari, S. (2020). *Talven 2019–2020 sää-, lumi- ja luonnonolosuhteiden poikkeuksellisuus ja vaikutukset poronhoitoon*. Luonnonvara- ja biotalouden tutkimus 58/2020. Helsinki: Luonnonvarakeskus.

Kumpula, J., Kurkilahti, M., Helle, T. & Colpaert, A. (2014): Both reindeer management and several other land use factors explain the reduction in ground lichens (*Cladonia* spp.) in pastures grazed by semi-domesticated reindeer in Finland. *Regional Environmental Change*. 14, 541–559.

Landauer, M., Rasmus, S. & Forbes, B.C. (2021). What drives reindeer management in Finland towards social and ecological tipping points? *Regional Environmental Change*. 21, 32.

Landbruksdirektoratet (2017). *Gjennomgang av kriseberedskapen i reindriften*. Rapport nr. 30/2017.

Landbruksdirektoratet (2020). *Gjennomgang av beitekrisen i reindriften 2020*. Rapport nr. 45/2020.

Lawrence, R. & Larsen, R.K. (2017). The politics of planning: assessing the impacts of mining on Sami lands. *Third World Quarterly*. 38(5), 1164–1180.

Maijala V., Kylmämaa L., Majuri K. & Mustonen J. (2013). Porojen talviruokinnan – Hyvien toimintatapojen opas. *Poromies*. 5/2013, Supplement.

Maijala V. & Nieminen M. (2001). Winter feeding of reindeer – experiences and results from Finland. *Rangifer Report*. No. 5, 16.

Pekkarinen, A.J., Kumpula, J. & Tahvonen, O. (2015). Reindeer management and winter pastures in the presence of supplementary feeding and government subsidies. *Ecological Modelling*. 312, 256–271.

Persson, A.-M. (2018). *Status of supplementary feeding of reindeer in Sweden and its consequences*. Umeå: Department of Wildlife, Fish and Environmental Studies, Swedish University of Agricultural Sciences, M.Sc. thesis 2018:15.

Rantamäki-Lahtinen, L. (2008). *Porotalouden taloudelliset menestystekijät.* MTT: n selvityksiä 156.

Riseth, J.A., Tømmervik, H. & Bjerke, J.W. (2016). 175 years of adaptation: North Scandinavian Sami reindeer herding between government policies and winter climate variability (1835–2010). *Journal of Forest Economics.* 24, 186–204.

Risvoll, C. & Hovelsrud, G.K. (2016). Pasture access and adaptive capacity in reindeer herding districts in Nordland, Northern Norway. *The Polar Journal.* 6, 87–111.

Rockström, U. & Åhman, B. (2017). *Råd och rekommendationer vid utfodring av renar.* Uppsala: Gård- och djurhälsan.

Rognmo, A., Markussen, K.A., Jacobsen, E., Grav, H.J. & Skytte Blix, A. (1983). Effects of improved nutrition in pregnant reindeer on milk quality, calf birth weight, growth, and mortality. *Rangifer.* 3, 10–18.

Rönnegård, L., Forslund, P. & Danell, Ö. (2002). Lifetime patterns in adult female mass, reproduction and offspring mass in semidomestic reindeer (*Rangifer tarandus tarandus*). *Canadian Journal of Zoology.* 80, 2047–2055.

Saarni, K. & Nieminen, M. (2011). *Tukipolitiikan vaikutukset Suomen poronhoitoon.* Riista- ja kalatalous, Tutkimuksia ja selvityksiä 10/2011.

Säkkinen, H., Timisjärvi, J., Eloranta, E., Heiskari, U., Nieminen, M. & Pukka, M. (1999). Nutrition-induced changes in blood chemical parameters of pregnant reindeer hinds (*Rangifer tarandus tarandus*). *Small Ruminant Research.* 32, 211–221.

Salmi, A. K., Niinimäki, S., & Pudas, T. (2020). Identification of working reindeer using palaeopathology and entheseal changes. *International Journal of Paleopathology.* 30, 57–67.

Skjenneberg, S. & Slagsvold, L. (1968). *Reindriftens og dens naturgrunnlag.* Oslo/Bergen/Tromsø: Universitetsforlaget.

Skuterud, L., Ytre-Eide, M.A., Hevrøy, T.H. & Thørrin, H. (2016). Caesium-137 in Norwegian reindeer and Sámi herders – 50 years of studies. In: Persson, B.R., Holm, E., Garcia-Tenorio, R. & Manjón, G. (eds) *II International Conference on Radioecological Concentration Processes: International Conference 50 years later.* Department of Applied Physics II, University of Seville, Seville, Spain.

Staaland, H. & Sletten, H. (1991). Feeding reindeer in Fennoscandia: The use of artificial food. In: Renecker L.A. & Hudson R.J. (eds) *Wildlife production: Conservation and sustainable development.* Fairbanks, Alaska: Agricultural and Forestry Experiment Station, University of Alaska Fairbanks, 227–242.

Statistics Sweden (SCB/SSR). (1999). *Svensk rennäring.* Halmstad: Svenska samernas riksförbund, Jordbruksverket, Sveriges lantbruksuniversitet, Statistiska centralbyrån.

STFS 2019:1. *Föreskrifter om ändring i Sametingets föreskrifter (STFS 2016:1) om katastrofskadeskydd.* Kiruna: Sametinget.

Tryland, M., Nymo, I.H., Romano, J.S., Mørk, T., Klein, J. & Rockström, U. (2019). Infectious Disease Outbreak Associated With Supplementary Feeding of Semi-domesticated Reindeer. *Frontiers in Veterinary Science.* 6, 126.

Turunen, M. & Vuojala-Magga, T. (2014). Past and present winter feeding of reindeer in Finland: herders adaptive learning of feeding practices. *Arctic.* 67, 173–188.

Turunen, M., Oksanen, P., Vuojala-Magga, T., Markkula, I., Sutinen, M.-L. & Hyvönen, J. (2013). Impacts of Winter Feeding of Reindeer on Vegetation and Soil in the Sub-Arctic: Insights from a feeding experiment. *Polar Research.* 32, 18610.

Turunen, M.T., Rasmus, S., Bavay, M., Ruosteenoja, K. & Heiskanen, J. (2016) Coping with difficult weather and snow conditions: Reindeer herders' views on climate change impacts and coping strategies. *Climate Risk Management.* 11, 15–36.

Turunen, M., Rasmus, S., Järvenpää, J. & Kivinen, S. (2020). Relations between forestry and reindeer husbandry in northern Finland: Perspectives of science and practice. *Forest Ecology and Management.* 457, 117677.

Uboni, A., Åhman, B. & Moen, J. (2020). Can management buffer pasture loss and fragmentation for Sami reindeer herding in Sweden? *Pastoralism.* 10, 13.

Vuojala-Magga, T., Turunen, M., Ryyppö, T. & Tennberg, M. (2011). Resonance Strategies of Sami Reindeer Herders in Northernmost Finland during Climatically Extreme Years. *Arctic.* 64, 227–241.

White, R.G. & Luick, J.R. (1984). Plasticity and constraints in the lactational strategy of reindeer and caribou. *Symposia of the Zoological Society of London.* 51, 216–232.

Wiklund, E., Malmfors, G., Finstad, G.L. Åhman, B., Skuterud, L., Adamczewski, J. & Garner, K. (2018). Meat quality and meat hygiene. In: Tryland, M. & Kutz, S.J. (eds.) *Reindeer and caribou: Health and disease.* Boca Raton: Taylor & Francis. 353–382.

13 Health and diseases of semi-domesticated reindeer in a climate change perspective

Morten Tryland, Birgitta Åhman and Javier Sánchez Romano

Introduction

Climate change, in combination with loss of land, has increased the need for supplementary winter feeding of semi-domesticated reindeer in Fennoscandia (Chapter 12). Feeding a reindeer requires that sufficient good quality feed is provided, while also keeping the right timing and conditions to avoid health and disease problems. Even though reindeer are gregarious, feeding may contribute to even higher animal density and more nose-to-nose contact between animals, thus facilitating transmission of infectious agents that may cause disease in reindeer. Poor hygienic conditions in feeding corrals may further increase exposure to pathogens, especially for newborn calves and other immunologically naïve animals.

Climate change affects all life in an ecosystem, including insects and other arthropod populations. Some are associated with reindeer as ectoparasites, with the subsequent harassment and stress, and may cause diseases and secondary infections. Other arthropods are temporary blood-sucking parasites that may also act as vectors for parasites, bacteria and viruses that can cause disease in reindeer.

In this chapter, we present some common conditions directly related to feed and feeding regimes, as well as the most relevant infectious agents and insect vectors that may cause or contribute to disease in semi-domesticated reindeer in Fennoscandia; we also address how they may be affected by climate change.

Health challenges and diseases associated with feeding

Disease conditions related to feed and feeding regimes

Several of the health problems associated with feeding are related to the change of diet and the fact that the rumen microorganisms need time to adapt to the new diet. Other problems are associated with unsuitable feedstuffs and poor hygienic quality of the feed.

DOI: 10.4324/9781003118565-18

Ruminal acidosis

In ruminal acidosis, digestion stops because the rumen content has become too acidic. It is a serious and relatively common condition when reindeer are shifted from natural pasture to grain-based commercial pelleted feed.

In a well-functioning reindeer rumen, the pH varies between 6 and 7 (Nilsson et al. 2006). Grain-based feeds for reindeer contain easily digestible carbohydrates (primarily starch) that lower the rumen pH when fed in large amounts. This affects the rumen microorganism population and may result in the growth of lactic acid-producing bacteria such as lactobacilli, making the rumen pH drop to 4–5 (Åhman et al. 2018). This can cause metabolic and life-threatening acidosis. Ruminal acidosis usually occurs within three weeks of the start of feeding (Åhman et al. 2018). Reindeer with ruminal acidosis are often lethargic, having poor appetite, increased thirst, decreased or ceased rumen contractions and sometimes diarrhoea (Rehbinder & Nikander 1999). The rumen content becomes liquid, which causes a typical sloshing sound from the stomach, thus the common name "*skvalpmage*" or "rippling belly".

It is difficult to treat severe cases of rumen acidosis, while early or milder cases may be cured by changing to a lichen diet or by providing a liquid energy mix with bicarbonate that can neutralize the rumen content.

Diarrhoea

Diarrhoea in reindeer is relatively common at the onset of feeding with grain-based pellets but has also been observed in reindeer that have been fed high-fibre diets (e.g., Josefsen et al. 2007). Diarrhoea may be only transient, but severe cases must be treated. Diarrhoea can also be caused by bacterial infections in the digestive tract, discussed later in this chapter.

Wet belly

A seemingly unique condition in reindeer that is associated with feeding is wet belly syndrome. Affected reindeer start to perspire, making the haircoat wet under the belly and often down the legs (Figure 13.1), sometimes also in the neck region (Åhman et al. 2002). Additional signs may be that the reindeer have an extraordinary appetite and that they get cold and curl up when lying down. The condition has been reported since the 1960s, when feeding experiments with reindeer started (e.g., Persson 1967).

The reason for wet belly is unknown. Hay or high-fibre diets containing straw have been suggested as causes of wet belly (Jacobsen & Skjenneberg 1977). However, Åhman et al. (2002) showed that wet belly may also occur when reindeer are fed a "natural" diet (e.g., lichens, bilberry shrubs and willow leaves) with no grass forage. Thus, although linked to feeding, wet belly seems not to be associated with any special feedstuff. Nonetheless, it is usually effective to change the feed.

Figure 13.1 Reindeer with wet belly. The condition is characterized by wet fur under the belly and down the hind legs and by an extraordinarily strong appetite.
Photo: Svenska Samernas Riksförbund/Gård & Djurhälsans bildarkiv för renens hälsa.

Bloat

Bloat, or ruminal tympany, occasionally affects reindeer during feeding (Åhman et al. 2018). It happens when the rumen is filled with gas or foam, and the animal is unable to burp up the gas. This causes a high and rapidly increasing pressure in the abdomen that may obstruct breathing and blood circulation, and the animal can die. The probable reason is the intake of large amounts of concentrates.

Accumulation of grass

The digestive system of reindeer is not adapted to handling large amounts of fibre (Hofmann 1989). Accumulation of undigested grass in the rumen is thus a well-known condition in reindeer that are fed mainly with hay or grass silage (Åhman et al. 2018). When grass is not digested, the reindeer do not get enough energy, remains hungry and continues to eat, but may die from emaciation although its rumen is full of forage. In most cases, the condition can be reversed if the animal is provided with more easily digested feed.

Infections and diseases associated with increased animal density and hygienic conditions

Feeding of reindeer will increase animal-to-animal contact and may also contribute to unfavourable hygienic conditions facilitating transmission of reindeer pathogens, especially in corrals and when the animals are fed over extended periods of time. Wet conditions and accumulation of faeces and mud may create a situation that results in infections affecting young calves (Foster 2010; Wikström 2014). Reindeer are sometimes infected with opportunistic bacteria, such as staphylococci, streptococci, *Escherichia coli* and *Trueperella pyogenes*, which may cause localized or generalized infectious diseases, but also with specific bacterial pathogens, such as *Salmonella* sp., *Moraxella* sp., *Listeria* sp., *Mycoplasma* sp. and others (Josefsen et al. 2019). Some selected infectious diseases in reindeer that may be associated with feeding are discussed below.

Infectious keratoconjunctivitis: Alphaherpesvirus and secondary bacterial infections

The eye disease infectious keratoconjunctivitis (IKC) was first thoroughly described in Scandinavian reindeer in Sweden (Bergman 1912). IKC in reindeer has been described as a multi-factorial disease, and a plethora of microorganisms have been identified in the eyes of affected reindeer, including the alphaherpesvirus cervid herpesvirus 2 (CvHV2), *Chlamydia* spp. and *Moraxella* spp. (Sánchez Romano et al. 2018; 2019; Tryland et al. 2009). CvHV2 is enzootic in most reindeer populations in Fennoscandia (das Neves et al. 2010), and a clinical trial with this virus demonstrated that it had the capacity to cause severe IKC in reindeer with no previous exposure and immunity to the virus (Tryland et al. 2017). However, this finding does not exclude the possibility of other pathogens being involved in the pathogenesis of IKC in reindeer (Sánchez Romano et al. 2018; 2019).

IKC may be seen in single animals in a herd but can also appear as large outbreaks, affecting tens or hundreds of animals, primarily calves and yearlings. Such outbreaks have been associated with stress and supplementary feeding (Tryland et al. 2009; Sánchez Romano et al. 2019). IKC can affect one or both eyes, and it often starts with increased lacrimation and fur discoloration under the eyes that can rapidly progress to purulent secretions, corneal and periorbital oedema, conjunctivitis and keratitis. The most severe cases may result in corneal ulcers, eye ruptures and permanent blindness (Tryland et al. 2009; 2017).

The lack of an effective treatment for CvHV2 together with its enzootic status make it difficult to control viral IKC outbreaks, but broad-spectrum antibiotics have been used to control secondary bacterial infections (Sánchez Romano et al. 2019; Tryland et al. 2009). Single cases are often slaughtered at an early stage by the herders, whereas outbreaks are often dealt with by a veterinarian.

Parapoxvirus (contagious ecthyma)

Contagious ecthyma (CE) is a disease caused by viruses from the genus *Parapoxvirus* (family *Poxviridae*). The disease has been diagnosed in semi-domesticated reindeer under natural herding conditions in Sweden, Finland and Norway (Tryland et al. 2019). The parapoxvirus Orf (ORFV) is distributed worldwide and affects small ruminants and a wide range of wild ruminant species; it may also cause painful cutaneous lesions in people handling infected animals. Another parapoxvirus, Pseudocowpoxvirus (PCPV), which has cattle as its main reservoir, has been associated with several CE outbreaks in Finland (Tryland et al. 2019).

Contagious ecthyma in reindeer is usually characterized by proliferative "cauliflower-like" lesions in the skin around the mouth and nostrils, as well as in the oral mucosa (Tryland et al. 2019). CE starts with the entry of the virus through small skin or mucosal lesions. The initially small and pink proliferative lesions develop to larger masses, often covered by thick, black crusts. Animals in the later stages of the disease may be unable to eat, and reduced body condition and emaciation can be observed. Currently, there is no specific treatment against CE, but supportive therapy and antibiotics can be used to control secondary bacterial infections.

Necrobacillosis

Necrobacillosis is caused by the bacterium *Fusobacterium necrophorum*, which is an obligate anaerobic rod that is part of the normal ruminal microbiota of ruminants, including reindeer (Aagnes et al. 1995). It is thus present in the environment but not able to penetrate intact skin or mucosal membranes. Abrasions and lesions, caused by external factors, viral infections or the eruption of new teeth in young animals, may pave the way for opportunistic infection by this bacterium. Although *F. necrophorum* is regarded as the primary pathogen, other opportunistic bacteria may contribute to the disease, such as *Trueperella pyogenes* and *Staphylococcus aureus* (Josefsen et al. 2019). In the past, necrobacillosis was mainly recorded as an infection of the feet, but also of the oral mucosa (Skjenneberg & Slagsvold 1968; Rehbinder & Nikander 1999). During recent disease outbreaks in semi-domesticated reindeer, the oral form has dominated, being associated with corralling and supplementary feeding (Wikström 2014; Tryland et al. 2019b). However, necrobacillosis may also affect the rumen and cause severe lesions, disease and mortality in animals showing no clinical signs in the mouth (Figure 13.2).

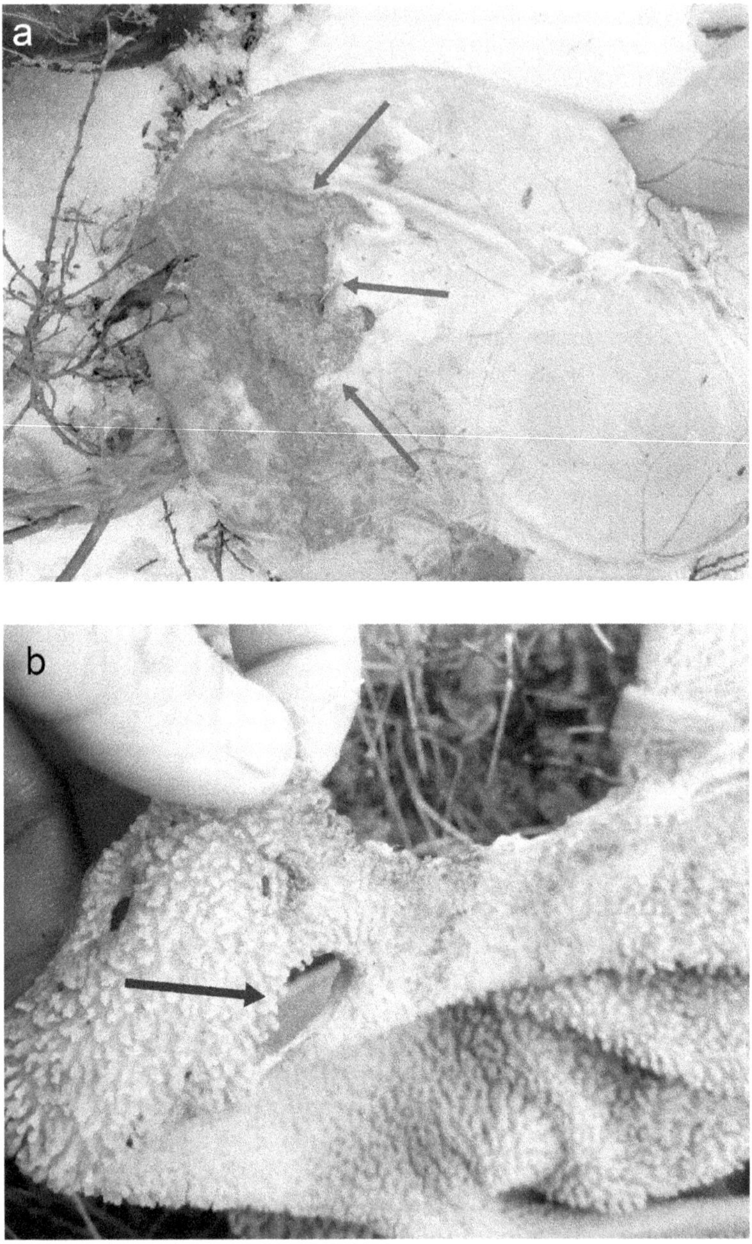

Figure 13.2 Necrobacillosis in reindeer can usually be identified as affecting the distal part of the legs (digital necrobacillosis) or the oral mucosa (oral necrobacillosis). However, the infection may also affect internal organs such as the rumen or liver, with no visible necrotic lesion on the oral mucosa. (a) the outside (serosa) of the rumen was necrotized and rumen content was leaking out in the abdomen. (b) An inspection of the inside of the rumen (mucosa) confirmed necrotic lesions.

Photo: Svenska Samernas Riksförbund/Gård & Djurhälsans bildarkiv för renens hälsa.

Pasteurellosis

Pasteurellosis in reindeer refers to a disease caused by the bacterium *Pasteurella multocida* (Josefsen et al. 2019), which is found as a commensal organism in the throat and the upper respiratory tract of many animal species. The potential pathogen is transmitted via direct contact between animals and is known to cause disease when associated with stress, such as heat, drought, insect harassment and animal transport, especially affecting young animals (Nordkvist & Karlsson 1962). In reindeer, pasteurellosis is usually characterized by per-acute haemorrhagic septicaemia, and animals can be found dead with no previously recorded clinical signs. With a more prolonged disease course, nasal discharge and coughing are common, and the calves display weakness and apathy and often die (Josefsen et al. 2019). Pasteurellosis in reindeer was first described to appear as large epizootics among semi-domesticated reindeer in Norway and Sweden in 1912–1914 (Horne 1915). Several outbreaks of pasteurellosis have been reported in Fennoscandia during the past century. An outbreak in a slaughter corral in Norway in 2010 caused septicaemia and killed 44 reindeer calves (Mørk et al. 2014).

Altered distribution of arthropods, arthropod vectors and associated reindeer pathogens

Some arthropods such as ticks, mosquitoes and midges feed on reindeer as temporary blood-sucking parasites. Through this blood-sucking activity, arthropods may act as vectors and transmit parasites, bacteria and viruses between hosts. The development and activity of arthropods depend on climatic conditions, such as wind, precipitation and temperature, which are factors that are expected to change in the northern regions (IPCC 2021). Thus, climate change may impact the epidemiology of vector-borne diseases (Wittmann & Baylis 2000; Ogden & Lindsay 2016). The generally low temperatures in Arctic regions are, in fact, close to minimum requirements for insect locomotion, and thus largely determine their activity patterns (Strathdee & Bale 1998). Snow cover, bird migration and plant flowering have a great effect on the seasonal development of insects and other arthropods (Strathdee & Bale 1998; Høye et al. 2007). They may also be quick responders to climatic conditions, having a short generation time, which also affects their role as vectors for infectious diseases for *Rangifer* (Boggs 2016).

Deer ked (Lipoptena cervi)

The deer ked (*Lipoptena cervi*) is a blood-sucking ectoparasite that is mainly associated with moose (*Alces alces*) and roe deer (*Capreolus capreolus*) in Fennoscandia (Välimäki et al. 2010). It has, however, expanded its range northwards during the last five years and is now affecting the reindeer herding area in Finland (Kynkäänniemi et al. 2020). The adults feed on the host's blood and reproduce in its fur, potentially causing hair breakage and heat loss (Härkönen et al. 2010). Even though the potential impact of deer ked in semi-domesticated reindeer has still not been thoroughly studied, there are indications of acute behavioural disturbances and an increase in potential stress due to deer ked infestation, which may pave the way for the appearance of several of the infectious diseases mentioned above (Kynkäänniemi et al. 2014).

Reindeer pathogens transmitted by ticks

The tick (arthropod, class Arachnida) *Ixodes ricinus* is the primary arthropod vector of zoonotic diseases in Europe (Gilbert 2010; Heyman et al. 2010). This species mainly feeds on rodents, hares and cervids, but also on birds, livestock, pets and humans. The distribution of *I. ricinus* is expanding towards higher latitudes and upwards in elevation from coastal to inland areas (Medlock et al. 2013; Jore et al. 2014; Mysterud et al. 2017). The conditions for ticks in higher altitudes may improve with predicted climate change scenarios, increasing the impact of ticks as disease vectors (Gilbert 2010). In Norway, ticks are now emerging on reindeer in Nordland county, sometimes combined with undiagnosed diseases. A recent study (2018–2019) indicated that ticks are now present in almost all the northern municipalities in Sweden, and not only in coastal regions (Jaenson et al. 2012).

Anaplasmosis (Tick-borne fever)

Anaplasma phagocytophilum is a worldwide tick-borne bacterium that causes tick-borne fever in sheep, cattle and wildlife; in humans the disease is called anaplasmosis (Stuen et al., 2013). In general, the clinical symptoms of tick-borne fever are high fever, anorexia and dullness. Anti-anaplasma antibodies have recently been detected among reindeer in Nordland county, Norway, suggesting exposure and disease in the population (unpublished data). An experimental inoculation of reindeer with blood from infected sheep caused fever five days post inoculation, followed by dullness and anorexia, as well as a severe reduction of white blood cells by 40–85%, demonstrating that reindeer are susceptible to *A. phagocytophilum*.

Borreliosis (Lyme's disease)

Borreliosis (Lyme's disease) is caused by bacteria of the *Borrelia burgdorferi* sensu lato complex, which, in Eurasia, is transmitted by *I. ricinus* and *I. persulcatus*. Borreliosis is estimated to affect 65,000 people in Europe annually (Berger 2014). A study of ticks in Nordland county, Norway, revealed that 21% of the nymphs and 46% of adult ticks contained *B. burgdorferi* bacteria, suggesting that reindeer are most probably exposed (Hvidsten et al. 2015).

Babesiosis

The disease babesiosis is caused by a group of protozooans. Babesiosis is characterized by high fever and blood in the urine (haematuria), and high mortality. Studies in southern Sweden revealed that 4% of the ticks carried *Babesia* spp. (Karlsson & Andersson 2016), and 53% of the cattle were exposed to *B. divergens* (Andersson et al. 2017). Babesiosis is only rarely reported from Norway and Finland. Reindeer have been shown to be susceptible to babesiosis (Langton et al. 2003; Wiegman et al. 2015).

Reindeer pathogens transmitted by mosquitoes and midges

Setaria tundra

Setaria tundra (*Filarioidea*; *Onchocercidae*) is a mosquito-borne filaroid nematode of which there have been several outbreaks in semi-domesticated reindeer in Finland, characterized by peritonitis and decreased body condition (Kutz et al. 2019). Adult forms of *S. tundra* live in the peritoneal cavity of reindeer (definitive host). A larval stage (L1) is released into the bloodstream and ingested by mosquitoes, in which they develop into stage L3 infective larvae in a temperature-dependent process, taking approximately two weeks at 21 °C (Laaksonen et al., 2009), before being transmitted to new hosts during mosquito feeding.

Bluetongue virus (BTV)

Bluetongue virus (BTV; genus *Orbivirus*, family *Reoviridae*) is transmitted by biting midges (*Culicoides* spp.) but can also have oral and transplacental transmission (Backx et al. 2009). BTV causes an acute disease in naïve sheep, with fever, excessive salivation, oedema of the face and cyanosis of the tongue and lips (hence the name "bluetongue"). BTV also infects other domestic animals and most species of wild ruminants, although frequently with no clinical symptoms. Increased ambient temperatures may shorten the time from the uptake of the virus in a vector to when it is infective (Wittman et al. 2002) and facilitate virus replication, contributing to a longer season with infective vectors (Mullens et al. 1995). BTV recently expanded its distribution, appearing in Denmark in 2007, Sweden in 2008 and Norway in 2009, but never seemed to reach the reindeer populations in these countries (Tryland et al. 2019).

Schmallenberg virus (SBV)

Schmallenberg virus (SBV; genus *Orthobunyavirus*, family *Bunyaviridae*) was first reported in dairy cattle in 2011 in Germany and the Netherlands, where it caused fever, diarrhoea and reduced milk production. This virus may also cause congenital malformations in newborn lambs, goat kids and calves, as well as premature birth, stillbirth or the birth of mummified foetuses (Wernike et al. 2014). SBV is transmitted by biting midges (*Culicoides* spp.) or mosquitoes. In Sweden, a serological screening of roe deer, red deer, moose and fallow deer (*Dama dama*) revealed SBV-antibodies in samples obtained after the vector season of 2012, but no such antibodies were detected in 2015. A possible explanation for why SBV did not become established among wild cervids in

Sweden, in contrast to more southern parts of Europe, is the occurrence of a vector-free season during winter (Malmsten et al. 2017). Serological screenings of wild and semi-domesticated reindeer in Norway and Finland did not demonstrate exposure of reindeer to SBV (Tryland et al. 2019).

Reindeer pathogens transmitted by other insect vectors

Rumenfilaria andersoni

Rumenfilaria andersoni (*Splendidofilariinae*; *Onchocercidae*) is a lymphatic-dwelling filaroid nematode that can be found in the lymphatic vessels of reindeer and other cervids, and as microfilariae larvae in peripheral blood. This parasite is transmitted by blood-sucking insects, but the actual vector, geographical distribution and health impact in reindeer remain unknown (Grunenwald et al. 2016; Laaksonen et al. 2015).

Onchocerca spp.

The parasite *Onchocerca skrjabini* (=*Onchocerca tarsicola*) is transmitted by blackflies (*Simuliidae*) and biting midges (*Ceratopogonidae*), which transport the subcutaneous microfilaria from one host to another. The infection has been described in reindeer in Sweden and Finland (Kutz et al. 2019; Laaksonen et al. 2017). The adult parasites are often found in connective tissues surrounding the tendons of the tibio-tarsal and radio-carpal joints (Rehbinder et al. 1979) and severe infections have been associated with granulomatous nodules in most organs and severe haemorrhagic tarsitis in Finnish reindeer (Kutz et al. 2018).

Concluding remarks

In this chapter, we have described selected health conditions, diseases and infections in semi-domesticated reindeer that, directly or indirectly, will or may be affected by climate change. Climate change will affect ecosystems and may change reindeer herding. To track these changes and their potential impact on reindeer health, it is necessary to gather data on health and disease parameters from reindeer continuously, to be able to support the reindeer herding industry in the future.

References

Aagnes, T.H., Sørmo, W. & Mathiesen, S.D. (1995). Ruminal microbial digestion in free-living, in captive lichen-fed, and in starved reindeer (*Rangifer tarandus tarandus*) in winter. *Applied and Environmental Microbiology*. 61(2), 583–591.

Åhman, B., Finstad, G.L. & Josefsen, T.D. (2018). Feeding and associated health problems. In: Tryland, M. & Kutz, S.J. (eds) *Reindeer and caribou: Health and disease*. Boca Raton: Taylor & Francis. 135–156.

Åhman, B., Nilsson, A., Eloranta, E. & Olsson, K. (2002). Wet-belly in reindeer (*Rangifer tarandus tarandus*) in relation to body condition, body temperature and blood constituents. *Acta Veterinaria Scandinavica*. 43, 85–97.

Andersson, M.O., Vichová, B., Tolf, C., Krzyanowska, S., Waldenström, J. & Karlsson, M.E. (2017). Co-infection with *Babesia divergens* and *Anaplasma phagocytophilum* in cattle (Bos taurus), Sweden. *Ticks and Tick-borne Diseases*. 8, 933–935.

Backx, A., Heutink, R., van Rooij, E. & van Rijn, P. (2009). Transplacental and oral transmission of wild-type bluetongue virus serotype 8 in cattle after experimental infection. *Veterinary Microbiology*. 138(3–4), 235–243.

Berger, S. (2014). *Lyme disease: Global status. 2014 edition*. Los Angeles, CA: GIDEON Informatics Inc. P.

Bergman, A.M. (1912). Smittosam hornhinneinflammation, dermatitis infectiosa, hos ren. *Skand Vet Tidskr*. 2, 145–177.

Boggs, C.L. (2016). The fingerprints of global climate change on insect populations. *Current Opinion in Insect Science*. 17, 69–73.

das Neves C.G., Roth S., Rimstad E., Thiry E. & Tryland M. (2010). Cervid herpesvirus 2 infection in reindeer: a review. *Veterinary Microbiology*. 143(1), 70–80.

Fernández-Pacheco, P., Fernández-Pinero, J., Agüero, M. & Jiménez-Clavero, M.A. (2008). Bluetongue virus serotype 1 in wild mouflons in Spain. *The Veterinary Record*. 162(20), 659–660.

Foster, A. (2010). Common conditions of reindeer. *In Practice*. 32(10), 462–467.

Gilbert, L. (2010). Altitudinal patterns of tick and host abundance: a potential role for climate change in regulating tick-borne diseases?. *Oecologia*. 162, 217–225.

Grunenwald, C.M., Carstensen, M., Hildebrand, E., Elam, J., Laaksonen, S., Oksanen, A. & Gerhold, R.W. (2016). Epidemiology of the lymphatic-dwelling filarioid nematode Rumenfilaria andersoni in free-ranging moose (*Alces alces*) and other cervids of North America. *Parasites & Vectors*. 9, 450.

Härkönen, L., Härkönen, S., Kaitala, A., Kaunisto, S., Kortet, R., Laaksonen, S. & Ylönen, H. (2010). Predicting range expansion of an ectoparasite – the effect of spring and summer temperatures on deer ked (*Lipoptena cervi*) performance along a latitudinal gradient. *Ecography*. 33, 906–912.

Heyman, P., Cochez, C., Hofhuis, A., Van Der Giessen, J., Sprong, H., Porter, S.R., Losson, B., Saegerman, C., Donoso-Mantke, O., Niedrig, M. & Papa, A. (2010). A clear and present danger: tick-borne diseases in Europe. *Expert Review of Anti-infective Therapy*. 8(1), 33–50.

Hofmann, R.R. (1989). Evolutionary steps of ecophysiological adaptation and diversification of ruminants: a comparative view of their digestive system. *Oecologia*. 78, 443–457.

Horne, H. (1915). Rensyke (Hæmorrhagisk septikæmi hos ren). *Norsk Veterinærtidsskrift*. 27, 41–47, 73–84, 105–119.

Høye, T.T., Post, E., Meltofte, H., Schmidt, N.M. & Forchhammer, M.C. (2007). Rapid advancement of spring in the High Arctic. *Current Biology*. 17(12), R449–R451.

Hvidsten, D., Stordal, F., Lager, M., Rognerud, B., Kristiansen, B.E., Matussek, A., Gray, J. & Stuen, S. (2015). Borrelia burgdorferi sensu lato-infected *Ixodes ricinus* collected from vegetation near the Arctic Circle. *Ticks and Tick-Borne Diseases*. 6(6), 768–773.

IPCC (2021). *Climate Change 2021: The Physical Science Basis. Contribution of Working Group I to the Sixth Assessment Report of the Intergovernmental Panel on Climate Change*. Cambridge: Cambridge University Press.

Jacobsen, E. & Skjenneberg, S. (1977). Resultater og erfaringer fa foring av rein med høy, surfor, "sauenøtt" og "komplettfor m/halm". *Forskning og forsøk i landbruket.* 28, 651–660.

Jaenson, T.G.T., Jaenson, D.G.E., Eisen, L., Peterson, E. & Lindgren, E. (2012). Changes in the geographical distribution and abundance of the tick Ixodes ricinus during the past 30 years in Sweden. *Parasites & Vectors.* 5, 8.

Jore, S., Vanwambeke, S.O., Viljugrein, H., Isaksen, K., Kristoffersen, A.B., Woldehiwet, Z., Johansen, B., Brun, E., Brun-Hansen, H., Westermann, S., Larsen, I.L., Ytrehus, B. & Hofshagen, M. (2014). Climate and environmental change drives *Ixodes ricinus* geographical expansion at the northern range margin. *Parasites & Vectors.* 7, 11.

Josefsen, T.D., Sørensen, K.K., Mørk, T., Mathiesen, S.D & Ryeng, K.A. (2007). Fatal inanition in reindeer (*Rangifer tarandus tarandus*): Pathological findings in completely emaciated carcasses. *Acta Veterinaria Scandinavica.* 49, 27.

Josefsen, T.D., Oksanen, A. & Gjerde, B. (2014). Parasitter hos rein i Fennoskandia, en oversikt. *Norsk Veterinærtidskrift.* 126, 185–201.

Josefsen, T., Mørk, T., Nymo, I.H. (2019). Bacterial infections and diseases. In: Tryland, M. & Kutz, S.J. (eds) *Reindeer and caribou: Health and disease.* Boca Raton: Taylor & Francis. 237–271.

Karlsson M.E. & Andersson, M.O. (2016). Babesia species in questing *Ixodes ricinus*, Sweden. *Ticks and Tick-borne Diseases.* 8, 10–12.

Kutz, S., Laaksonen, S. Åsbakk, K. & Nilssen, A.C. (2019). Parasitic infections and diseases. In: Tryland, M. & Kutz, S.J. (eds) *Reindeer and caribou: Health and disease.* Boca Raton: Taylor & Francis. 177–235.

Kynkäänniemi, S.M., Kettu, M., Kortet, R., Härkönen, L., Kaitala, A., Paakkonen, T., Mustonen, A.M., Nieminen, P., Härkönen, S., Ylönen, H. & Laaksonen, S. (2014). Acute impacts of the deer ked (*Lipoptena cervi*) infestation on reindeer (*Rangifer tarandus tarandus*) behavior. *Parasitology Research.* 113(4), 1489–1497.

Kynkäänniemi, S.M., Kortet, R. & Laaksonen, S. (2020). Range expansion and reproduction of the ectoparasitic deer ked (*Lipoptena cervi*) in its novel host, the Arctic reindeer (*Rangifer tarandus tarandus*), in Finland. *Parasitology Research.* 119, 3113–3117.

Laaksonen, S., Solismaa, M., Kortet, R., Kuusela, J. & Oksanen, A. (2009). Vectors and transmission dynamics for *Setaria tundra* (Filarioidea; Onchocercidae), a parasite of reindeer in Finland. *Parasites & Vectors.* 2(1), 1–10.

Laaksonen, S., Oksanen, A. & Hoberg, E. (2015). A lymphatic dwelling filarioid nematode, *Rumenfilaria andersoni* (Filarioidea; Splendidofilariinae), is an emerging parasite in Finnish cervids. *Parasites and Vectors.* 8, 228.

Laaksonen, S., Oksanen, A., Kutz, S., Jokelainen, P., Holma-Suutari, A. & Hoberg, E. (2017). Filarioid nematodes, threat to arctic food safety and security. *Game Meat Hygiene: Food Safety and Security.* 2, 213–223.

Langton, C., Gray, J., Waters, P.F. & Holman, P.J. (2003). Naturally acquired babesiosis in a reindeer (*Rangifer tarandus tarandus*) herd in Great Britain. *Parasitology Research.* 89, 194–198.

Malmsten, A., Malmsten, J., Blomqvist, G., Näslund, K., Vernersson, C., Hägglund, S., Dalin, A.M., Ågren, E.O. & Valarcher, J.F. (2017). Serological testing of Schmallenberg virus in Swedish wild cervids from 2012 to 2016. *BMC Veterinary Research.* 13(1), 84.

Medlock, J, Hansford, K, Bormane, A, Derdakova, M, Estrada-Peña, A, George, J-C, Golovljova, I, Jaenson, T, Jensen, J-K, Jensen, P, Kazimirova, M, Oteo, JA, Papa, A, Pfister, K, Plantard, O, Randolph, S, Rizzoli, A, Santos-Silva, MM, Sprong, H, Bortel, W. (2013). Driving forces for changes in geographical distribution of *Ixodes ricinus* ticks in Europe. *Parasites & Vectors.* 6(1), 1–11.

Mørk, T., Sunde, M., Josefsen, T.D. (2014). Bakterieinfeksjoner hos rein [Bacterial infections in reindeer]. *Norsk Veterinærtidsskrift*. 126(2), 222–228.

Mullens, B.A., Tabachnick, W.J. Holbrook, F.R. & Thompson, L.H. (1995). Effects of temperature on virogenesis of bluetongue virus, serotype 11 in Culicoides variipennis sonorensis. *Medical and Veterinary Entomology*. 9(1), 71–76.

Mysterud, A., Jore, S., Østerås, O. & Viljugrein, H. (2017). Emergence of tick-borne diseases at northern latitudes in Europe: a comparative approach. *Scientific Reports*. 7, 16316.

Nilsson, A., Åhman, B., Murphy, M. & Soveri, T. (2006). Rumen function in reindeer (*Rangifer tarandus tarandus*) after sub-maintenance feed intake and subsequent feeding. *Rangifer*. 26, 73–83.

Nordkvist, M. & Karlsson, K.A. (1962). Epizootisk förlöpande infektion med *Pasteurella multocida* hos ren. *Nordisk veterinaermedicin*. 14, 1–15.

Ogden, N.H. & Lindsay, L.R. (2016). Effects of climate and climate change on vectors and vector-borne diseases: ticks are different. *Trends in Parasitology*. 32(8), 646–656.

Persson, S. (1967). Studier av renarnas energibehov. *Kungliga Lantbruksstyrelsen, Meddelanden – Serie B. Lantbruksavdelningen. Rennäringen*. (75), 23.

Rehbinder, C., Christensson, D. & Forssell, I. (1979). Investigations on parasitic visceral granulomas in reindeer. *Nordisk veterinaermedicin*. 31(6), 282–283.

Rehbinder C. & S. Nikander. (1999). *Ren och rensjukdomar*. Uppsala: Studentlitteratur.

Sánchez Romano, J., Mørk, T., Laaksonen, S., Ågren, E., Nymo, I.H., Sunde, M. & Tryland, M. (2018). Infectious keratoconjunctivitis in semi-domesticated Eurasian tundra reindeer (*Rangifer tarandus tarandus*): microbiological study of clinically affected and unaffected animals with special reference to cervid herpesvirus 2. *BMC Veterinary Research*. 14(1), 15.

Sánchez Romano, J., Leijon, M., Hagström, Å., Jinnerot, T., Rockström, U.K. & Tryland, M. (2019). Chlamydia pecorum associated with an outbreak of infectious keratoconjunctivitis in semi-domesticated reindeer in Sweden. *Frontiers in Veterinary Science*. 6, 14.

Skjenneberg, S. & Slagsvold, L. (1968). *Reindriften og dens naturgrunnlag*. Oslo: Universitetsforlaget. (In Norwegian. English translation: Anderson, C.M. & Luick, J.R. eds. (1979). *Reindeer husbandry and its ecological principles*. Juneau, Alaska: U.S. Department of the Interior Bureau of Indian Affairs.

Strathdee, A.T. & Bale, J.S. (1998). Life on the edge: insect ecology in arctic environments. *Annual Review of Entomology*. 43(1), 85–106.

Stuen, S., Granquist, E.G. & Silaghi, C. (2013). *Anaplasma phagocytophilum* – a widespread multi-host pathogen with highly adaptive strategies. *Frontiers in Cellular and Infection Microbiology*. 3, 31.

Tryland, M., Das Neves, C.G., Sunde, M. & Mørk, T. (2009). Cervid herpesvirus 2, the primary agent in an outbreak of infectious keratoconjunctivitis in semi-domesticated reindeer. *Journal of Clinical Microbiology*. 47(11), 3707–3713.

Tryland, M., Sánchez Romano, J., Marcin, N., Nymo, I.H., Josefsen, T.D., Sørensen, K.K. & Mørk, T. (2017). Cervid herpesvirus 2 and not Moraxella bovoculi caused keratoconjunctivitis in experimentally inoculated semi-domesticated Eurasian tundra reindeer. *Acta Veterinaria Scandinavica*. 59(1), 23.

Tryland, M., Das Neves, C.G., Klein, J., Mørk, T., Hautaniemi, M. & Wensman, J. (2018). Viral infections and diseases. In: Tryland, M. & Kutz, S.J. (eds) *Reindeer and Caribou – Health and Disease*. Boca Raton: CRC Press. 273–303.

Tryland, M., Nymo, I.H., Sánchez Romano, J., Mørk, T., Klein, J. & Rockström, U. (2019). Infectious disease outbreak associated with supplementary feeding of semi-domesticated reindeer. *Frontiers in Veterinary Science*. 6, 126.

Välimäki, P., Madslien, K., Malmsten, J., Härkönen, L., Härkönen, S., Kaitala, A., Kortet, R., Laaksonen, S., Mehl, R., Redford, L., Ylönen, H. & Ytrehus, B. (2010). Fennoscandian distribution of an important parasite of cervids, the deer ked (*Lipoptena cervi*), revisited. *Parasitology Research*. 107(1), 117–125.

Wernike, K., Conraths, F., Zanella, G., Granzow, H., Gache, K., Schirrmeier, H., Valas, S., Staubach, C., Marianneau, P., Kraatz, F., Höreth-Böntgen, D., Reimann, I., Zientara, S. & Beer, M. (2014). Schmallenberg virus – two years of experiences. *Preventive Veterinary Medicine*. 116(4), 423–434.

Wiegmann, L., Silaghi, C., Obiegala, A., Karnath, C., Langer, S., Ternes, K., Kämmerling, J., Osmann, C. & Pfeffer, M. (2015). Occurrence of Babesia species in captive reindeer (*Rangifer tarandus*) in Germany. *Veterinary Parasitology*. 211(1–2), 16–22.

Wikström, E. (2014). Dödlighet hos renkalvar vid kalvning i hägn. SLU, Fakulteten för veterinärmedicin och husdjur, Inst. f. husdjurens utfodring och vård, Uppsala. Master thesis available at https://stud.epsilon.slu.se/6615/.

Wittmann, E.J. & Baylis, M. (2000). Climate change: effects on culicoides-transmitted viruses and implications for the UK. *The Veterinary Journal*. 160(2), 107–117.

Wittmann, E.J., Mello, P.S. & Baylis, M. (2002). Effect of temperature on the transmission of orbiviruses by the biting midge, Culicoides sonorensis. *Medical and Veterinary Entomology*. 16(2), 147–156.

Part V

Prospects and synthesis

14 Tipping points and regime shifts in reindeer husbandry

A systems approach

Jon Moen, Bruce C. Forbes, Annette Löf and Tim Horstkotte

Introduction

Change is pervasive in both nature and society. Some changes are slow and gradual while others are fast or discontinuous. The occurrence of fast and abrupt shifts has been the focus of much discussion in recent years as they challenge our understanding of system dynamics, pose massive governance challenges and are associated with high risks for human well-being (e.g., Galaz et al. 2016; Ratajczak et al. 2018; Turner et al. 2020; Lenton 2020). Many examples of abrupt changes and regime shifts have been documented in ecological systems, including rapid collapses of fish, mammal and bird populations, and changes in disturbance regimes such as wildfires and insect outbreaks (Ratajczak et al. 2018; Biggs et al. 2018). Sometimes abrupt change, whether in nature or society, is attributed to delayed policy responses and slow implementation rates (Martin et al. 2020), or changes in human behaviour (e.g., Gladwell 2000). While considerable effort has been devoted to predicting future abrupt changes in various systems (e.g., Scheffer 2009), most methods rely on statistical analyses of long-time series, which are seldom available (but see, e.g., Galaz et al. (2009) on the potential of 'mining' digital information in order to detect early warning signals). Increasing our understanding of abrupt change in different systems at different scales alongside societies' abilities to navigate such dynamics, or transform if needed (Moore et al. 2014), therefore remains a topical task. In this chapter, we adopt a systems approach to reindeer husbandry. The chapter is intended to be a synthesis, drawing on the previous chapters and other research, to identify and discuss the prevalence of tipping points, the potential for regime shifts and alternative states of reindeer husbandry as a coupled social-ecological system and livelihood.

Terminology of change – tipping points

The terminology pertaining to abrupt changes is diverse and sometimes confusing. Abrupt changes can, in general, be defined as changes that occur over shorter time periods relative to typical rates of changes for a given system

DOI: 10.4324/9781003118565-20

(Ratajczak et al. 2018). Many different concepts have been used to describe this, partly because it is a developing research field and partly because the concepts originate from different disciplines. Some of the common concepts include regime shifts, critical transitions, tipping points, punctuated equilibria, alternative stable states, thresholds, collapses, surprises and state shifts (see Milkoreit et al. 2018 for an overview). All of these concepts emphasize different aspects of change, but some general themes emerge: the existence of multiple states of a system, internal feedbacks that regulate these different states, non-linearity of changes and potential difficulties in reversing them.

In this chapter, we will use the concept of tipping points and adhere to the definition of Milkoreit et al. (2018: 9):

> A point or threshold at which small quantitative changes in the system trigger a non-linear change process that is driven by system-internal feedback mechanisms and inevitably leads to a qualitatively different state of the system, which is often irreversible. This new state can be distinguished from the original by its fundamentally altered (positive and negative) state-stabilizing feedbacks.

This shift between system states across a tipping point is sometimes called a regime shift (Biggs et al. 2018).

The non-linear changes and state shifts described above are characteristic of a system, which can be defined as a 'set of things – people, cells, molecules, or whatever – interconnected in such a way that they produce their own pattern of behavior over time' (Meadows 2008). While the concept of a system may resonate well within natural sciences, some researchers within social sciences and humanities are less attracted, claiming, for instance, that system boundaries are seldom defined (see, e.g., Byrne 1998; Olsson et al. 2015). This is, of course, true, but the focus in systems analysis is on the interactions between the agents or entities in the system and not the borders themselves.

Another helpful heuristic is to distinguish between tipping points and transformations. The former are typically considered innate to system dynamics while the latter entail specific focus on the component of human agency and intent (Löf 2010). Although natural resource management in many regards has moved beyond rigid command-and-control approaches, and although it is well known that policy interventions rarely generate the intended results, tipping points and transformative potential go hand in hand. For example, recognizing the risk of undesirable alternative system states can induce transformative action otherwise considered too costly, while other tipping points can actively be sought to push and change system dynamics along what is perceived as more sustainable pathways (Westley et al. 2011).

To understand and describe abrupt changes, it is important to identify key external drivers (i.e., pressures outside the defined system, such as climate or policy change, which are two relevant examples in relation to reindeer

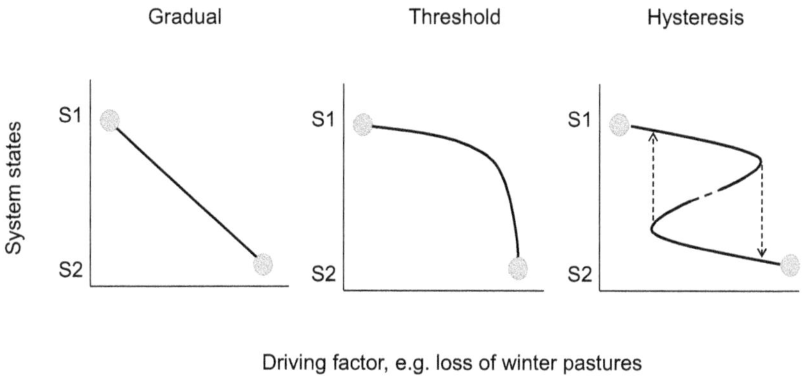

Figure 14.1 S1 and S2 are different system states, for instance, a system based on natural pastures vs a system based on supplementary feeding or ranching. The line represents the response to a driving factor. A gradual response suggests a direct and immediate impact, while a threshold suggests that the system can buffer changes in the driving factor to some extent. A hysteresis effect suggests that there is a threshold, but also that the shift to the alternative state changes the feedbacks in the system so that it is very difficult to reverse the process.

husbandry). Ratajczak et al. (2018) give a good overview, and here we will only touch briefly on the types of relationships between external drivers and system responses. The simplest gradual responses (except no response, of course) are linear or unimodal changes (Figure 14.1). Such responses also imply that the system may recover along the same trajectory if the driver is relaxed or changed. The system may also be able to buffer a change in a driver until a threshold is reached, beyond which the system rapidly or abruptly shifts into a new state. A threshold response may also include hysteresis, where the conditions have changed so much that it is very difficult for the system to recover to the original state even if the driver of change is relaxed. Ecological examples of hysteresis effects that are relevant for reindeer pastures and have lasted for centuries include harvesting trees for firewood in treeline areas which changed the microclimate and prevented trees re-establishing when logging ceased (Karlsson et al. 2007), and long-term grazing and trampling effects in historical reindeer milking pens which changed nutrient regimes and prevented shrubs from invading the herbaceous vegetation even when these sites were no longer in use (Egelkraut et al. 2018).

Van Ginkel et al. (2020) consider different kinds of tipping points. It is clear that the ecological part of a social-ecological system can change abruptly in response to different drivers, for instance as a result of reinforcing feedback loops, such as where global warming melts permafrost causing emissions of greenhouse gases from degrading organic soils which in turn causes stronger global warming (e.g., Hollesen 2015). There may also be shifts originating from the social part of the SES, for instance where a new behaviour, idea or technology is adopted by a large majority, or when adaptation actions fail to meet policy or management objectives (van Ginkel et al. 2020).

It has been suggested that primary industries, such as forestry, agriculture and reindeer husbandry, may be especially at risk of crossing tipping points (Yletinen et al. 2019). This is because they are, to varying degrees, affected by human activities, the ecosystems that they rely on may be severely changed, and they are often dependent on anthropogenic inputs that are holding the system close to critical thresholds (see, e.g., Rist et al. 2014). Maintaining these systems in a production state has been compared to Sisyphus continuously pushing a boulder up a mountain (Rist et al. 2014). Using the lenses of tipping points and regime shifts, this chapter focuses on reindeer husbandry as a social-ecological system, highlighting the inseparability of humans, reindeer and the environment, and conceptually explores the macro-level of emergent phenomena, such as abrupt changes to the livelihood. It is important to note that pastoral systems such as reindeer herding differ from other primary industries in that they have inbuilt mechanisms to reduce vulnerability and deal with different forms of uncertainty, for example, through seasonal and local mobility (Galvin et al. 2009; Horstkotte et al. 2014). However, this assumes the existence of alternative pastures that are both accessible and diverse (Löf 2013) and makes herding practices totally dependent on competing for land users' consideration (Chapters 4 and 8).

Social tipping points

Much of the research on tipping points has focused on ecological or social-ecological systems (SES; Biggs et al. 2018) or on earth system changes (e.g. Lenton 2020). There are fewer studies on social tipping points, at least under that umbrella term. There is much research in social sciences and the humanities about, for instance, behaviour, values, norms, institutions, stability and change that is highly relevant, but it is not possible to summarize it in this chapter. However, we will briefly introduce the subject of social tipping points and relate them to concepts of transformation.

The concept of social tipping points became well known through the popular book by Gladwell (2000). He argues that tipping points in social contexts are governed by three different rules, or agents of change, which he called the Law of the Few, the Stickiness Factor and the Power of Context. In short, this means that whether or not change comes about is often dependent on very persuasive or connected individuals (cf. network theory) who can package information so it attracts people's attention, plus the time has to be right to receive the information or idea. A typical example might be the rapid spread of demonstrations for increased action to combat climate change by the Fridays for the Future movement, which was initiated by a single teenager (Greta Thunberg). These conditions are, however, very difficult to govern, or plan for. Governance scholars highlight different institutional capacities and responses that can increase the ability to navigate abrupt change. For example, the importance of monitoring early warning signals through innovative approaches (see Galaz et al. 2009), preparing for change so that when a window of opportunity arises the institutional side is ready to act (Gelcich et al. 2010), or simply recognizing transformation as a policy option which widely differs from typical incremental responses (Löf 2010; Westley et al. 2011). Transformation in this context thus entails a component of intentionality and gains recognition through, for example, the literature on sustainability transitions (Avelino et al. 2016).

While the literature is vast and spans many different disciplines and fields, much seems to converge around highlighting the role of *ideas* (and essentially politics) in explaining both unwanted stability (or negative resilience) and the potential for transformative change. In particular, the materiality and structural dimensions of ideas, manifested for example in policy discourses (Schmidt 2011), socio-technological systems (Sovacool et al. 2020), epistemic communities and coalitions between different actors (Meijerink 2005) and, more broadly, the formal and informal institutionalization of certain ideas and perspectives (Otto et al. 2020), are key in understanding transformation and transformative potential (for a discussion see also Chapters 7 and 8).

In summary, a tipping point can be connected to, or caused by, changes in both the biophysical environment and socio-economic conditions. Changing environmental conditions can cause an existing portfolio of policies or behaviour to fail (van Ginkel et al. 2020), and changing values or norms can cause changes in the environment through resource use and extraction patterns, for instance in terms of public perceptions of sustainable land use (Chapter 8).

Risks of future regime shifts in reindeer husbandry

Many of the trends and external factors described in the chapters in this book may push reindeer husbandry across a tipping point so that it enters into a new regime or state, one which differs from how it is practised today. The 'ideal' or traditional state of reindeer husbandry is based on free-ranging animals, the use of natural pastures, and governed by traditional knowledge, although the exact details vary between different regions and countries due, for instance, to different forms of governance, quality and amount of pasture, local history and pressures from other land users (see Chapter 1 for a general description). Alternative states that have been raised and discussed in earlier chapters include reliance on supplementary feeding to compensate for losses of pastures (Chapter 12), fenced herds to protect from predation (Chapter 6), becoming a meat-processing industry based on more centralized herding practices (e.g., Landauer et al. 2021) or a total loss of reindeer husbandry. All of these alternative states are seen as undesirable by the herders (e.g., Axelsson Linkowski et al. 2020; Landauer et al. 2021). We will discuss some of the drivers and state shifts on a general level, although we also acknowledge that trends and threats vary over the reindeer husbandry area in Fennoscandia.

The loss of pastures due to other land use (Chapter 4) is, of course, a key driver that is affecting, and will continue to affect, reindeer husbandry in many ways. Herders have been forced to adopt a number of strategies to at least partly compensate for the losses (see Uboni et al. 2020 for Sweden and Landauer et al. 2021 for Finland). These include changes in harvest strategies (towards calf slaughter) and herd structure (towards more females in the herd), introductions of modern machinery and equipment to allow the remaining pastures to be used more efficiently, and increased supplementary feeding to compensate for lost pastures (see Chapter 12). Herders have also been forced to abandon traditional rotational grazing strategies that would allow lichens to recover (Axelsson Linkowski et al. 2020). All of these strategies come with increased financial and psychosocial costs (see more below).

The gradual loss of pastures over decades (Chapter 4) may, from a systems perspective, be seen as a 'slow variable'. Slow variables determine the underlying structure of the system, whereas the dynamics of the system arise from interactions and feedbacks between fast variables (Biggs et al. 2012), such as the number of animals slaughtered, economics or weather. A shift from a system based on natural pastures to a system based on feeding, i.e., a form of ranching, may be an example of a regime or state shift. While feeding becomes more common in most of the reindeer herding area in Fennoscandia, such a shift has not yet occurred in most districts. However, in some herding districts, for instance, in the majority of the herding districts in Finland, feeding is a common practice with majority of the reindeer being fed either in the field or in feeding pens, and it has been so since the 1970s (Nieminen 2010; Turunen & Vuojala-Magga 2014; Landauer et al. 2021). In a participatory study with two herding districts in Sweden, herders mapped the key factors influencing their decision-making process with regard to supplementary feeding, including their interrelations, among which parallel land use and governance-related factors alongside herding economy were identified as important in influencing choices about using supplementary feeding (Horstkotte et al. manuscript). However, the study primarily showed the lengths to which herders would go in order to prevent becoming locked into a system based on supplementary feeding.

Predators may also have significant effects on reindeer husbandry (see Chapter 6 for details). Due to successful conservation policies, numbers of large carnivores have increased over the reindeer husbandry area during recent decades from a previous low caused by long-term human persecution. The effects of carnivores on reindeer husbandry are dependent on the management of both the carnivore populations and of coping strategies adopted by the herders, such as supplementary feeding to decrease losses or avoidance of certain areas. Predation may cause direct losses due to mortality, but also more indirect effects by disturbing herds and calving areas, reducing grazing time and increasing energy expenditure, with a consequent decline in the condition of the reindeer. From a systems perspective, losses due to female mortality may cause herds to collapse. The loss of an adult female creates a gap in calf production until that female has been replaced by a new female. High losses due to predation may thus make it impossible to maintain reindeer numbers and have severe effects on the economy as the potential for harvests declines (Åhman et al. 2014). Coping with predators has also led to a severe physical and mental burden being placed on the herders and may cause young herders to question the long-term sustainability of reindeer husbandry as a livelihood (Vuojala-Magga 2012). In Finland, the compensation scheme for losses to predators may lead to more time being spent finding carcasses to receive compensation than is available for actual herding practices (Heikkinen et al. 2011). In Sweden, where compensation is paid based on numbers of predators and not on confirmed losses, one herder still commented, in relation to predation losses, that 'I am in the business of herding reindeer, not in the business of feeding predators' (J. Moen, *pers. comm.*).

Reindeer husbandry is seen by Sámi herders both as a tradition and as a way of life rather than primarily as an economic enterprise (e.g., Karlsson 2015; Nordin 2007, see also Chapter 8 for a longer discussion). Even so, there is a need for a sustainable economy, and both pasture losses and mortality as a result of large carnivores cause tangible costs. These costs in reindeer herding could be either purely financial or more indirect, such as increased workload or stress (e.g., Uboni et al. 2020). Financial costs include, for instance, costs for supplementary feed, transport of reindeer, use of helicopters for gathering the herds and buying, maintaining and running a fleet of vehicles. The increased use of technologies, such as GPS, drones and GIS, in reindeer husbandry has further required a large monetary input into the herding enterprises. On top of this, transaction costs, i.e., costs associated with defining, establishing and maintaining rights, have increased greatly as consultation processes with other land users, such as forestry, mining and wind power companies, increase (Bostedt et al. 2015).

The increase in costs has consequences for the sustainability of reindeer husbandry and for reindeer herders. For instance, some older reindeer herders find it difficult to motivate young people to take up reindeer husbandry because it is so difficult to meet the costs; this could lead to a demographic tipping point with very few herders and loss of tradition and culture (Lépy et al. 2018; Landauer et al. 2021). Reindeer herders also show significantly higher levels of anxiety and depression compared to both urban and other rural people (Kaiser et al. 2010). The strongest factors related to emotional disorders are work-related, caused for example by large losses due to predators, extreme weather conditions, financial pressure and conflict with competing land users. A combination of the direct and indirect costs may be one reason for an observed change in the number of reindeer enterprises with different herd sizes. In Sweden, both large (more than 500 reindeer) and small enterprises (less than 100 reindeer) have increased since at least the mid-1990s, while the number of medium-sized enterprises has decreased (see Appendix 2 in Uboni et al. 2020 for data sources). This might indicate that enterprises need to have more reindeer than previously to be economically sustainable. However, the right to own and herd reindeer is connected to membership of a herding district and the ownership of an earmark. Therefore, in order not to lose this right and to give the younger generation a chance to establish as reindeer herders in the future, some enterprises still maintain a small reindeer herd, often in their children's name (e.g., Nordin 2007; Karlsson 2015).

It is difficult for herders and herding enterprises to increase revenue to counteract the increasing financial costs. Pastures are used as a common resource with internal rules regulating access to them within the herding districts, and all the available pastures are used (see Chapter 7; Axelsson-Linkowski et al. 2020). If a herder would like to increase the size of his or her herd, there will be a zero-sum game in which someone else would have to reduce their herd size (Karlsson 2015). The space for larger herds (or new herds) is further reduced by the encroachment of other land users (see Chapter 4). Overall, diminished economy, loss of hope for the future of reindeer herding by the younger generation and reduced physical space for reindeer husbandry may all push the system towards a tipping point.

Some authors have described abrupt changes in SES as collapses, i.e., a complete loss of the system (e.g. Diamond 2005). Cumming and Peterson (2017) defined a system collapse on the basis of four criteria: (i) loss of identity, (ii) rate of loss, (iii) losses of social-ecological capital and (iv) long-term consequences of those losses. It may be illustrative to summarize the potential risks that traditional reindeer husbandry faces in light of these criteria. The loss of identity (i) means that key actors, system components and interactions disappear. One such loss of identity could be the shift from a pastoral system based on natural pastures to a system more akin to ranching, based on supplementary feeding (Chapters 4 and 12; Helle & Jaakkola 2008; Turunen & Vuojola-Magga 2014). This means that important system components, such as pastures, are lost, and that feedbacks affecting both practical herding decisions and economic interactions are changed. The rate of loss or change (ii) may also be high in certain areas, certainly within one generation of herders. For instance, the cumulative impacts of other forms of land use have restricted the usefulness of traditional ecological knowledge (TEK) and the transmission of that knowledge from one generation to the next (Axelsson Linkowski et al. 2020). This is further exacerbated by climate change, with what is considered 'normal' shifting faster than the practice-based knowledge (Löf et al. 2012; Löf 2013). Further, the success of large carnivore conservation since the 1990s has increased the predation pressure on some herds to a point where Åhman et al. (2014) suggested that one of the sub-herds in the Njaarke herding district had already collapsed, as the losses of female reindeer were so substantial that herd size could no longer be maintained (Chapter 6). Losses of social-ecological capital (iii) can, for instance, be connected to the loss of TEK between generations as described above. This is especially critical in a pastoral society that is as much rooted in culture and tradition as it is in the monetary aspects of herding. Finally, consequences are certainly long-lasting (iv); for instance if people give up their livelihood, or if the younger generation does not take up reindeer herding as a livelihood (e.g., Karlsson 2015). Further, the rights to the land (for herding, fishing and hunting) are tied to reindeer herding and the continued use of the land (Brännström 2017; Labba 2017). A collapse of reindeer husbandry could mean that the usufructuary rights of the reindeer herders become further questioned by other land users, as has already happened in several court cases in Sweden (Sasvari & Beach 2011).

An external 'shock' that may change the trajectory of Sámi reindeer husbandry into a new regime would be a shift in recognition and implementation of Indigenous rights. These usufructuary rights exist in parallel with other property rights holders and land owners such as the state, private individuals or companies. Legislation in relation to conflicts between land owners and reindeer herding districts in all three countries is generally weak, and many conflicts have ended up in the courts (e.g. Sasvari & Beach 2011; Brännström 2018, see also Chapter 8). While recent legal developments relating to Sámi rights in Sweden could potentially remedy some of these structural imbalances (Allard & Brännström 2021), the Swedish state's record consistently demonstrates empty talk over actual implementation (Mörkenstam 2019). Moreover, juridification (seeking recognition through the courts when the political system fails) as a strategy is costly, as the courts only try the exact questions that lie before them, and their rulings may therefore not lead to a long-term solution of the conflict (Brännström 2017). The only way to take responsibility for the legal development in the conflict with other land users is through the legislative bodies in the countries. The UN Human Rights Council has also repeatedly criticized the Nordic countries for not sufficiently protecting Indigenous rights (e.g., UN General Assembly, 9 August 2016). Should existing legislation be implemented and complemented with new legislation in which Indigenous rights are strengthened, land use and resource extraction would probably take on new trajectories within the Sámi reindeer husbandry areas in all three countries – as would reindeer husbandry.

In summary, it is clear that there are several routes by which traditional reindeer husbandry can shift into an alternative state. Loss of pastures, increased predation pressure, encroachment by other land users, increasing emotional stress and lack of hope for the future may all drive reindeer husbandry towards a more ranch-like form with supplementary feeding, fencing, and more stationary herds. On top of this, climate change will cause additional stresses on both reindeer and reindeer husbandry as a system (Chapter 5). The system is able to buffer against some of these drivers (e.g., Uboni et al. 2020), but the adaptive space and buffering abilities are limited. Where there is a chance of buffering, state shifts are characterized by thresholds, where change may happen suddenly once the threshold is passed. For instance, loss of quality and quantity of pastures can be buffered by supplementary feed – but only to a certain point, beyond which economy, lack of workforce, or emotional stress may force people to give up herding. Loss of animals to predation may reach a point where it is not possible to maintain herd sizes, as described for Njaarke herding district above. Several of these state shifts may also exhibit hysteresis effects, i.e. difficulties in shifting back. For instance, if people give up herding, traditional knowledge will be lost (e.g., Axelsson Linkowski et al. 2020), and if pastures are not used, grazing rights may be lost.

So, what can be done to reduce the risk of unwanted state shifts? Reindeer herders have pushed their potential to adapt their livelihood to internal and external changes and continue to do so. Despite the characterization of reindeer husbandry as a 'traditional' livelihood, this does not imply stasis of what reindeer herders define as the identity of their livelihood. However, the cultural markers of free-ranging animals relying on natural grazing grounds provide the basis for the identity and require intact social-ecological relationships within the herding districts, as well as in their interaction with the external society. These system qualities need to be strengthened as they innately provide resilience – such as maintaining and increasing diversity of pastures, securing access to pastures during crises with regard to forage availability and increasing mobility (e.g., through a landscape perspective where migration routes and safe passages are protected). Several of these strategies will demand structural, institutional and legislative changes, but also discursive changes in terms of how we imagine what sustainability is, what constitutes the best use of forested areas and whether herders are considered to be one of many stakeholders – or as the rights holders they really are according to the law.

References

Åhman, B., Svensson, K. & Rönnegård, L. (2014). High female mortality resulting in herd collapse in free-ranging domesticated reindeer (Rangifer tarandus tarandus) in Sweden. *PLoS ONE*. 9, e111509.

Allard, C. & Brännström, M. (2021). Girjas Reindeer herding community v. Sweden: Analysing the merits of the Girjas case. *Arctic Review*. 12, 56–79.

Avelino, F., Grin, J., Pel, B. & Jhagroe, S. (2016). The politics of sustainability transitions. *Journal of Environmental Policy and Planning*. 18, 557–567.

Axelsson Linkowski, W., Fjellström, A.-M., Sandström, C., Westin, A., Östlund, L. & Moen, J. (2020). Shifting strategies between generations in Sami reindeer husbandry: the challenges of maintaining traditions while adapting to a changing context. *Human Ecology*. 48, 481–490.

Biggs, R., Schlüter, M., Biggs, D., Bohensky, E.L., BurnSilver, S., Cundill, G., Dakos, V., Daw, T.M., Evans, L.S., Kotschy, K., Leitch, A.M., Meek, C., Quinlan, A., Raudsepp-Hearne, C., Robards, M.D., Schoon, M.L., Schultz, L. & West, P.C. (2012). Towards principles for enhancing the resilience of ecosystem services. *Annual Review of Environment and Resources*. 37, 421–448.

Biggs, R., Peterson, G.D. & Rocha, J. (2018). The regime shifts database: a framework for analyzing regime shifts in social-ecological systems. *Ecology and Society*. 23(3), 9.

Bostedt, G., Widmark, C., Andersson, M. & Sandström, C. (2015). Measuring transaction costs for pastoralists in multiple land use situations: reindeer husbandry in northern Sweden. *Land Economics*. 91, 704–722.

Brännström, M. (2017). *Skogsbruk och renskötsel på samma mark: En rättsvetenskaplig studie av äganderätten och renskötselrätten*. PhD. diss. Umeå University.

Brännström, M. (2018). Samiska markrättigheter i förändring? Hovrättens dom i Girjas-målet väcker frågor om innebörden av rättigheter till fast egendom. *Juridisk Publikation*. 1, 25–47.

Byrne, D. S. (1998). *Complexity theory and the social sciences: An introduction*. London: Routledge.

Cumming, G.S. & Peterson, G.D. (2017). Unifying research on social-ecological resilience and collapse. *Trends in Ecology & Evolution*. 32, 695–713.

Diamond, J. (2005). *Collapse: How Societies Choose to Fail or Succeed*. New York: Viking Press.

Egelkraut, D., Aronsson, K.-Å., Allard, A., Åkerholm, M., Stark, S. & Olofsson, J. (2018). Multiple feedbacks contribute to a centennial legacy of reindeer on tundra vegetation. *Ecosystems*. 21, 1545–1563.

Galaz, V., Crona, B., Daw, T., Bodin, Ö., Nyström, M. & Olsson, P. (2009). Can web crawlers revolutionize ecological monitoring. *Frontiers in Ecology and Environment*. 8, 99–104.

Galaz, V., Österblom, H., Bodin, Ö. & Crona, B. (2016). Global networks and global change-induced tipping points. *International Environmental Agreements: Politics, Law and Economics*. 16(2), 189–221.

Galvin, K.A. (2009). Transitions: Pastoralists living with change. *Annual Review of Anthropology*. 38, 185–198.

Gelcich, S., Hughes, T.P., Olsson, P., Folke, C., Defeo, O., Fernández, M., Foale, S., Gunderson, L.H., Rodríguez-Sickert, C., Scheffer, M., Steneck, R.S. & Castilla, J.C. (2010). Navigating transformations in governance of Chilean marine coastal resources. *Proceedings of the National Academy of Sciences*. 107, 16794–16799.

Gladwell, M. (2000). *The Tipping Point: How Little Things Can Make a Big Difference*. New York: Little Brown Publisher.

Heikkinen, H.I., Moilanen, O., Nuttall, M. & Sarkki, S. (2011). Managing predators, managing reindeer: contested conceptions of predator policies in Finland's southeast reindeer herding area. *Polar Record*. 47, 218–230.

Helle, T.P. & Jaakkola, L.M. (2008). Transitions in herd management of semi-domesticated reindeer in northern Finland. *Annales Zoologici Fennici*. 45, 81–101.

Hollesen, J., Matthiesen, H., Møller, A.B. & Elberling, B. (2015). Permafrost thawing in organic Arctic soils accelerated by ground heat production. *Nature Climate Change*. 5, 574–578.

Horstkotte, T., Sandström, C. & Moen, J. (2014). Exploring the multiple use of boreal landscapes in Northern Sweden: The importance of social-ecological diversity for mobility and flexibility. *Human Ecology*. 42(5), 671–682.

Kaiser, N., Sjölander, P., Edin Liljegren, A., Jacobsson, L. & Salander Renberg, E. (2010). Depression and anxiety in the reindeer-herding Sami population of Sweden. *International Journal of Circumpolar Health*. 69, 383–393.

Karlsson, H., Hörnberg, G., Hannon, G. & Östlund, L. (2007). Long-term vegetation changes in the northern Scandinavian forest limit: a human impact-climate synergy?. *Holocene*. 17, 37–49.

Karlsson, A.M. (2015). *Decision making for sustainable reindeer herding – values, goals, and decisions*. Ph.D. diss. Swedish University of Agricultural Sciences.

Labba, K. (2017). *Renskötselns interna organisering*. Ph.D. diss., University of Tromsö.

Landauer, M., Rasmus, S. & Forbes, B.C. (2021). What drives reindeer management in Finland towards social and ecological tipping points? *Regional Environmental Change*. 21, 32.

Lenton, T.M. (2020). Tipping positive change. *Philosophical Transactions of the Royal Society B*. 375, 20190123.

Lépy, É., Heikkinen, H.I., Komu, T. & Sarkki, S. (2018). Participatory meaning making of environmental and cultural changes in reindeer herding in the northernmost border are of Sweden and Finland. *International Journal of Business and Globalization.* 20, 203.
Löf, A. (2010). Exploring adaptability through learning layers and learning loops. *Environmental Education Research.* 16, 281–295.
Löf, A. (2013). Examining limits and barriers to climate change adaptation in an Indigenous reindeer herding community. *Climate and Development.* 5(4), 328–339.
Löf, A., Sandström, P., Baer, K., Stinnerbom, M. & Sandström C. (2012). *Renskötsel och klimatförändring: Risker, sårbarhet och anpassningsmöjligheter i Vilhelmina norra sameby.* (Political Science Department Research Report 2012:4). Umeå: Umeå University.
Martin, R., Schlüter, M. & Blenckner, T. (2020). The importance of transient social dynamics for restoring ecosystems beyond ecological tipping points. *Proceedings of the National Academy of Sciences.* 117, 2717–2722.
Meadows, D. (2008). *Thinking in Systems: a Primer.* London: Chelsea Green Publisher.
Meijerink, S. (2005). Understanding policy stability and change. The interplay of advocacy coalitions and epistemic communities, windows of opportunity, and Dutch coastal flooding policy 1945–2003. *Journal of European public policy.* 12, 1060–1077.
Milkoreit, M., Hodbod, J., Baggio, J., Benessaiah, K., Calderón-Contreras, R., Donges, J.F., Mathias, J.-D., Rocha, J.C., Schoon, M. & Werners, S.E. (2018). Defining tipping points for social-ecological systems scholarship – an interdisciplinary literature review. *Environmental Research Letters.* 13, 033005.
Moore, M.L., Tjörnbo, O., Enfors, E., Knapp, C., Hodbod, J., Baggio, J.A., Norström, A., Olsson, P. & Biggs, D. (2014). Studying the complexity of change: toward and analytical framework for understanding deliberate social-ecological transformations. *Ecology and Society.* 19(4), 54.
Mörkenstam, U. (2019). Organised hypocrisy? The implementation of the international indigenous rights regime in Sweden. *The International Journal of Human Rights.* 23, 1718–1741.
Nieminen, M. (2010). Why supplementary feeding of reindeer in Finland? *Rangifer Report.* 14, 41.
Nordin, Å. (2007). *Renskötseln är mitt liv.* Miscellaneous Publications no 10, Vaartoe, Umeå: Umeå University.
Olsson, L. Jerneck, A., Thoren, H., Persson, J. & O'Byrne, D. (2015). Why resilience is unappealing to social science: theoretical and empirical investigations of the scientific use of resilience. *Scientific Advances.* 1, e1400217.
Otto, I.M., Donges, J.F., Cremades, R., Bhowmik, A., Hewitt, R.J., Lucht, W., Rockström, J., Allerberger, F., McCaffrey, M., Doe, S.S.P., Lenferna, A., Morán, N., van Vuuren, D.P. & Schellnhuber, H.J. (2020). Social tipping dynamics for stabilizing Earth's climate by 2050. *Proceedings of the National Academy of Sciences.* 117, 2354–2365.
Ratajczak, Z., Carpenter, S.R., Ives, A.R., Kucharik, C.J., Ramiadantsoa, T., Stegner, M.A., Williams, J.W., Zhang, J. & Turner, M.G. (2018). Abrupt change in ecological systems: inference and diagnosis. *Trends in Ecology and Evolution.* 33, 513–526.
Rist, L., Felton, A., Nyström, M., Troell, M., Sponseller, R.A., Bengtsson, J., Österblom, H., Lindborg, R., Tidåker, P., Angeler, D.G., Milestad, R. & Moen, J. (2014). Applying resilience thinking to production ecosystems. *Ecosphere.* 5, 73.
Sasvari, A. & Beach, H. (2011). The 2011 Swedish supreme court ruling: a turning point for Saami rights. *Nomadic Peoples.* 15, 130–135.

Scheffer, M. (2009). *Critical Transitions in Nature and Society*. Princeton: Princeton University Press.

Schmidt, V.A. (2011). Speaking of change: why discourse is key to the dynamics of policy transformation. *Critical Policy Studies*. 5, 106–126.

Sovacool, B.K., Hess, D.J., Amir, S., Geels, F.W., Hirsh, R., Medina, L.R., Miller, C., Palavicino, C.A., Phadke, R., Ryghaug, M., Schot, J., Silvast, A., Stephens, J., Stirling, A., Turnheim, B., van der Vleuten, E., van Lente, H. & Yearley, S. (2020). Sociotechnical agendas: Reviewing future directions for energy and climate research. *Energy Research & Social Science*. 70, 101617.

Turner, M.G., Calder, W.J., Gumming, G.S., Hughes, T.P., Jentsch, A., LaDeau, S.L., Lenton, T.M., Shuman, B.N., Turetsky, M.R., Ratajczak, Z., Williams, J.W., Williams, A.P. & Carpenter, S.R. (2020). Climate change, ecosystems and abrupt change: science priorities. *Philosophical Transactions of the Royal Society B*. 375, 20190105.

Turunen, M. & Vuojala-Magga, T. (2014). Past and present winter feeding of reindeer in Finland: herders' adaptive learning of feeding practices. *Arctic*. 67, 173–188.

Van Ginkel, K.C.H., Botzen, W.J.W., Haasnoot, M., Bachner, G., Steininger, K.W., Hinkel, J., Watkiss, P., Boere, E., Jeuken, A., Saiz de Murieta, E. & Bosello, F. (2020). Climate change induced socio-economic tipping points: review and stakeholder consultation for policy relevant research. *Environmental Research Letters*. 15, 023001.

Vuojala-Magga, T. (2012). Adaptation of Sámi reindeer herding: EU regulations and climate change. In: Tennberg. M. (ed.) *Governing the uncertain*. Dordrecht: Springer. 101–122.

Westley, F., Olsson, P., Folke, C., Homer-Dixon, T., Vredenburg, H., Loorbach, D., Thompson, J., Nilsson, M., Lambin, E., Sendzimir, J., Banerjee, B., Galaz, V. & van der Leeuw, S. (2011). Tipping Toward Sustainability: Emerging Pathways of Transformation. *Ambio*. 40, 762.

Uboni, A., Åhman, B. & Moen, J. (2020). Can management buffer pasture loss and fragmentation for Sami reindeer herding in Sweden? *Pastoralism*. 10, 23.

Yletinen, J., Brown, P., Pech, R., Hodges, D., Hulme, P.E., Malcolm, T.F., Maseyk, F.J.F., Peltzer, D.A., Perry, G.L.W., Richardson, S.J., Smaill, S.J., Stanley, M.C., Todd, J.H., Walsh, P.J., Wright, W. & Tylianakis, J.M. (2019). Understanding and managing social-ecological tipping points in primary industries. *BioScience*. 69, 335–347.

15 Pathways for action

The need for Sámi self-determination

Åsa Larsson Blind

Sámi reindeer herders: rights holders and knowledge holders

Sámi reindeer herding is a traditional livelihood of the Sámi people. This needs to be the starting point in any discussion on policy or governance concerning Sámi reindeer herding. It means that Sámi reindeer herders are not an interest group or a stakeholder amongst others (Sarkki et al. 2021). Rather, Sámi reindeer herders are rights holders and knowledge holders. Reindeer herding is imbedded in and interlinked with the Sámi culture and therefore also in the Sámi society. In that sense, Sámi reindeer herding is a societal structure. It is thus not possible to divide the practical aspects of reindeer herding from the surrounding Sámi societal structures. The often-used phrase, "reindeer herding is not an occupation, but a way of life", may sound like a cliché, but contains a lot of truth. This is well expressed by Retter (2020) when describing the Sámi culture's relationship to nature:

> Nature is the foundation for making a living, it is also the frame for our living conditions, surroundings, the experience of the past, knowledge and way of life. Sámi culture is therefore dependent on healthy and productive ecosystems. The foundational values have their origin in Sámi worldview and understanding of human life. The Sámi society is closely connected to the values and indigenous knowledge that are transferred from generation to generation.

This brings forward the importance of knowledge. Knowledge builds societies. Academic knowledge has helped to build the societies of the Nordic countries. In the same way, there needs to be made space for Indigenous knowledge to continue to build the structures of the Sámi society, in the same way as it has done in pre-colonial times. The two parallel knowledge systems, academia and Indigenous knowledge, existing side by side create the possibility to utilize the strengths and richness of both systems. For this to be effective, both knowledge systems need to be respected in their own capacity to create a knowledge base of "best available knowledge".

DOI: 10.4324/9781003118565-21

The way the academic research system is structured and the basis on which academic research accumulates "academic knowledge" is well-known. It is important to understand that Indigenous knowledge is based on a different structure. The Ottawa Principles of Indigenous knowledge (2015) defines this system as

> A systematic way of thinking and knowing that is elaborated and applied to phenomena across biological, physical, cultural and linguistic systems. Traditional Knowledge is owned by the holders of that knowledge, often collectively, and is uniquely expressed and transmitted through indigenous languages. It is a body of knowledge generated through cultural practices, lived experiences including extensive and multigenerational observations, lessons and skills. It has been developed and verified over millennia and is still developing in a living process, including knowledge acquired today and in the future, and it is passed on from generation to generation.

The challenges of including Indigenous knowledge in decision making are noted in several chapters in this publication (Chapter 7, 8 and 9). Incorporation Indigenous knowledge in decision-making processes is a way to increase validity and legitimacy in the processes, and in their outcomes. The inclusion of Indigenous knowledge is also brought forward as an important step in meaningful recognition of reindeer herders' institutions as a means for empowerment and self-determination. But as discussed in Chapter 7, even though the need for implementation of Indigenous knowledge into policies is identified, so far it has been insufficiently executed in current governance systems.

As Indigenous knowledge is vital to understanding Sámi reindeer herding, and thus vital to any decisions on state policies and management systems affecting reindeer herding, this poses challenges for all outside actors. This is the challenge for researchers aiming at studying reindeer herding or tapping into the knowledge of reindeer herders. There are of course researchers studying the reindeer as an animal, but as soon as the research includes aspects of the living conditions of that reindeer, the social system of reindeer herding needs to be considered. This is also the challenge for state authorities aiming at developing well-functioning governance models that will result in expected and desired outcome to satisfy all involved actors.

Many studies bring forward the need to incorporate Indigenous knowledge and academia, but fewer studies give guidance on how to accomplish that demanding task. When research projects are limited to specific aspects of Sámi reindeer herding, the implications of such a selective focus and the consequences and possibilities to draw conclusions from those results remain uncertain and even disputable. When the separate parts added together might make up a different holistic meaning than the different parts by themselves, what implications does that have on research and the possibilities to utilize the results?

As Indigenous knowledge is a systematic way of thinking and knowing, there is no way of picking bits and pieces. Without the necessary cultural insight, it is difficult to understand choices made within Sámi reindeer herding and strategies chosen for the future. This gap of knowledge can result in management systems leading to unexpected, unplanned results. Laakso (2008) described this phenomenon as a "shadow field" of reindeer herding management that is invisible to outsiders. This highlights the difficulty for outside management systems to predict the outcome and suggest appropriate solutions to current and future challenges without the inclusion of Indigenous knowledge holders of Sámi reindeer herding communities.

It is thus necessary to include the knowledge holders in the processes to ensure that knowledge is used in the correct way and to interpret the meaning correctly. The other way around, it is important to interpret the results of academic research through the lens of reindeer herding to get practical meaning. This makes it evident that only the meaningful participation and inclusion of Indigenous peoples in these processes open the possibilities to find good solutions for the ways forward. The reindeer herding Sámi thus need to be included in any processes affecting them.

The potential of Indigenous knowledge for better land use management

Indigenous peoples' societies are amongst the ones contributing the least to climate change drivers, yet Indigenous peoples around the world are amongst those most affected by the consequences of the climate crisis. One of the major crises the world is facing in relation to the climate change is the rapid loss of biodiversity. In a World Bank report from 2008 it is stated that "Traditional Indigenous Territories encompass up to 22 percent of the world's land surface and they coincide with areas that hold 80 percent of the planet's biodiversity". This has been reiterated by the EU in the European Parliament resolution from 2018 on violation of the rights of Indigenous peoples in the world, including land grabbing. In line with this, the IPBES global assessment report on biodiversity and ecosystem services from 2019 showed that nature generally is declining less rapidly in Indigenous peoples' land than in other lands. Indigenous peoples, like the Sámi, have successfully preserved biodiversity in their areas through the continuation of their traditional livelihoods. This shows that Indigenous livelihoods and traditional ways of life benefit biodiversity. It also shows that Indigenous knowledge entails a valuable asset in promoting biodiversity and has much to teach current industrialized land use regimes that result in a rapid decrease of biodiversity. Indigenous knowledge that has successfully preserved and restored biodiversity across the globe should therefore be seen as a self-evident asset in discussions on how to design land use management systems to halt biodiversity loss. One could boldly say that promoting Indigenous livelihoods, culture and ways of life should categorize as a climate action in itself. It would also be natural for states and other actors to partner up with Indigenous peoples to design initiatives to stop biodiversity loss and combat climate change. However, the recognition of Indigenous knowledge as being equally valid as academic science as a way of knowing remains lacking. It is a waste of potential to exclude Indigenous peoples from processes where they could clearly contribute with valuable insights to enhance the quality in decisions on land use management. A way forward to preserve biodiversity and restore key habitats would be for science and Indigenous knowledge to work in partnership while also being restitutive and rights based (Ogar et al. 2020).

Indigenous peoples have been advocating for environmental issues, sustainability and climate action for decades. For Sámi, productive and healthy ecosystems both on land, including water systems, and in marine areas are central prerequisites to Sámi culture and identity. Impacts on ecosystems such as pollution, climate change and changed land use are therefore a threat to Sámi culture and should be prevented (Resolution, Tråante 2017). Yet, Indigenous peoples are still struggling to be recognized as legitimate parties in the decision-making forums on these issues, both nationally and internationally. Despite Indigenous peoples' willingness to contribute to the solution and the value of Indigenous knowledge confirmed also by academic science, Indigenous peoples find themselves on the outside of land use decisions. This is also the case for reindeer herding Sámi.

Existing colonial structures + green washing of economic development = Green colonialism

Instead of recognizing Indigenous knowledge as a foundation, different initiatives are launched as "new" approaches to land management, seemingly aiming to promote sustainable solutions. One of these initiatives is the "Nature-based Solutions (NbS)" that is promoted as a new approach of solutions inspired and supported by nature, which are cost-effective, simultaneously provide environmental, social and economic benefits and help build resilience. The UN Environment Program (UNEP) describes NbS as "locally appropriate actions that address challenges such as climate change and provide human well-being and biodiversity benefits by protecting, sustainably managing and restoring ecosystems". NbS are projected as a new and innovative way to put people and nature together. However, for Indigenous peoples it simply coincides with our traditional ways of living that protect the environment, restore ecosystems, provide food and economies while respecting nature. Another example of a "new approach" is Circular Economy (CE). This is a model for production and consumption, which involves using existing materials and products for as long as possible and aims at tackling global challenges like climate change, biodiversity loss, waste and pollution. Circular Economy is discussed on multiple levels. For example, it is included in the EU Green Deal, the EU Commission's program to respond to the challenge to tackle climate and environmental challenges, and Sweden has adopted a national strategy for Circular Economy.

The promoted strategies speak of inclusiveness, just transition and a focus on sustainability. But despite this, the invisibility – again – of the perspectives that Indigenous peoples can contribute with to all the discussions and actions are remarkable. The global and national problems that are sought to be resolved through these initiatives also affect Indigenous peoples, and these "new" approaches contain nothing new to Indigenous peoples such as the reindeer herding Sámi. This shows, though, that Indigenous peoples and their knowledge are made invisible time and time again when it comes to land use decisions and in the search for more sustainable land use.

This clearly indicates that the implementation of the Sámi right to self-determination is the way for Sámi reindeer herders to access their right to participate in land use decisions on their territories and to continue to use their Indigenous knowledge of reindeer herding.

This would empower Sámi reindeer herders to raise their concerns and promote cultural appropriate solutions in their own capacity with their own voice. Most importantly, it would pave the way towards effective participation with their Indigenous knowledge for the best for the Sámi society and livelihoods.

Indigenous peoples also experience that these promoted approaches serve as a green washing of the discourse of climate-friendly initiatives, sustainability and climate actions. In reality, these promoted approaches seem to be a way of letting industries driven by the main goal of economic growth continue their extractive land use on Indigenous lands. The industrialization that created the global society of today was developed on the notion of economic growth as the main driver. We now see the consequences of this development and the world finds itself in a global crisis. In the name of societal development and economic growth, the Sámi have been forced to give up pasture lands, hunting grounds and fishing areas resulting in the undermining of the Sámi culture. Lands have been claimed by industries such as forestry and mining, creating wealth on national level without recognition of the Sámi rights to their territories. Now, in the name of climate action, investments in "green energy" result in more mines and other industry development on Sámi lands. This time, land claims go under the expression of "Green mining" and that these mines are necessary for the majority of society's green shift to a climate friendly, more sustainable society. Also, wind power developments are rapidly established as "green energy" investments.

Of course, action needs to be taken to combat climate change. The problem for reindeer herding Sámi, however, is that the same colonial power structures are still in force that have led to the current situation of diminishing pastures due to other land use encroachments and that kept Sámi excluded from their rights as a people. The great investments made in wind power developments as "green energy" are made without the notion of climate justice and proportionality, thus not respecting Sámi rights and without recognizing the great consequences Sámi reindeer herding communities are left tackling.

For reindeer herders, this new wave of development in the name of climate action and green shift looks just the same as other industrial development of the past. A mine is still a mine, and the permit process is still made without the recognition of Sámi reindeer herding property right to their territories and the right to Free, Prior and Informed Consent (FPIC). In the eyes of the Sámi people, this is to be categorized as Green colonialism, since the same non-participatory power relations are intact, and the colonial structures continue to keep Sámi Indigenous knowledge on land use management invisible. The same colonial structures that exclude Sámi from their rights as a people still apply. Only this time, it becomes even harder to dispute, since opposing Sámi reindeer herders are rendered climate unfriendly – in addition to the already used label of backwards striving when opposing "necessary" development.

This situation has unfortunately forced Sámi reindeer herders to take their issues to court and even to the international arena. In 2013, the Vapsten reindeer herding community reported the mining project of Rönnbäcken, Sweden, to the UN Committee on the Elimination of Racial discrimination, CERD (54/2013). In 2020, CERD gave its conclusion that Sweden is violating the International Convention on the Elimination of all forms of Racial Discrimination, because the Swedish Mining Act does not recognize the property rights inherent in Sámi reindeer herding. CERD recommended Sweden to revise the Mining Act in order to prevent the same violation to occur again (UN Doc. CERD/C/102/D/54/2013). This brings the current colonial structures into light, showing that national legislation is systematically discriminating against Sámi reindeer herders in Sweden. In 2018, Fovsen Njaarke reindeer herding district also submitted a case to CERD (67/2018), regarding the Fosen wind power development in Norway. CERD replied a few months later, urging Norway to halt the project due to its extensive impact on the reindeer herding Sámi culture. However, Norway ignored the CERD decision and permitted the windmill company to continue with the constructions parallel to the national court proceedings. In October 2021, the Supreme Court in Norway issued its verdict in the case (HR-2021-1975-S). A unanimous court ruled that the wind power development is violating Sámi Indigenous rights, and the concession and the expropriation permit were rejected. This is of course a victory for the reindeer herding community, but the damage is done as the windmill park already is constructed and great values of pasture lands destroyed.

These examples pinpoint the difficulty for Sámi reindeer herders to access their rights within the current national systems. It also highlights the need for change that paves the way for inclusion of Sámi reindeer herders in decisions-making processes that affect them. A change focused on rights to FPIC and climate justice, contrary to only including reindeer herders' opinions in a late stage of the permit process. But as long as the Sámi right to self-determination is not implemented, there will be an unbalance of power relations under which the Sámi depend on the benevolence of outside decision makers. For these colonial structures to be broken, Sámi self-determination needs to be implemented.

Strong Sámi institutions

The Sámi people have had their societal structures interrupted and suppressed by colonization (Chapter 1). The Sámi people have therefore been denied the possibility to develop their society on their own terms and based on their own knowledge and priorities. The societal development in Sápmi as a colonized territory has been shaped according to the knowledge base and priorities of the majority national state. In order to move towards a change of the current management system that excludes Indigenous ways of knowing, there is a need to secure the resources and capacity building for the Sámi society to establish and strengthen their own institutions. There is a need to strengthen Sámi institutions on all levels, from the local *siidas* to the national and cross-border organizations. Indigenous knowledge can then be used as a knowledge base for the Sámi society to express their own priorities, restore suppressed societal structures and regain societal strength. Only then, the Sámi will have the possibility to fulfil the place as equal partners relative to states and regional authorities in decision-making processes and negotiations of present and future resource management within the Sámi areas.

Strengthening Sámi institutions is also crucial in relation to research institutions. Research has the potential of finding ways forward. But for research to truly contribute to a pathway forward, it needs to be scrutinized for its perspective and foundational outsets and be based on the best knowledge available, including both academic and Indigenous knowledge. Through research projects, there are possibilities to test solutions that otherwise would not have been accepted; thus, research can create pathways forward in issues that are seen as locked by long-term conflicts.

However, there are still challenges in bringing forward the Sámi reindeer herding viewpoints even through research. To be Indigenous is to constantly be perceived as political. This means that research on Indigenous issues can be seen as political. Studies on Indigenous issues often identify a need to change power imbalance or to change structures that seldom are in line with state policy. Consequently, research in the field of Indigenous issues, in this case Sámi reindeer herding, and the researchers choosing this field of research, may run the risk to be seen as political. With this also comes a risk that the research outcomes are seen as "pro-indigenous" and thus biased, rather than being seen as knowledge-based in their approach and unbiased in their intentions by raising an important, but too often, invisible Indigenous perspective.

Strong Sámi institutions are also an imperative foundation for the implementation of Sámi self-determination. An important part of self-determination entails control over knowledge production processes, thus increased influence over research, and the politics of what is perceived as important research issues (Chapter 8)

Full and effective participation

The Sámi people have the right to take part in decisions that affect them. There is an urgent need for structures that include Sámi reindeer herders in decision-making processes. But for this to be useful, these structures need to be designed in a way that the participation is full and effective. As has been shown by Turi (2016), participation without it being meaningful does not necessarily mean influence. Without the meaningful, full and effective participation that ensures influence in decision-making processes, Sámi reindeer herding will be subject to priorities set by the surrounding societies, resulting in the continued degradation of living conditions for reindeer herding societies and upholding of current unbalanced power structures. For the participation to be effective, there need to be a co-design on the structures of the processes and also a respect for the internal decision-making structures of the Sámi society in reaching positions in important issues and processes of choosing mandated representatives.

This publication brings forward many examples for the consequences of lacking influence for reindeer herding communities and the result of the current unbalance in power relations between reindeer herding and state authorities. This includes the current unsustainable situation due to the limited space available for adaptation actions to cope with climate change (Chapter 5), decreasing flexibility fragmentation of pasture lands progresses and pasture resources decline due to other land users' extractive resource exploitation (Chapter 4), which decreases the adaptive capacity to respond to extreme weather events and climate change. The described current conditions for reindeer herding, brought forward through the results of this publication, paints a clear picture of the lack of consideration of the needs of reindeer herding in current land and resource management structures. It is thus clear that the current state regulations fail to respect and include the Indigenous rights of Sámi reindeer herding and a call for a change that will secure these rights to be respected. The Sámi people also need to be included in the designing of the management structures to sufficiently include and integrate Indigenous knowledge in a meaningful way. Once again, this shows the need for the implementation of Sámi self-determination.

Self-determination as a means to establish a sustainable management system

In the discussion of self-determination, there is often an emphasis on the current conflicts as examples of what self-determination would be utilized to influence, such as wind power developments, mines, predator policies or climate change adaptation. This is expected but at the same time a hindrance for taking steps forward towards implementing the right to self-determination. It is exactly in these types of land use issues where the current decision makers are resisting Sámi reindeer herders' influence (Chapter 8). Since the current decision making is made without significant influence by the Sámi reindeer herding community, there is a tendency to transfer the current dominant view on the right to decide on how a new future system possibly could be conceived and implemented. This sets limitations on moving forward. The current unbalanced situation where local Sámi communities are involved too little and too late in the processes, causing protests and appeals, is not fostering a real discussion based on a constructive outset. Shifts in power balance are never easy and the transfer of power seldom happens without resistance.

The colonial view on the Sámi people as not being capable of managing their own affairs tends to show up in between the lines in state authorities' and policy's discussions. For instance, Sámi reindeer herding constantly needs to prove itself capable of good management according to the conditions or goals defined by these authorities. If decisions following Sámi positions step out of already determined directions and are perceived as negative for other competing interests, these perceived negative consequences are used as an argument for keeping the decision-making power within state control. Thus, reindeer herding is only considered capable of good management if complying with state authority management regimes and goals. This results in the continuation of current power relations and top-down management system. It also disregards the knowledge of Sámi reindeer herding and violates the right to self-determination. What is seen as good management is measured against the outside view and priorities. Even when done in the name of self-determination, it is an example of quite the opposite: a state's top-down management system based on ongoing colonial structures.

Between the lines of management regulations and authority argumentation lies an assumption that, without state governance, resources would be misused by Sámi reindeer herders – assuming that Sámi reindeer herders would use as much as they can of available resources, despite the need. This indicates little or no understanding of the sustainable foundation of traditional Indigenous livelihoods. That assumption is rather based on the approach of resource exploitation inherent in an industrialized society. On the contrary, the foundation for Sámi reindeer herding is based on the notion that one should use only what you need, live off on the abundance and not drain the resource of its capacity to sustain also in coming years and for coming generations (Sámi Traditional Ecology, Resolution from the 21st Saami Conference, Tråante 2017).

A way forward is to shift focus from the specific issues debated to the design and form of Sámi self-determination. The right to self-determination should be Interpreted procedurally rather than substantially. Mörkenstam (2015) interprets Indigenous peoples' right to self-determination to imply recognition of Indigenous peoples as having a standing equal to nation states, "i.e. *as if* they were sovereigns". The meaning of self-determination in political practice would thus be the outcome of negotiations between two (or more) equal political entities. The approach to self-determination as procedural might be a way to overcome what is discussed as a risk: current governing systems will face aggravated levels of tensions and conflict because the wrong questions about addressing self-determination and power inequalities are posed under the present dominant understanding of what reindeer herding is or can or should be.

Conclusion

The broad work of the ReiGN project shows why the rights-based approach and the emphasis on self-determination of Sámi reindeer herding are necessary.

The outcome of the project shows the need for a better-designed governance system, well adapted to Sámi reindeer herding that is equipped to tackle the current and future challenges of land use conflicts and climate change adaptation.

For Sámi reindeer herding, the current situation is far from sustainable and there is no more time left to waste. There is an urgent need to start using the knowledge within reindeer herding communities to find sustainable solutions for land use and resource management and cultural acceptable adaptation measures to cope with the changing climate. There is an unused potential that is held back by rigid state management systems and the lack of implementation of Sámi self-determination.

Unfortunately, there are still colonial structures in the state's approach towards the Sámi people and towards Sámi reindeer herding, in particular, upholding that other people than the Sámi reindeer herders themselves know what is best for reindeer herding. Sámi reindeer herders are still not included in the decision-making processes, as should be the natural working method: for the expert knowledge holders and rights holders to be involved in decisions that concern them. As long as this very outdated, but sadly not yet erased, view is imbedded in state policies on reindeer herding management, we will not reach a truly functional governance model for reindeer herding. It will not be possible to make use of the extensive knowledge base of Indigenous knowledge and, most importantly, the Nordic states will not have fulfilled their obligation to respect the rights of the Sámi. The path forward to find sustainable solutions for the future is thus the way through the implementation of Sámi self-determination.

The only way to interrupt colonial structures is for the national states to stop colonizing. As long as the current, historically inherited structures are not interrupted, all discussions on land use, resource use and development in Sápmi will be conducted in the shadow of unequal power relations. This maintains top-down, exclusive decision making that is not equipped to tackle the current and future land management challenges in Sápmi. As outlined in the introduction, the pathway forward in solving present and future challenges for Sámi reindeer herding, whether talking about climate change adaptation, land use encroachment or development of governance models, is the implementation of Sámi self-determination. Only when being recognized as equal partners, the Sámi will be able to bring their knowledge to the table on their own terms, and in their own capacity contribute to solving the challenges society is facing through cultural acceptable solutions based on best available knowledge.

References

European Commission. (2019). Communication from the Commission to the European Parliament, The European Council, The Council, The European Economic and Social Committee and the Committee of the Regions: *The European Green Deal*. Document ref. COM (2019) 640. Brussels. https://eur-lex.europa.eu/resource.html?uri=cellar:b828d165-1c22-11ea-8c1f-01aa75ed71a1.0002.02/DOC_1&format=PDF

European Parliament resolution of 3 July 2018 on violation of the rights of indigenous peoples in the world, including land grabbing (2017/2206(INI)). www.europarl.europa.eu/doceo/document/TA-8-2018-0279_EN.pdf

International Covenant on Civil and Political Rights, ICCPR. (1966). www.ohchr.org/en/professionalinterest/pages/ccpr.aspx

IPBES. (2019). Summary for policymakers of the global assessment report on biodiversity and ecosystem services of the Intergovernmental Science-Policy Platform on Biodiversity and Ecosystem Services. Díaz, S. M., Settele, J., Brondízio, E., Ngo, H., Guèze, M., Agard, J et al. (eds.). IPBES Secretariat, Bonn: Germany.

Laakso, A.M. (2008). The shadow field of reindeer management: A case study from Finland. *Acta Borealia*, 25(2), 138–159.

Mörkenstam, U.(2015). Recognition *as if* sovereigns? A procedural understanding of indigenous self-determination. *Citizenship Studies.* 19(6–7), 634–648.

Ogar, E., Pecl, G., & Mustonen, T. (2020). Science must embrace traditional and indigenous knowledge to solve our biodiversity crisis. *One Earth*, 3(2), 162–165.

Ottawa Principles of Indigenous knowledges (2015). www.arcticpeoples.com/knowledge#indigenous-knowledge

Resolution from the 21st Saami Conference, Tråante. (2017). Original title; Sámi Árbevirolaš Ekologiija Cealkámuš Sámiid 21. Konferánssas, Tråante 2017 https://static1.squarespace.com/static/5dfb35a66f00d54ab0729b75/t/5e1862053ff49c5358ccf4be/1578656263176/SÁMI_ÁRBEVIROLAŠ_EKOLOGIIJA.pdf

Retter, G.-B. (2020). Å skrive sorg. In: Björlykhaug, K.I. & Vetlesen, A.J. (eds.). *Det går til Helvete. Eller?: Om kjærlighet, sorg og raseri i natur-og klimakrisens tid.* Fornebu: Dinamo Forlag.

Sarkki, S., Heikkinen, H.I. & Löf, A. (2021), Reindeer Herders as Stakeholders or Rights-Holders? Introducing a Social Equity-Based Conceptualization Relevant for Indigenous and Local Communities. In: Nord, D.C. (ed.) *Nordic Perspectives on the Responsible Development of the Arctic: Pathways to Action.* Cham: Springer. 271–292.

Sobrevila, C. (2008). *The role of indigenous peoples in biodiversity conservation: The natural but often forgotten partners.* The World Bank, no. 44300.

Turi, E. I. (2016). *State Steering and Traditional Ecological Knowledge in Reindeer-Herding Governance: Cases from western Finnmark, Norway and Yamal, Russia.* PhD. diss. Umeå University.

16 Final reflections

Jon Moen, Tim Horstkotte, Øystein Holand and Jouko Kumpula

We live in a time of rapid change. Owing to the all-encompassing impacts of human civilization on ecosystem processes and society in the Anthropocene, transformation and adaptation are required to avoid exceeding planetary boundaries. Yet, multiple crises currently cause a 'perfect storm' of challenges for humanity. The effects of ongoing climate change are already upon us: polar ice caps are shrinking, and extreme weather events lead to flooding, heat waves, and wildfires. Loss of biodiversity is so fast and ubiquitous that scientists warn of a sixth global mass extinction, driven by the cumulative effects of land use and climate change. Human population growth has resulted in increased consumption of the planet's resources and degradation of the ecosystems that provide them. Land use change and globalization likely contributed to the spread of the COVID-19 pandemic in 2020.

Certainly, there are attempts to address these challenges, as declared in international policies. The Paris Agreement on climate change aims to limit global warming to well below an increase of 2 °C in the mean global average compared to pre-industrial levels, preferably even below 1.5 °C. The Convention on Biological Diversity (CBD) attempts to reduce loss of biodiversity by engaging participating countries to develop strategies for sustainable use and for biodiversity conservation. More generally, the United Nations Sustainable Development Goals (SDGs) formulate goals for desirable futures from environmental to societal perspectives. While these global agreements have been signed by most countries, they still seem to fall short of the defined targets due to insufficient political will and meaningful actions. But even while seemingly inadequate, global agreements inevitably will affect national politics and policies and finally will have local consequences.

All global goals require local implementation and thus must be locally accepted and implemented. Attempts to reduce the use of fossil fuels require that the provisioning of energy be transformed to renewable energy sources, such as wind, hydro, tidal, wave, solar and bioenergy sources. Such transformations will require extensive infrastructures that will themselves claim large land resources. Indeed, bioenergy requires land where raw materials for bioenergy can be produced. Further, new mines for mineral extraction need to be opened to advance the electrification and digitalization of our society.

DOI: 10.4324/9781003118565-22

Such forms of adaptation, mitigation, and transformation to global challenges are always local and will thus affect local people and enterprises. Global goals may involve opposite needs and, consequently, also exacerbate conflicts over land use and rights to resources. An example of this is the so called Nimbyism (not in my back yard), where local residents protest against development – because it is close to them, but not because they oppose the development itself. Clearly, policies and actions are needed to distribute the societal burdens and to compensate those that are negatively affected.

Northern landscapes have long been seen by the populated south of Fennoscandia as peripheral areas with abundant natural resources to extract, hunt or harvest, while at the same time being sparsely populated, with few visible traces of traditional land use to the untrained eye. Today, wind farms are being built; mineral resources are mapped and mines are being opened; and heated debates are taking place on how forests should be managed in northern Fennoscandia and elsewhere in the Arctic. While these areas are already severely affected by climate change, they are also seen as part of the solution: 'resource peripheries' that are shaped to meet the needs of more heavily populated areas. This will inevitably affect ecosystems and their biota, local and Indigenous peoples, local enterprises and land use patterns. As this new wave of 'green colonialism' is sweeping through the Arctic, it may also re-focus perceptions of democracy and self-determination of 'the local' in response to 'the global'.

Reindeer pastoralism in Fennoscandia is affected by all of these processes, which are interwoven globally, regionally and locally. It is through this integrative lens that we developed the various chapters of this book, as outlined in Chapter 1. Indeed, a systems approach is necessary to understand this complex social-ecological system and to identify potential tipping points beyond which the system changes its identity and operation (Chapter 14).

Climate change is challenging traditional use of pastures, e.g., through changes in snow and ice conditions (Chapter 5) and is affecting reindeer health and diseases (Chapter 13). Competing land use is reducing the quantity and quality of pasture lands (Chapter 4), and decisions about policies shaping the future of reindeer pastoralism are increasingly taken by organizations and actors far removed from the local (Chapter 8). These changes will influence the production potential and the viability of reindeer pastoralism (Chapter 10) – and hence could drain the herders' economy (Chapter 11), especially as the need for supplementary feeding increases (Chapter 12). Knowing the reindeer and their behaviour is an essential part of the herders' traditional knowledge and is particularly needed when the effects of different changes and disturbances on reindeer pastoralism are evaluated or need to be communicated to other land users (Chapter 3). Also, maintaining the genetic diversity of the herd (Chapter 2) is a prerequisite for maintaining welfare of the reindeer and their adaptive capacity in an evolutionary sense. At the same time, genetic diversity also improves options to develop selection practices (Chapter 10) and to compose resilient herds in balance with the pasture resources.

Several chapters in this book emphasize the need for a more holistic view of land use and associated livelihoods, where multiple actors, effects, pressures and alternatives have to be considered. Chapter 8 shows that interactions between reindeer herding communities and other actors are governed in a fragmented system of sectoral silos and weak instruments. To a significant degree, this stalemate is caused by current governing systems that have been created to address problems and solutions as defined by other actors than the reindeer herders themselves. Chapter 6 discusses this situation with regard to the conservation success of large carnivore resurgences in the Nordic countries. With negative effects on reindeer pastoralism, it is evident that conservation goals need to be balanced with livelihood needs and human welfare.

Several factors have contributed to the lack of a more holistic regional land use planning. There are strong status and power imbalances between the actors, where other forms of land use often represent national or general interests in resource development, fuelled both by economic interests and, in part, by the green transition to a carbon-neutral economy. Further, Chapter 8 shows that framing reindeer pastoralism as an economically oriented production system (as opposed to a way of life; Chapters 7 and 15), working under a market economy, may be responsible for the current top-down governing system (Chapter 9).

Another factor is a lack of recognition and implementation of Indigenous rights. In contrast to Norway, Sweden and Finland have not ratified the indigenous and Tribal Peoples Convention (ILO 169), repeatedly receiving criticism by the United Nations for not respecting these rights. Among other things, this means that Sámi reindeer herders are seen as 'stakeholders' like any others, and not as 'rightsholders', even though their grazing rights are part of the constitutions of the Nordic countries.

In light of these multifaceted challenges, what pathways and actions are possible? Recognition of herders' customary rights, as well as of traditional knowledge, as an evidence base for national laws and international agreements, and a reversal of colonial influences of knowledge invalidation, have been identified as pathways to escape social-ecological traps (Chapter 7). Inclusion of herders' knowledge through collaborative processes that respect the integrity and complementarity of both scientific and traditional knowledge systems can increase the validity and relevance of decision-making processes and become an important resource to address the challenges of climate change, sustainable land use and the transformation to a carbon-neutral society. Further, Chapter 6 argues that the inclusion of traditional knowledge is vital in the sustainable management of large carnivores to achieve both conservation and livelihood goals.

In Chapter 15, the vice-president of the Saami Council explicitly affirms the significance of self-determination and the right to participate in and influence decisions that affect Sámi reindeer herders and the Sámi society. The chapter also raises questions about the role of participatory research. Many of the authors in this book have long and fruitful experiences in participating with reindeer herders in various research projects and have benefitted from the exchange of academic and Indigenous knowledge. But as stated in Chapter 15:

> The two parallel knowledge systems, academia, and Indigenous knowledge, existing side by side creates the possibility to utilize the strengths and richness of both systems. For this to be effective, both knowledge systems need to be respected in their own capacity to create a knowledge base of 'best-available-knowledge'.

It is our hope that we can continue to co-create the 'best-available-knowledge', and that this book has contributed with insights that address the challenges of global environmental change to reindeer pastoralism in Fennoscandia. We also hope that these insights will make an impact in policy-making for an increased well-being of people and reindeer in the North on our path towards a transformation to a just and sustainable society.

Index

Note: Page numbers in *italics* indicate figures and in **bold** indicate tables on the corresponding pages.

For Product Safety Concerns and Information please contact our EU representative GPSR@taylorandfrancis.com
Taylor & Francis Verlag GmbH, Kaufingerstraße 24, 80331 München, Germany

www.ingramcontent.com/pod-product-compliance
Ingram Content Group UK Ltd.
Pitfield, Milton Keynes, MK11 3LW, UK
UKHW020316270926
475702UK00050B/917
* 9 7 8 0 3 6 7 6 3 2 6 7 0 *